Ast

Man and the changing environment

Holt, Rinehart and Winston
New York Chicago San Francisco Atlanta Dallas
Montreal Toronto London Sydney

Robert G. Franke
Iowa State University

and

Dorothy N. Franke
Des Moines Area Community College

Man and the changing environment

to David, Daniel, and Sara

Quotation on page 155 from THE PEOPLE, YES by Carl Sandburg, copyright, 1936, by Harcourt Brace Jovanovich, Inc.; renewed, © 1964, by Carl Sandburg. Reprinted by permission of the publishers.

Copyright © 1975 by Holt, Rinehart and Winston
All rights reserved.

    Franke, Robert G.
Man and the changing environment.

    Includes bibliographies.
1. Biology. 2. Ecology. I. Franke, Dorothy N., joint author. II. Title.
QH308.2.F7   574   74-30330
    ISBN: 0-03-084714-1

Printed in the United States of America.
  6 7 8 9   038   9 8 7 6 5 4 3

Chapter frontispieces
1. Courtesy of U.S. Forest Service. Photo by P. Freeman Helm.
2. Courtesy of Mount Wilson and Palomar Observatories.
3. From Z. Burian, *Prehistoric Man*, Artia, Prague, 1961.
4. From Z. Burian, *Prehistoric Man*, Artia, Prague, 1961.
5. Courtesy of San Diego Zoo. Photo by R. Garrison.
6. *Des Moines Register*, Des Moines, Iowa.
7. Courtesy of National Audubon Society. Photo by G. M. Haist.
8. Courtesy of Agency for International Development, Washington, D.C.
9. Courtesy of Agency for International Development, Washington, D.C.
10. Courtesy of U.S. Department of Interior, Bureau of Reclamation.
11. From *Arts in Society*. Photo by M. L. Brisson, University of Wisconsin–Green Bay.
12. Courtesy of World Health Organization. Photo by P. N. Sharma.
13. Courtesy of U.S. Department of Interior, Bureau of Reclamation.
14. Courtesy of U.S. Department of Interior, National Park Service.
15. Courtesy of U.S. Forest Service. Photo by L. J. Prater.

Preface

Interest in environmental problems is intensifying, especially since Earth Week, 1971. Many dedicated specialists, such as economists, engineers, and scientists, are putting their expertise and energy in this direction foremost. More lay persons, also, are looking for ways to express their concern. In many communities recent developments involving environment, such as recycling centers or picketing local industries that heavily pollute, have been initiated by energetic lay people. This action has brought results, although too often for every sign of progress, new environmental problems are uncovered. Whether expert or lay person, anyone concerned with "doing something" about environmental problems soon realizes that solutions need hard work and dedication and the expertise of many specialists.

One specialist who is always consulted sooner or later is the biologist, specifically the ecologist. *Biology* is the study of life; *ecology* is the division of biology that includes the study of the environment and all organisms in it — plants, animals, and man. Hence the biologist–ecologist is aware of the principles by which life-forms have existed in their environments for millenia. He can provide insight into what must be done to set right once more the relationship between earth and man.

Until the last few years, when the interest of expert and lay person in man's environmental problems became intense, ecological principles remained the domain of the biologist. Now, with environmental crises threatening our resources, our life style, and perhaps our very lives or our children's lives, the basic laws by which plants and animals relate to their environment must be made public knowledge. No senator can responsibly decide a public issue or lay person responsibly vote without having been exposed to such basic biological information.

The widespread dissemination of basic biological ideas is difficult. Beside television, the most effective medium is probably the classroom. School boards and administrators throughout the country are sensing the opportunities for environmental education. Slowly, curriculum changes are being made. Higher education is rallying, too, and many are providing environmental biology courses for a general education audience in which some basic biological principles, together with a description of our environmental problems, are the subject. More courses of this kind, and more qualified teachers, are needed. Vital, too, is the development of more appropriate teaching materials. This book was written to help meet this need.

Man and the Changing Environment is basically a biology text. The broad overview of biology and environmental problems is constantly stressed; and, as often as possible, the relationship of biology and the environmental problems to man is emphasized. The book is not intended for advanced biology courses, such as ecology, but for students in introductory biology courses. Consequently, *Man and the Changing Environment* should be useful wherever an introductory biology course emphasizing environmental problems is taught, from university to community college.

The book begins with the description of biological evolution, introducing the student to the process that relates organisms and environment. It continues with a description of man as one of these organisms, eventually discussing man's difficulties in managing his existence in the environment. In an attempt to maintain continuity from chapter to chapter and not lose sight of the overview, some subjects have been omitted, and others treated generally. Only use of the text will make clear if the careful inclusion or omission of material has been appropriate.

Many people have contributed a great deal to this book. Some are research scientists whose work produced the material that has been recorded here. Others have been concerned writers who took the time to dig out and collate significant information. Information has been generously borrowed from all these people. Those who deserve special thanks are Dr. Clark Bowen, whose vision led to the creation of the biology program at Iowa State University, including the course "Environmental Biology"; Dr. Fred Smith, Chairman of the Department of Botany and Plant Pathology, who encouraged the writing of the book in many ways; Dr. Lawrence Mitchell, who read and commented on many of the chapters; Drs. Marilyn Bachmann and David Ehrenfeld, who read, criticized, and suggested many improvements in organization, emphasis, and wording; and the Iowa State University Library personnel, who secured many references.

Dr. Robert O. Richards, Department of Sociology and Anthropology, Iowa State University, deserves special thanks for his interest, enthusiasm, and criticism during the final stages of manuscript preparation.

Finally, we thank Lillian Johnson, who in addition to her full-time work still found time to help in innumerable ways. It is no overstatement to say that without her contribution this book could not have been written.

Robert G. Franke
Dorothy N. Franke

Contents

Preface   v
1. Evolution, Organisms, and Environment   3
2. From Cell to Ecosystem   31
3. The Arrival of Man   61
4. The Distribution of Life   85
5. Behavior and Survival   119
6. The Growth of the Human Population   155
7. Characteristics of Animal Populations   189
8. The Control of the Human Population   211
9. Feeding the World   237
10. Air Pollution   275
11. Water: Maintenance of Quality and Quantity   299
12. Biocides   341
13. The Use of Energy, Minerals, and Soil   367
14. Man and Living Resources   397
15. Approaching Solutions to Environmental Problems   421
    Index   437

Chapter 1

But I am not equipped to philosophize and this book will not attempt it.

This only will it assert: a man's philosophy is likely to be truer and deeper if he has a knowledge of biology and organic evolution. Heaven knows we are artificial enough these days in our town-bred existence. A little feeling of oneness with the rest of life may even make us as intelligent and as sensitive as the men who, thrilled and awed and elevated at touching the skirts of creative understanding, painted the cave walls of Lascaux and Altamira two hundred long centuries ago.

The Story of Life
H. E. L. Mellersch

Evolution, organisms, and environment

Today man is living in a critical relationship with his environment. He is overrunning the world with too many people, overusing it by rapidly removing its resources, and abusing it through neglect and thoughtlessness. This book is concerned primarily with these crucial problems. Continued existence for man on earth — to say nothing of a better existence — depends on identifying our environmental problems, understanding their causes, and discovering ways to solve them.

Of these tasks, perhaps the easiest is the *identification* of environmental problems. You do not have to be an environmentalist to recognize smog in Los Angeles or the ravages of strip mining in Appalachia. Once identified, the *solution* of environmental problems is most challenging, because problem resolution requires an understanding of causes.

To understand the causes of ecological problems — where man "went off-base" — we shall first examine the type of relationship that man has had with his environment in the past. Man, always changing, evolving, has lived on the earth for millions of years. Why has he endured? What kind of relationship did he have which proved so life-sustaining? What biological principles help explain his past success? How significant would this information be in assuring man of a future? Of course, these questions are answered best by examining the clues left by ancient man. However, insight can also be gained by a close look at how other species relate to their environments. Thus, we shall look into the life-sustaining nature of the relationships that organisms maintain with their environments. In examining the adaptations of ancient man and of other organisms to their environment, you will soon appreciate a major biological truth: life is a product of its interaction with the environment through time. The identity of any plant or animal — from its survival in the past, to its appearance and activities today, to its potential for survival in the future — is intimately related to the environment.

The specific factor that binds organisms (including man) with their environments is revealed when their relationships are examined closely. That factor is change. Environments change and organisms change, and as a consequence the relationship between them also changes. In fact, the significance of change in the relationship leads to a second biological truth: the strongest bond between organisms and environment appears to be the influence that each has on the other for change.

So, the book is about the change in plants and animals (man in particular) as the organisms relate to their environments. In this sense it is a book about *biological evolution*, the change through time of living things and the relationships of the organisms with their surroundings.

At the same time the book is about *ecology*. Ecology is the study of how plants and animals interact in their environments. The origin of the word, from the Greek "study of the house" (*oikos*, house; *logus*, study), emphasizes the complex subject of ecology. Ecology is a study of how the occupants in their "house" *interact*, not just study of the house itself. This is a book about ecology in that we examine the complex interactions of plants and animals in the natural environment, especially those that involve man, which produce change.

We shall begin by examining examples of change in plants and ani-

5   Evolution, organisms, and environment

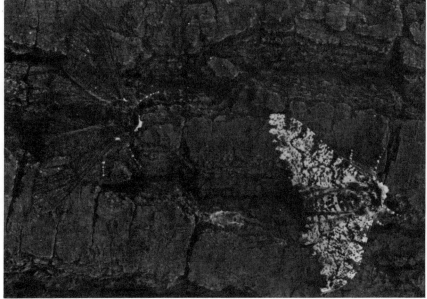

Figure 1–1
White and dark forms of the peppered moth *Biston betularia*. Top: At rest on a light background in an unpolluted countryside. Bottom: On a soot-covered tree trunk near Birmingham, England. (H. B. D. Kettlewell, Department of Zoology, Oxford, England.)

mals. We will note how the environment appears to influence these changes and then look at explanations for the changes. Finally, we shall examine the evidence that biologists accept as proof of change. Out of this should come better insight into man's nature, his current difficulties in relating to the environment, and his prospects for the future.

Changing organisms in changing environments

A few examples that can be found today will illustrate the changes that some organisms have undergone as a result of interaction with their changing environments. We shall look at the change in a species of moth (the peppered moth) over only a few decades of changing environment, the changes in species of pasture plants in grazed and ungrazed fields through only a few generations, and the result of changes in a plant species of yarrow in California which apparently developed over generations in response to the diverse environmental conditions at various altitudes. These simple illustrations hint at the awesome dynamics behind environmental change and organismic transformation.

Examples of changes in organisms

Extensive studies of the peppered moth (*Biston betularia*), especially in Great Britain, reveal a most interesting change in the color of the moth's populations over a few decades in areas surrounding industrial towns. Before the mid-1800s, most representatives of the species that were collected were light in color. When moths rested on the trunks of lichen-covered trees, they were often difficult to see (Fig. 1–1, top). Then in 1948 collections of this insect made near Manchester, a thriving industrial town, differed greatly from previous collections. It was a dark melanic form (Fig. 1–1, bottom). In the following decades throughout England, notably near industrial areas, increasing numbers of the dark form were found. Today, in some areas *only* this form occurs, the light variety seemingly having been replaced.

The change of color in the peppered moth over a few decades illustrates change in a species over a very short time. Other examples can be cited, such as the changes in pasture plants that grow in heavily grazed areas. A mixture of grass and *Trifolium* clover seed was planted in a pasture in Maryland. After sowing, a fence was built down the center of the field. During the next 3 years the pasture on one side of the fence was cut several times for hay; the other side was heavily grazed. Then an investigator dug up plants from each side of the fence and planted them in his experimental plot. Most of the plants from the grazed pasture remained short; most of the plants from the other side of the fence grew tall. Seed was collected from many of the plants and sowed. In general, plants derived from the collected seed were found to exhibit the same height characteristic as the parent plants. In other words, the plants and their descendants from the grazed side of the fence appeared to be permanently changed.

Yarrow plants of the genus *Achillea* illustrate changes in a species over many generations. *Achillea* are very widespread in the Northern Hemisphere. Among the many structural variations seen within the several species is height. Generally, this characteristic appears to be correlated with the altitude at which the yarrow grows: the greater the altitude, the shorter the plant. Figure 1–2 illustrates the mean heights of plants collected at various altitudes across California.

Biologists experimented with the plants collected at these altitudes. Clonal plants — plants made from cuttings from an original plant — were grown from tall specimens originally collected at the coast (Fig. 1–3). These clonal plants were planted at three different altitudes: at sea level, at about 5,000 feet above sea level, and at about 10,000 feet. Clones of the coastal plants when planted at the midelevation produced shorter plants than at sea level. When planted at the highest elevation all clones but one died.

A similar experiment was done with clones of plants collected at the midelevation. These plants when collected were generally shorter than those found at sea level. However, when they were planted at sea level, the plants grew larger than at 5,000 feet. At 10,000 feet, some of the plants died; those that did grow were shorter than the same plants grown at the 5,000-foot level.

These observations suggested that yarrow plants had developed a survival ability that was, in some way, correlated with their particular heights and respective environments in which they were annually collected. Presumably, this adaptation to environment took a long time to come about, because clones transplanted to atypical altitudes showed only a limited tendency (if any at all) to adapt quickly, and sometimes they even died. What can explain this tendency of some plants to adapt to new environments and others to die? For that matter, what can explain the natural occurrence of tall plants at sea level, short plants at high elevations, and intermediate plants at moderate elevations?

How organisms change

Some generalizations about the changes observed in plants and animals in nature are illustrated in the changes described in the three organisms cited above:

1. Changes in organisms appear to accompany changes in the environment.
2. The relationship between a plant or animal species and its environment is not static but appears to be dynamic, in that any change in one may induce change in the other.
3. Organisms appear to better their survival chances because of an innate tendency to change over generations.
4. Man, directly or indirectly, can strongly influence the changes.
5. If the changes in the environment are too severe, individual organisms may die.

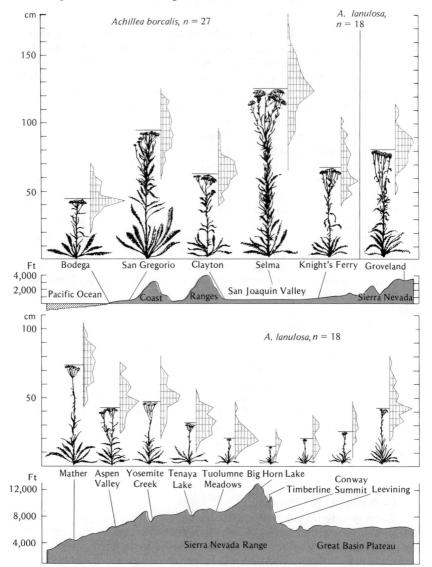

Figure 1–2
Variation in yarrow (genus *Achillea*). Top: Ranges of variation in height in collections of yarrow plants from sea level to the Sierra Nevada Mountains in California (arrows point to average heights). Bottom: Approximate sites of collection within a 200-mile range across California at latitude 38°N. (From Clausen, Keck, and Heisey, Publication 581, Carnegie Institution of Washington, Washington, D.C.)

The changes in the moth and the pasture plants as a result of the influence of environment is clear, mainly because the changes occurred in the obvious characteristics of color and height over only a few generations.

The change in the yarrow is not so clearly recognized from the observations cited, although the suggestion is implicit that the slight variations seen in members of the population when transplanted to foreign environments may be the cause of the unique heights of the populations found growing naturally at different elevations. In brief, the yarrow plants, the peppered moth, and the pasture plants are all illustrations of change in species over time within changing environments.

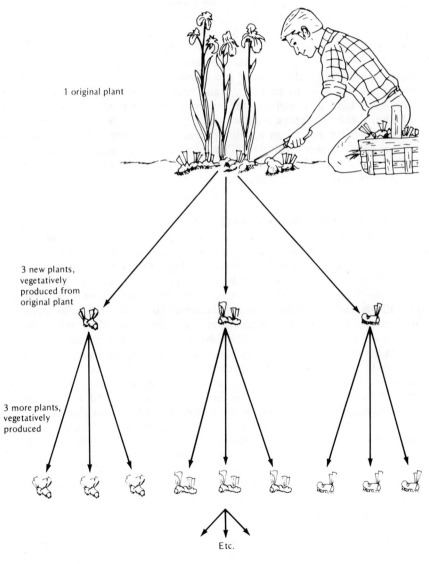

Figure 1–3
How "clones" are produced.

The important question for us to ask now is: How can these aspects of change in populations through time be explained? However, before that question can be answered, more must be said to specify the type of change that we are discussing. Two types of change occur in plants and animals: (1) the change to an individual in its lifetime, and (2) the change to a group of similar interbreeding individuals, a *population*. The first is often easy to explain. Inadequate fertilizer will produce a field of deformed plants; a poor diet may produce a bowlegged baby. Often adverse changes in individual organisms can be corrected by improving their environment. Furthermore, changes in individuals have significance only while a particular organism is alive, because the alterations to the body of the organism are not genetic and cannot be transferred to offspring.

The evolution of a population is less readily explained. Changing the environment during the lives of the members of the population of a particular species will not suddenly transform the individual organisms into a different kind of organism that warrants a new name. Such change is apparently subtle, always requires many generations, is a genetic phenomenon, and often results in a new *species* of organisms. It is the process known as *biological evolution*. This population change, a result of biological evolution, is the type of change in organisms that will be discussed in the remainder of the chapter.

The synthetic theory of evolution

How can such aspects of change in populations be explained? To answer this question we shall examine an explanation for the origin of new kinds of organisms, or *species,* resulting from populational changes. This explanation is known as the *synthetic theory of evolution.* An understanding of the theory explains changes in populations through time. The theory also illustrates the intimacy and balance of the relationship between all organisms and their environments and should make clear why disruption of either a population of organisms or their environments profoundly affects the other — often with serious consequences, and even death to the organisms involved. For man, an understanding of the synthetic theory of evolution can lead to a clearer understanding of how *Homo sapiens* evolved to his present position in the hierarchy of life and what can happen to him as a species if his present environment is disrupted carelessly.

The synthetic theory of evolution is a broad explanation for the origin of diverse types of organisms on earth throughout time. The theory is constantly being updated as new information regarding genetics and heredity becomes known. The basis for the theory is not recent. In fact, Charles Darwin, a British naturalist, first published the basis of the theory in 1859 in his famous book *The Origin of Species* (Fig. 1–4). In his book Darwin stated the following ideas:

>   1. *Organisms tend to overproduce.* Plants and animals always produce an abundance of offspring. For example, Darwin once calculated that 19 million elephants could theoretically descend from one pair

of elephants in 750 years if each generation of offspring survived to reproduce.

2. *Of the many offspring born, only a few survive.* Darwin pointed out that the number of creatures in the wild normally stay about the same. He concluded that this is so because only a few of them survive after birth. Such a mathematically possible profusion of elephants is, therefore, highly unlikely. Darwin observed that those offspring most likely to die are those least able to survive in their en-

Figure 1–4
Charles Darwin, 1809–1882. (Radio Times Hulton Picture Library.)

vironment. Their deaths are a result of the intense competition among the offspring for basic needs, such as food and space. Darwin called such competition a *struggle for existence*. The victory of the best-adapted he called *the survival of the fittest*.

3. *Individuals within a population show variation.* Darwin pointed out that some members of a population survive and others die because individuals in that population vary. Such variation appears to be *innate* in the individuals, that is, a characteristic they are born with. The plants or animals that survive and reproduce often pass the survival traits to some of their offspring. Consequently, a selection for survival traits goes on through the generations. This process is known as *natural selection*.

4. *Great amounts of time are needed for natural selection to produce new species.* If enough time is available, natural selection will often yield a population of organisms greatly different from its predecessors. Darwin's evidence for this came mainly from the comparison of modern animals with fossils of extinct animals in South America. Contemporary animals were unlike the fossil remains; yet they often resembled the fossil forms in many basic ways. Presumably, those ancient forms changed through many generations by the process of natural selection to give rise to the contemporary types. Thus, Darwin became increasingly convinced that great amounts of time were normally necessary for evolution to occur.

Natural selection as expressed in *The Origin of Species* provided an explanation for the creation of new kinds of living things which tied together many observations. Nevertheless, Darwin's idea of natural selection was severely attacked by critics, mainly because it suggested that the same process also produced man. This suggestion challenged a literal Biblical interpretation of man's creation because it meant that man arose from a nonhuman form. If man is closely akin to other forms of life, he is not unique. Historically this has been a difficult idea for many people to accept.

As the decades passed, new observations in biology were added to Darwin's original ideas as expressed in his book. The new and old ideas eventually formed the synthetic theory, which today provides a convincing explanation for the origin of new kinds of organisms and explains the origin of the complex intimacy and interdependence between plants and animals and their environments that we see today.

Gene theory

One of the new ideas added to Darwin's theory of natural selection was the concept of *gene*, an element of the sex cells that serves as a transmitter of heredity characteristics. Darwin could not explain the origin of variation among individuals in a population; the gene concept could. Gene theory accounted for the origin of characteristics of an organism and their transmission to offspring. From it developed several new biological ideas which helped explain how organisms and environment interact through time to produce new species. These ideas included *mutationism* and *gene recombination,* the significance to species change of the *isolation* and the *migration* of population fragments, and the *size* of a population.

In the early 1900s the gene, carrier of characteristics that unfold as a plant or animal develops, was found to undergo relatively permanent changes, *mutations,* arising from causes other than those of normal gene recombination. This discovery was very important to evolutionists, because a changed or mutated gene may produce a new characteristic in some offspring. It appeared to evolutionists as if a source had been found of individual variation essential as raw material to Darwin's natural selection.

An additional factor found to influence species formation was *genetic recombination.* Genetic recombination is the assorting of the genes (1) in the production of each egg and sperm, and (2) upon fusion of the eggs and sperms, the *gametes,* in fertilization. Recombination, together with mutation, produces the differences in individuals in a population. In turn, it is the differences that are "worked on" by natural selection to produce populations of different characteristics.

With regard to the interaction of organisms and environment, three major conditions prevail that influence the fate of gene changes resulting from mutation and genetic recombination and in turn allow natural selection to operate. One influence is the *migration* into the population of individuals from nearby different populations of the same species. The individual organisms that mix into the original population introduce forms or combinations of genes not occurring there or add to the percentage of genes already in the population. Migration of members out of a population also may change the number of genes making up a population.

Another influence is *population size* in a given environment. In a large population any effect of a change in the percentages of genes due to mutation or migration, for example, is obscured by the large percentage of unchanged genes. However, in a small population, any new gene may represent a great proportional change within the total number of genes. Thus, the effects of the gene change will be more readily observed in the smaller population. This rapid change in the percentages of genes in a small population, which in time may affect the appearance of a population, is known as *genetic drift.*

A third influence on the evolution of new species is the *isolation* of breeding segments of the population. When this occurs, the forces of natural selection operate within each isolated population segment and may eventually transform, or *evolve,* the separated populations into increasingly unique groups. This is because isolation prevents any mixing of the genes between isolated sectors of the total population. Eventually, natural selection may differentiate the isolated populations so much that individuals of one can no longer reproduce with individuals of the other. At that point, two new species have evolved where originally there was one.

A classic example of the development of two species from (presumably) one is that of squirrels living on the rim of the Grand Canyon. Both species have almost identical characteristics, except that those on the northern rim of the canyon are white-tailed and black-bellied, whereas those on the southern rim have gray tails and white undersides. Between the two populations is the deep, impassable canyon. Because the two species are so much alike, biologists think the two types of squirrels probably were one species until the two isolated populations could no longer interbreed, owing to the enlarging of the Grand Canyon.

Thus, *geographical isolation* has augmented natural selection working on individual variations within a population. Isolation of other kinds may also occur. For example, two distant parts of a large population may develop different reproductive timings; that is, one segment of the population may produce eggs and sperms at one time, whereas members of the same population but in another area may produce sperm and eggs at a slightly different time. Thus, fertilization *between* the two populations be-

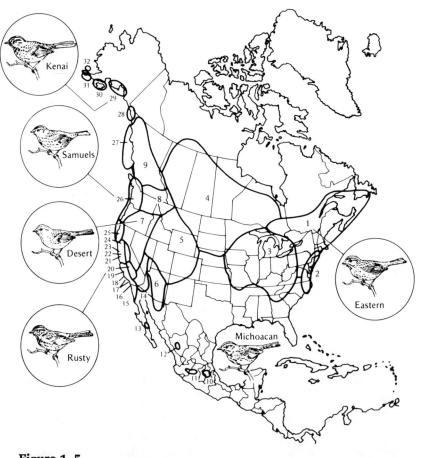

Figure 1-5
Distribution of subspecies of song sparrow (*Melospiza melodia*). Only six subspecies found some distance from each other are illustrated. Note their similar appearance. (From G. Ledyard Stebbins, *Processes of Organic Evolution*, 2nd ed., © 1971. Reprinted by permission of Prentice-Hall, Inc., Englewood Cliffs, N.J.)

comes unlikely. This isolation is called *seasonal isolation*. Such a situation is found in several species of frogs, for example.

Other types of isolation are known. For example, populations may have varied or incompatible mating patterns (*behavioral isolation*) or structural differences in their sex organs (*mechanical isolation*).

Consequences of the synthetic theory

Observations of plant and animal populations in nature in various stages of evolutionary development can be observed. In fact, the races of man are subpopulations that may have occurred because conditions on earth at one time prevented easy mixing of human genes. Perhaps if isolation had continued in the past, the races would have lost their interfertility. Today, populations of birds and other animals slightly different from each other but still capable of interbreeding (Fig. 1–5) are common [these are known as *demes* (Fig. 1–6) or, if more unique, as *subspecies*]. On the other hand, some of these populations, if left undisturbed, may eventually be unable to interbreed. If so, *speciation* has occurred, and two new species have been created (Fig. 1–7). Thus, the synthetic theory, involving isolation, genetic drift, migration, and genetic variation, appears to explain the origin of new species and contributes to understanding the delicate and intimate relationships of organisms and their environments.

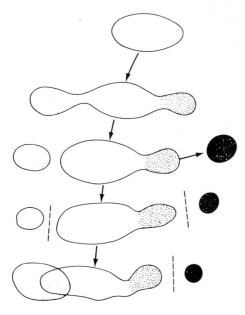

First stage.
A single population in a homogeneous environment.

Second stage.
Differentiation of environment, and migration to new environments produces racial differentiation of races and subspecies (indicated by different kinds of shading).

Third stage.
Further differentiation and migration produces geographic isolation of some races and subspecies.

Fourth stage.
Some of these isolated subspecies differentiate with respect to genic and chromosomal changes which control reproductive isolating mechanisms.

Fifth stage.
Changes in the environment permit geographically isolated populations to exist together again in the same region. They now remain distinct because of the reproductive isolating barriers which separate them, and can be recognized as good species.

Figure 1–6
Sequence of events that leads to the production of different races, subspecies, and species, starting with a homogeneous group of populations. (From G. Ledyard Stebbins, *Processes of Organic Evolution,* 2nd ed., © 1971. Reprinted by permission of Prentice-Hall, Inc., Englewood Cliffs, N.J.)

16   Chapter 1: Evolution, organisms, and environment

Figure 1-7
Miniature pig bred by scientists at the Hormel Institute of the University of Minnesota for use in research. The miniature pig (with the registered trademark PIGmeePIG) can be more easily accommodated in the research laboratory than a regular pig as this picture shows. (Courtesy of J. H. Belknap, Hormel Institute, University of Minnesota, Austin, Minn.)

In summary, the synthetic theory includes the following ideas:

1. The change of species is a change in those characteristics determined by genes: function, morphology, and behavior of organisms.
2. The basic materials used in evolutionary change are the inheritable gene variations produced by mutation and genetic recombination.
3. The population is the living unit that changes and may produce a new species.
4. A change in a population is basically due to the shift of percentages of genes in the population from generation to generation.
5. Several factors affect the frequencies of genes in a population, including mutation, migration, genetic drift, and natural selection.

These aspects of the synthetic theory most directly explain the origin of new species through the millenia that life has been on earth. Most important for our concern, however, is the understanding provided by the synthetic theory for the close relationship between types of organisms and their environments. Evolution of life-forms through thousands of years has given rise to unique kinds of plants and animals, all of which are closely related to the kind of environment in which they are found. The synthetic theory explains how this relationship came about. An understanding of the theory also brings home the realization that a balance exists between organism and environment which has evolved over many years and which, if altered suddenly, could result in serious consequences for the organisms involved. This information relates significantly to modern man — surrounded with environmental crises that are the result of sudden alterations in his surroundings.

Evidence that populations change

The major theme of this book is the relationship of plants and animals to their environments. Consequently, understanding the synthetic theory of evolution is essential to understanding what the book is about. This is because the synthetic theory not only explains how plants and animals have evolved into the relationships they now have with their environments, but also it helps explain the extraordinarily complicated and intimate interrelationships that can be observed in any natural environment.

Understanding the synthetic theory also explains why disruptions to the environment or the populations in the environment seriously upset the very balance that characterizes their existence. Consequently, and more to the point, an understanding of the synthetic theory gives insight into the types of crises that face man today as a result of the disruption of his relationship with environmental conditions.

To emphasize the significant role of the synthetic theory in explaining the complexities of the natural world today and to better understand our contemporary problems with that world, we shall look next at how changes in populations of organisms and their environments seem to illustrate the synthetic theory in action. The first changes that we shall explain are those of the peppered moth, the pasture plants, and the yarrow cited earlier. Your understanding of the synthetic theory requires that the causes of change in the peppered moth population should be looked for in the environmental conditions in which the moth lives as well as in the moth population. This can be better understood by glancing at Figure 1.1. Notice how obviously the light moths stand out against the dark tree trunk but how well disguised they are against the light trunk. Here is the clue to why the color of the population of moths changed over just a few decades. In the early half of the 1800s England was experiencing the Industrial Revolution, a movement greatly dependent on the burning of coal. By the middle of the 1800s soot proved to be a common air pollutant of England, coating buildings, streets, and even trees. As the trees gradually darkened, white moths resting on the trunks became increasingly obvious and predators had an easier time discerning them. Subsequent experiments have shown that birds are natural predators of the moths and can very easily discern white moths on dark backgrounds. Thus, natural selection eventually led to a change in the percentage of dark forms in the moth population.

Today, many of the industrial areas in England are burning different fuels to reduce air pollution. It will be interesting to see if the color of the population of moths in the industrial areas changes back to a lighter form in the next few decades. Our understanding of the way in which life-forms evolve suggests that it should.

An understanding of the synthetic theory also explains the development of the short-growing pasture plants described earlier. In this case, in the 3 years during which cattle grazed the pasture, tall plants were prevented from reproducing. Thus, the short plants less apt to be eaten were allowed to survive and reproduce. Eventually, only short plants would prevail. Recall that these short plants taken into the laboratory and allowed to reproduce tended to produce predominantly short offspring, which suggested that, in fact, the predominant types of genes that now prevailed in the population produce short plants.

Although the synthetic theory explains the changes observed in populations of peppered moths and pasture plants, not enough generations of either have passed to allow the production of new species. The interaction of the yarrow plants with their environment had presumably been going on for a greater period of time and through many more generations, and thus, it appears that new kinds of yarrow are further along in their development into unique types of living things.

How does the synthetic theory of evolution explain the changes in yarrow? Remember that yarrow clones planted at various altitudes do not show great ability to survive there. Usually a new environment, if it does not kill the plants, helps to evolve changes. The plants displaying these changes may be able to survive in that new environment and even produce offspring. In time, the plants as a group in each of these environments would tend to produce a population adapted to the unique environmental conditions in which they grow. These changes would occur in the population over many generations. The characteristic heights of plants at the various altitudes can be interpreted as changes that result from the interaction of innate or genetic potential of the plants and their environments. In time, this interaction would produce populations of plants with unique characteristics that relate to the new environment in which the populations grow. Such an explanation is consistent with the synthetic theory.

Observations from populations of organisms that exist in changing environments would appear, therefore, to offer some of the most convincing evidence that evolution is not to be viewed simply as a historical process of the past but as going on at this time. Experiments in the labortory tend to support this idea.

One organism whose body was deliberately altered by scientists in the laboratory is the darling of genetic research, the common fruit fly. Geneticists were interested in the number of generations needed to change a simple characteristic of the fly's body. They chose the number of bristles on the fly's abdomen. Breeding experiments began with the population of relatively bristly fruit flies, those with abdomens that had an average of 36 bristles. Then, by always breeding the least hairy flies, those with the lowest number of bristles, they produced in 30 generations a population of flies with comparatively bristleless abdomens (an average number of 25 bristles). The researchers bred a hairier breed of flies as well. They mated only flies with relatively large numbers of abdominal bristles. In 20 generations they produced fruit flies with an average of 56 bristles per abdomen, probably some of the hairiest flies alive. Like the fruit fly's abdomen, changes in parts of other organisms have been developed deliberately. In all such experiments in the laboratory, the interaction between the genetic potential for change in the organism and the environment artificially altered by man is the cause of the evolution of a population of creatures with "new" characteristics.

In addition to field and laboratory data, evidence exists from geology and paleontology which offers proof that the synthetic theory operated in the past as well as now. Geological investigations appear to support operation of the synthetic theory in the past. *Geology* includes the study of the inanimate environment and is, thus, a source of information regarding the changes in the earth's surface through past millenia (some of the details of these changes are described in Chapter 2). An understanding of the synthetic theory reveals that closely correlated with any changes in the earth's environment are changes in the types of plants and animals. Testimony of these changes in living forms is found in the abundant fossil evidence scattered around the surface of the earth. The study of fossils, *paleontology*, reveals the history of life and suggests that in the past unique forms of animals and plants existed which are no longer found and that many contemporary plants and animals seem to have common fossil ancestors. From geology comes an understanding of the past changes in the earth's environment, and from paleontology comes an understanding of forms of life that lived in those environments. Changes in the earth are well-correlated with the changes that have occurred in populations of past organisms (Fig. 1–8). These changes are divisible into major periods of time called *geological periods*. The geological periods and their major kinds of plants and animals are shown in Table 1–1. This *geological time scale* is often divided into five major *eras,* which in turn are subdivided into *periods*.

Another area of biology which indicates that evolutionary processes have been working in the past to produce the present kinds of plants and animals is taxonomy. *Taxonomy* specializes in comparing organisms, grouping similar organisms, and naming the groups. Although taxonomists are sometimes only concerned with a label for a group, they are also frequently concerned with discernible similarities that might form the base for a natural grouping. The taxonomist relies heavily on the work of the comparative anatomist. For example, the discovery that the rat, elephant, and giraffe — and even the porpoise, which cannot turn its neck — have the same number of neck bones is significant to a taxonomist searching for evidence of natural relationships. From such information supplied by the comparative anatomist the taxonomist can conclude that

Table 1-1
Geological events and characteristic plants and animals that have occurred on the earth since the origin of life

Era	Period	Epoch	Time at beginning of each period (millions of years ago)	Geological events and climate	Biological characteristics
Cenozoic (Age of Mammals)	Quaternary	Recent	0.025	End of fourth ice age; climate warmer	Dominance of modern man; modern species of animals and plants
		Pleistocene	0.6–1	Four ice ages with valley and sheet glaciers covering much of North America and Eurasia; continents in high relief; cold and mild climates	Modern species; extinction of giant mammals and many plants; development of man
	Tertiary	Pliocene	12	Continental elevation; volcanic activity; dry and cool climate	Modern genera of mammals; emergence of man from man-apes; peak of mammals; invertebrates similar to modern kinds
		Miocene	25	Development of plains and grasslands; moderate climates; sierra mountains renewed	Modern subfamilies rise; development of grazing mammals; first man-apes, temperate kind of plants; saber-toothed cat

Table 1-1 (continued)

Era	Period	Epoch	Time at beginning of each period (millions of years ago)	Geological events and climate	Biological characteristics
		Oligocene	34	Mountain building; mild climates	Primitive apes and monkeys; whales; rise of most mammal families; temperate kind of plants; archaic mammals extinct
		Eocene	55	Land connection between North America and Europe during part of epoch; mountain erosion; heavy rainfall	Modern orders of mammals; adaptive radiation of placental mammals; subtropical forests; first horses
		Paleocene	75	Mountain building; temperate to subtropical climates	Dominance of archaic mammals; modern birds; dinosaurs all extinct; placental mammals; subtropical plants; first tarsiers and lemurs
Mesozoic (Age of Reptiles)	Cretaceous		130	Spread of inland seas and swamps; mountains (Andes, Himalayas, Rockies, etc.) formed; mild to cool climate	Extinction of giant land and marine reptiles; pouched and placental mammals rise; flowering plants (angiosperms); gymnosperms decline

Table 1-1 (continued)

Era	Period	Epoch	Time at beginning of each period (millions of years ago)	Geological events and climate	Biological characteristics
	Jurassic		180	Continents with shallow seas; Sierra Nevada Mountains	Giant dinosaurs; reptiles dominant; first mammals; first toothed birds
	Triassic		230	Continents elevated; widespread deserts	First dinosaurs; marine reptiles; mammal-like reptiles; cone trees dominant
Paleozoic (Age of Amphibians)	Permian		260	Rise of continents; widespread mountains; Appalachians formed; cold, dry, and moist climate; glaciation; red beds	Adaptive radiation of reptiles which displace amphibians; many marine invertebrates extinct; modern insects; cone trees (gymnosperms) appear
	Pennsylvanian*		310	Shallow inland seas; glaciation in Southern Hemisphere; warm, moist climate; cool swamp-forests	Origin of reptiles; diversification in amphibians; gigantic insects
	Mississippian*		350	Inland seas; mountain formation; warm climates; hot swamp lands	Amphibian radiation; insects with wings; sharks and bony fish; crinoids
(Age of Fishes)	Devonian		400	Small inland seas; mountain formation; arid land; heavy rainfall	First amphibians; mostly freshwater fish; lungfish and sharks; forests and land plants (ferns); wingless insects; corals

Table 1–1 (continued)

Era	Period	Epoch	Time at beginning of each period (millions of years ago)	Geological events and climate	Biological characteristics
	Silurian		425 to 430	Continental seas; relatively flat continents; mild climates; land rising; mountains in Europe	Fish with lower jaws; invasion of land by arthropods and plants
(Age of Invertebrates)	Ordovician		475	Oceans greatly enlarge; submergence of land; warm mild climates into higher latitudes	Ostracoderms (first vertebrates); trilobites abundant; land plants
	Cambrian		550	Lowlands; mild climates	Marine invertebrates and algae; all invertebrate phyla and many classes; abundant fossils; trilobites dominant
Proterozoic (Precambrian)			2,000	Volcanic activity; very old sedimentary rocks; mountain building; glaciations; erosions; climate warm moist to dry cold	Fossil algae 2.6 billion years old; sponge spicules; worm burrows; soft-bodied animals; autotrophism established
Archeozoic (Precambrian)			4,000 to 4,500	Lava flows; granite formation; sedimentary deposition; erosion	Origin of life; heterotrophism established

*The Pennsylvanian (Upper) and Mississippian (Lower) are often referred to as the Carboniferous period.
Source: C. P. Hickman, Integrated Principles of Zoology, The C. V. Mosby Company, St. Louis, 1966.

24   Chapter 1:   Evolution, organisms, and environment

Figure 1-8
The Grand Canyon wall is an excellent site to gain an appreciation of the changes that have occurred in the populations of organisms in an area through millenia. A trip into the Grand Canyon on mule or foot reveals fossils in the layers of rock that compose the canyon wall. Generally, only a few fossils of a limited number of types of organisms exist in the lower strata, whereas plentiful fossils of more abundant types of ancient organisms are found in the upper strata. Types and numbers of fossils appear correlated with past changes in the earth's surface. (Courtesy of U.S. Geological Survey. Photo by L. F. Noble.)

these diverse organisms most likely shared a common ancestor and that any taxonomic grouping should reflect their degree of relationship (Figs. 1-9 and 1-10).

Comparative biochemists and physiologists also contribute information which suggests that biological evolution has been active in the past to produce the organisms we see today. These scientists study like functions in organisms which structurally may be grossly different. For example, the chemical responsible for the food-making process of photosynthesis is essentially the same in all green plants, from the single-celled alga to the oak tree. This suggests that apparently diverse plants are related through ancient common ancestors that probably existed when the photosynthetic process came into being.

To these scientists similarities among organisms means that organisms are related. To them the relationship is a history best explained by the synthetic theory of evolution. This relationship can be viewed as a

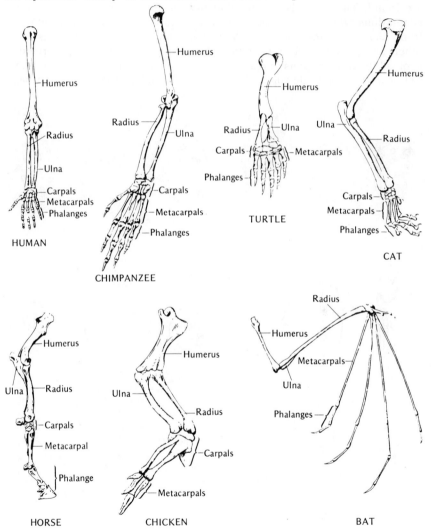

Figure 1-9
Note the basic similarity among the bones of the forelimbs of several vertebrates. To most biologists the remarkable similarities in basic structure suggest that these diverse animals had a common ancestor millenia ago. (Courtesy of E. Hackel, Department of Natural Science, Michigan State University, East Lansing, Mich.)

long sequence of diverse types of plants and animals coming out of the past, offspring of ancient populations and ever-changing environments, yielding the forms that populate the world today.

One other source of evidence comes from *biogeography*, study of the patterns of distribution of animal species around the earth. Biogeographers are well-informed about the special adaptation that plants' and animals have evolved to exist in particular environments in which they are found. They are also aware that despite the many unique structural adaptations which organisms have evolved to their specific environments, basic characteristics are still held in common with related organisms in other environments. For example, the white polar bear has probably evolved in Arctic regions in response to the snow and ice that prevail there. Yet despite the specific color adaptation of the polar bear, a bear is a bear.

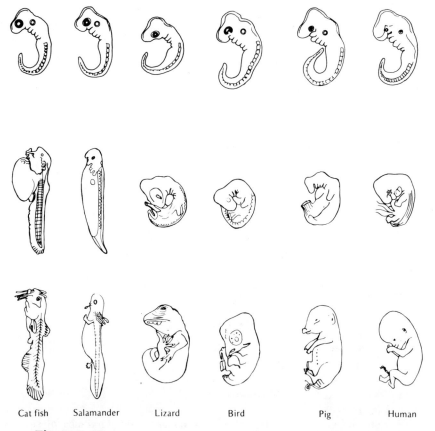

Cat fish   Salamander   Lizard   Bird   Pig   Human

Figure 1–10
Note the similarity among the early embryonic stages of various vertebrates. Such embryonic similarity probably indicates a common ancestry for all these higher animals.

Any other bear, although colored differently or found elsewhere, possesses basic characteristics in common with polar bears. This suggests that all bears had a common ancestral type which has diversified through the ages as bears spread into different environments. Climate and genetic potential have worked together to produce a unique kind of organism; but no matter what the environment, certain basic characteristics persist in all the diverse types of animals in this group.

Evidence from geology, paleontology, comparative biochemistry, morphology, physiology, and biogeography converge to support the concept that life from the past into the present has been constantly changing into new kinds of organisms and that these changes have been closely correlated with changes in the environment. This interrelation between changes in organisms and changes in the environment can best be explained by the synthetic theory of evolution. The theory provides for us today an understanding not only of the diversity of life-forms but also the interrelationships among kinds of organisms and between organisms and their environments which accounts for their continued existence today. In view of this understanding, it is interesting to contemplate the future as the processes of evolution continue, as climates change, and as the types of organisms familiar to us today evolve into still different forms. What significance do these ideas have to man today and to our descendants in the future?

The significance to man of evolution

The theme of this chapter is change — change in organisms and change in environments, and how the changes in each affect the other. But what significance does this theme have to our struggle for survival? For several reasons, an understanding of environmental and biological change as explained by the synthetic theory of evolution is significant to man.

First, the process of change that explains the evolution of contemporary plants and animals also is considered by biologists to explain how man came about. This story is explained more fully in Chapter 3, but it is important to realize from the start that man is an intergral part of the natural scheme of things and that he shares much common history with plants and other animals which inhabit the earth. This suggests that many of the rules by which other organisms have evolved and have managed to stay alive as species most likely apply to him also. The integration of man into the fabric of nature is essential for a continuance of a biologically sound existence on earth. It may well be that man's definition of himself as a being outside the natural structure and function of the natural world has accounted for some of the environmental atrocities which he has wrought and which now threaten his biological existence in many ways.

Second, reflection on the mechanism of change in populations in a changing environment makes clear that the emphasis in evolution is seldom on individuals but rather is on the members of a *population*. Here, too, may lie a lesson for man, in that many of the activities of individual segments of our population or of individuals, which today seem to be threatening our existence, may have to be curtailed in the interests of the population as a whole. More specifically, it may be that those who persist in polluting our environment so that human existence is threatened or made uncomfortable may have to be forced economically or politically to alter their actions so that the population as a whole is better off. This realization, if true, has important personal, sociological, and even political implications for the future.

Finally, an understanding of change in organisms in their environments suggests that man can aggressively and deliberately use the rules by which organisms change to make his existence more comfortable or easier — if he decides to. Such manipulation can even include evolving human beings who are more capable of surviving contemporary environmental problems. A knowledge of the process of species change can lead to an understanding of contemporary man's relationship to his environment and, more, give insight into the nature, cause, and cure of the environmental problems that confront us today. Helena Curtis emphasizes this point in *Invitation to Biology* (Worth Publishers, Inc., New York, 1972):

> *For those of us concerned with the present environmental crisis — and who among us can afford not to be? — a knowledge of the processes of evolution helps in our understanding of the roots of our present dilemma. The interrelationships of the living things . . . are the end result of evolutionary processes which have built up a network of intricate and often, to us, invisible adaptations and dependencies. We have been late in discovering that pulling a single thread in this web of life can bring about changes in its entire structure.*

Summary

In this chapter we have examined the relationship of organisms in their environment as a product of ages of evolution. The evolutionary process was explained by the synthetic theory of evolution, a synthesis of ideas, including Darwin's theory of natural selection, mutation and recombination of genes, isolation and the migration of population fragments, and genetic drift. Evidence that the synthetic theory is operating to produce new populations of plants and animals was interpreted in contemporary populations of the peppered moth and yarrow. Other evidence was interpreted from genetic experiments in the laboratory, from geology and paleontology, from comparative morphology and biochemistry, and from biogeography. An understanding of the synthetic theory as the mechanism by which organisms come about was cited as significant to man, because it suggests that man is an integral part of the natural scheme of living things; the rules by which plants and other animals relate to the environment also apply to him; evolution works on the population and not the individual; and man can choose to ensure a fruitful future for mankind rather than to be faced with the kinds of environmental crises that could lead to our demise.

Supplementary readings

DeBeer, G. R. 1964. *Atlas of Evolution.* Thomas Nelson & Sons, Ltd., London. 202 pp.

Eaton, T. H., Jr. 1970. *Evolution.* W. W. Norton & Company, Inc., New York. 270 pp.

Ehrlich, P. R., and R. W. Holm. 1963. *The Process of Evolution.* McGraw-Hill Book Company, New York. 347 pp.

Mellersh, H. E. L. 1958. *The Story of Life.* G. P. Putnam's Sons, New York. 263 pp.

Merrell, D. J. 1962. *Evolution and Genetics: The Modern Theory of Evolution.* Holt, Rinehart and Winston, Inc., New York. 420 pp.

Moody, P. A. 1970. *Introduction to Evolution,* 3rd. ed. Harper & Row, Publishers, Inc., New York. 527 pp.

Ross, H. H. 1966. *Understanding Evolution.* Prentice-Hall, Inc., Englewood Cliffs, N.J. 175 pp.

Simpson, G. G. 1964. *This View of Life: The World of an Evolutionist.* Harcourt Brace Jovanovich, Inc., New York. 308 pp.

Simpson, G. G. 1968. *The Meaning of Evolution: A Study of the History of Life and of Its Significance for Men,* rev. ed. Yale University Press, New Haven, Conn. 368 pp.

Stebbins, G. L. 1966. *Processes of Organic Evolution.* Prentice-Hall, Inc., Englewood Cliffs, N.J. 191 pp.

Stebbins, G. L. 1969. *The Basis of Progressive Evolution.* University of North Carolina Press, Chapel Hill, N.C. 150 pp.

Volpe, E. P. 1967. *Understanding Evolution.* William C. Brown Company, Publishers, Dubuque, Iowa. 160 pp.

Chapter 2

It is interesting to contemplate a tangled bank, clothed with many plants of many kinds, with birds singing on the bushes, with various insects flitting about, and with worms crawling through the damp earth, and to reflect that these elaborately constructed forms, so different from each other, and dependent upon each other in so complex a manner, have all been produced by laws acting around us. These laws, taken in the largest sense, being Growth and Reproduction; Inheritance which is almost implied by reproduction; Variability from the indirect and direct action of the conditions of life, and from use and misuse: a Ratio of Increase so high as to lead to a Struggle for Life, and as a consequence to Natural Selection, entailing Divergence of Character and the Extinction of less-improved forms. Thus, from the war of nature, from famine and death, the most exalted object which we are capable of conceiving, namely, the production of the higher animals, directly follows. There is grandeur in this view of life, with its several powers, having been originally breathed by the Creator into a few forms or into one; and that, whilst this planet has gone cycling on according to the fixed law of gravity, from so simple a beginning endless forms most beautiful and most wonderful have been, and are being evolved.

The Origin of Species
Charles Darwin

From cell to ecosystem

A lack of an understanding of the complex interaction of plants and animals may be jeopardizing the very existence of life on earth. Such a threat involves man for at least two reasons: he often is the disrupter of the complex interactions; and he, like other animals, is a species in the "web of life." Hence, it may be man himself whose existence is in question. To understand better why man has environmental problems today, to see why these problems threaten all life, and to gain insight into solutions to the problems, the patterns of nature must be comprehended.

One view of nature's pattern, the intricate and balanced relationship between environment and life-forms, was described in Chapter 1. The synthetic theory of evolution was said to describe the way in which this relationship came about, the process operating through millenia which has given rise to the diverse and numerous types of plants and animals found on the earth today.

In this chapter the rules that nature sets for the development of life-forms will again be examined, this time mainly from a historical frame rather than through a description of the evolutionary mechanism. In brief, we shall look at the early environment of the earth before life appeared, some major events up to the present time in the development of life from simple to complex, and the integrated product of the evolution of life forms and environment, the *ecosystem*.

The early environment

In the beginning there was only environment. No life existed 4.5 to 5 billion years ago. Then the earth, one of many small masses of gas and dust and ice broken away from a larger whirling, rotating conglomerate from outer space, underwent profound changes that resulted in some of the characteristics of the planet today — solid rocks, spacious seas, a gaseous atmosphere — and life.

Scientists have evidence and hypotheses that help sketch the sequence of changes in the early earth. The small dust clouds destined to become young planets, including the earth, settled into orbits around the sun. They were intensely hot as they rotated and condensed. Heavy materials, such as iron and sulfur, which made up the masses, sank into the planetary cores; light elements, such as oxygen, nitrogen, and carbon, layered on the surfaces. On planets most distant from the sun, such as Jupiter and Pluto, the hydrogen, carbon, and nitrogen solidified because of the cold, and on planets nearest the sun, the light gases soon dissipated because of the heat. But on the earth the great heat combined the light gases into small invisible molecules and thus formed the original atmosphere of the earth.

These early compounds were of several kinds. Hydrogen atoms combined to form hydrogen gas (H_2); hydrogen and oxygen formed water (H_2O); the carbon and hydrogen atoms yielded methane (CH_4); the nitrogen and the hydrogen atoms produced ammonia (NH_3); and the carbon and oxygen formed carbon dioxide (CO_2). Today, some of these atmospheric molecules still are found, such as CO_2 and H_2O. Others, such as CH_4 and

NH_3, have been lost. Still others, such as oxygen gas (O_2), appeared after the advent of life.

For the next billion to 1.5 billion years, the earth cooled. As a consequence, the water molecules condensed and formed the oceans. These shallow seas probably covered all the earth's surface until landmasses, composed of the earth's first rocks, were exposed. These rocks surfaced in a turbulent time, and as a result most of them quickly eroded. Their pieces, carried away to settle at the bottom of the vast sea, were compressed after more millions of years into new rocks, known as *sedimentary rocks*. Other original rocks, which poked through the vast sea surface, had a different destiny. Apparently subjected to extreme temperatures on the early, changing earth, they melted and fused into another kind of rock, known as *metamorphic rock*. Metamorphic rocks aged 3.6 to 3.3 billion years have been found. Curiously, sedimentary rocks dated near that point, too — about 3.2 to 2.7 billion years ago — are also found. This means that the original rocks which gave rise to these metamorphic and sedimentary rocks solidified earlier, about 3.5+ billion years ago.

The profound changes that the early earth was undergoing were to produce other consequences. *Stromatolites*, fossils of simple plants known as *blue-green algae*, are found in some sedimentary rocks. By the time sedimentary rocks were being squeezed into existence in the ocean's depths, cellular life in the form of simple plants already existed on earth. This suggests that the very beginnings of life were still earlier, before the sedimentary rocks were formed, probably when the oldest and original rocks were solidifying, 3.6 to 3.3 billion years ago.

Life from the environment: Oparin's hypothesis

But in a turbulent environment originating as a hot, swirling mass of gas and only slowly cooling to the point of allowing rocks to form, how could life originate? Scientists have been intrigued by this question for a long time, and hypotheses and myths prevail as to how life-forms could have come about. Today, an explanation most respected by biologists is that of A. I. Oparin, a Russian scientist who in 1936 published a hypothesis for life's origin which led to a great deal of laboratory experimentation.

Oparin stated a logical hypothesis. He said that life must have come out of the materials that were present in the early earth environment. These included kinds of matter — such as water, ammonia, carbon dioxide, and methane — and specific forms of energy — such as heat, lightning, and ultraviolet light. From reactions between the matter and energy came new combinations of matter. Eventually, from the small compounds found in the oxygen-less atmosphere came large compounds, such as those found in living systems today. In time, a wide variety of the compounds became distributed in the great sea that covered the earth. Most likely, Oparin presumed, these organic compounds were formed in the water that covered the earth's surface. The result, a vast "hot dilute soup" (as English scientist J. B. S. Haldane labeled it), contained sugar, alcohol, acids, and even amino acids, the building substances of proteins found in all living forms today.

From this soup containing diverse molecules to cells to whole organisms such as the human being is a giant step in evolution not easy for any scientist to explain. However, Oparin proposed that the proteins of the soup tended to attract each other and to form complexes known as *coacervates*. This clumping of proteins leading to coacervate formation was due mainly to the attraction of negatively and positively charged portions of the protein molecules and was a first step on the evolutionary path leading to complex plants and animals.

The development of the coacervate, often in the form of a small droplet, may well have marked the developmental step just previous to the coming of life itself, according to Oparin. Probably unique chemical reactions involving large organic compounds eventually occurred in the coacervate droplet, which enabled it to become considerably independent of the outside environment. At that point scientists believe life arrived on earth.

Many awe-inspiring stories have been told about the creation of life — some far more beautiful than Oparin's. What makes his "myth" of creation any more plausible than any other creation story? Oparin's theory has one advantage. Like all good scientific hypotheses, it is stated in such a way that it can be put to experimental test. Oparin's story conceptualized eons of time completely outside the experience of man, but laboratory replication of the processes Oparin grappled with permits the scientist to directly examine the possibility of his argument.

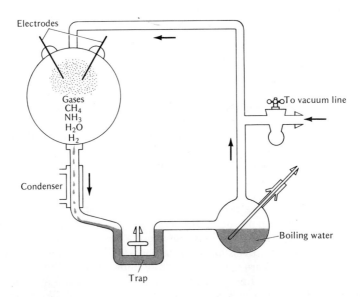

Figure 2-1
Apparatus used by Miller to simulate the primitive earth conditions. (Reprinted from *Journal of American Chemical Society*, vol. 77, p. 2351, 1955. Copyright © 1955 by American Chemical Society. Reprinted by permission of the copyright owner.)

Testing the hypothesis

Parts of Oparin's hypothesis have since been tested in the laboratory. In 1953 Stanley Miller, a young graduate student at the University of Chicago, simulated the early environmental conditions. From the small molecules of ammonia, water, carbon dioxide, and methane he produced several large molecules, including acetic acid (vinegar), lactic acid, urea, and amino acids (Fig. 2–1). He used an electric spark as the energy source, supposedly replicating the lightning in the early atmosphere. The experiment is important because it shows that the conditions on the young earth — even though no oxygen gas was present — could have produced several molecules basic even today in life systems. Amino acids, for example, form the basis of proteins. Since Miller's work, other researchers, using diverse energy sources, such as visible light, ultraviolet light, ionizing radiation, x rays, and even ultrasonic vibration, have synthesized from the simple original molecules five- and six-carbon sugars and other molecules essential in the metabolism of living systems.

Oparin's ideas received important support and elaboration when another researcher, Sydney Fox, tested Oparin's hypothesis. Fox showed that amino acids could be *polymerized,* attached to each other to form long chains. This discovery was significant because polymerized amino acids in certain sequences form proteins, essential structural and functional molecules in living systems. Fox heated the amino acids in a closed apparatus, constantly removing the water that formed as the amino acids joined

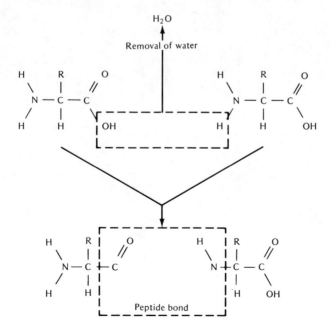

Figure 2–2
Formation of a peptide bond.

(Fig. 2–2). Fox's experiment suggests that the early environmental conditions on a solidifying earth, such as dry hot rock surfaces, could have induced polymerization.

Subsequent experiments showed that these polymerized molecules, which Fox called *proteinoids,* could be digested by protein enzymes, as well as used by bacteria as a food. This suggests that if proteinoids occurred on the early earth, they could have been eventually broken down by other molecules and the energy retrieved from this "food" to sustain early life processes.

Scientists may never know the exact role that polymerized amino acids played in the development of early life. Obviously, the orderliness they exhibit when joining may have been a major contribution. For example, in one experiment, a mixture of various amino acids (including one-third glutamic acid and one-third aspartic acid) was heated. The amino acids joined, but always in the following percentage composition: 13 percent glutamic acid, 55 percent aspartic acid, and 32 percent other amino acids. Apparently, the structure of amino acids influences to some degree the combinations they form. This experiment suggests that even on the early earth there existed some signs of orderliness in the structure of the early molecules.

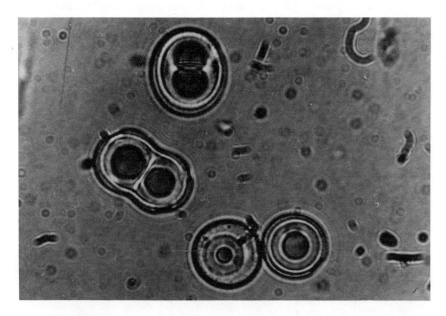

Figure 2–3
Early cells may have resembled the "microspheres" shown in this photograph. Microspheres have been aggregated from proteinoids in the laboratory in a manner consistent with Oparin's hypothesis for the origin of the cell. (Courtesy of S. W. Fox, Institute of Molecular and Cellular Evolution, University of Miami, Coral Gables, Fl.)

From environment to cell

Biologists have realized for nearly one and one-half centuries that almost all forms of life are composed of cells. Since the cell is such a basic structure of living things, it must have originated at a very early time on earth, before life differentiated. What kind of events in the early environment can be hypothesized as giving rise to the progenitor of the modern cell? Oparin hypothesized that large molecules, such as proteins, came together to form coacervates. From such aggregates came the cell, a biological creation so successful as to form the basic structural plan for all life-forms, including man (Fig. 2–3).

That coacervates may well have been the progenitor of the cell is suggested by their easy formation from proteins, or even carbohydrates, in simulated laboratory conditions probably present on the early earth, and their behavior after formation. For example, coacervates can "grow." Molecules in the coacervate most often carry a negative charge, and hence, positively charged molecules can be attracted to them and actually fuse with them. Frequently, the positively charged molecules will, in turn, attract other free negatively charged molecules floating outside the coacervate. Thus, the microdroplet will swell in a watery solution, becoming filled with numerous charged particles from outside the structure.

Another functional characteristic reminiscent of today's cells resides in the coacervate's outer boundary, or "membrane." The walls of aggregating molecules form a kind of primitive outer boundary similar to that of the modern cell. Like the outer membrane of today's cells, it helps to facilitate molecular activity within the coacervate by concentrating the reaction molecules. It also prevents uncontrolled passage of molecules in and out of the cell-like structure.

Early membranes and those of contemporary cells may have been structured alike, too. In addition to protein and/or carbohydrates, early membranes may have incorporated phosphorus-containing fat molecules, like membranes of many modern cells. These molecules are capable of a great number of structural variations which add stability to coacervates subjected to changing temperature or acidity. In the early environment this would have been significant, because coacervates made of just proteins and carbohydrates are comparatively easily disrupted.

Coacervates in transition to cells no doubt developed increasingly complicated functions. Life systems today are characterized not only by their component parts such as molecules or membranes, but by their elaborate interaction, *metabolism,* which maintains the life characteristics. Metabolism involves energy exchanges. Compounds come and go, yielding matter and energy useful to build other compounds *(catabolism)* or utilizing energy and materials to become larger or to replicate *(anabolism).* From the very first, anabolism and catabolism probably occurred together. By increasing in complexity these processes evolved into the metabolic systems seen in plants and animals today — chains of reactions driven by enzymes complex in structure and function.

How did metabolism begin? Experiments by Fox with coacervates made of proteinoids (which he called *microspheres*) give some suggestions. Fox and a fellow worker, Krampitz, discovered that certain amino acid polymers made in the laboratory when incorporated into microspheres had the ability to hasten the breakdown of *glucose,* a common six-carbon sugar found today in all living systems. Other sugars also could be broken down at a much more rapid rate. The synthesized amino acid polymers had the innate ability to function as *catalytic* molecules in living systems, that is, to hasten the interaction of other molecules and remain unchanged themselves. In the early history of life, many molecules probably had a catalytic effect on other compounds and hastened their reactions. These early catalysts were probably the first *enzymes.*

Most likely some of these early enzymes were more effective catalysts than others and, thus, persisted until today. For example, molecules with iron are thought to have been especially significant in the early days of life. Scientists believe this because iron was abundant on the early earth, and experiments in the laboratory show that iron, even when not attached to other elements, can cause many molecules to react when it is brought near them. Some of these reactions are found to be essential in the functioning of life systems. Another effective enzyme molecule in early life was one containing a particular combination of sulfur and hydrogen known as a *sulfhydryl group.* Probably in the past, many kinds of enzyme molecules with sulfhydryl groups proved more effective catalytically and, thus, persisted.

This explanation for the development of enzymes in the early life forms emphasizes once again the significant role of environment in effecting the evolution of the cell. The concept that only combinations and reactions of molecules which existed or functioned well in the given environment could persist to evolve into new forms and functions is consistent with the hypotheses of early workers such as Oparin, Fox, and Miller. It is also consistent with the theory of natural selection proposed by Darwin when describing the evolution of species of plants and animals. Thus, the role of environment is significant in the development of life at all levels. Indeed, whether studying molecules or man, one maxim always applies: life and environment are only different views of the same grand scene — nature spiraling into increasing complexity of forms and functions as the millenia go on.

To stress the role of environment in helping to shape early life is also to emphasize once again the orderliness of the first life and the comparative disorganization of the early milieu. In the early seas, probably, coacervates and unattached molecules floated side by side, the boundary between the organized and unorganized being faint and transient. Chance occurrence of appropriate quantities of matter and energy resulted in simple structures that eventually led to the awesome hierarchy of organic intricacy characterizing plants and animals today.

Two other kinds of molecules probably aided in the first steps of that journey, the transformation of coacervates into cells. One was an energy-storing molecule; the other, a coding molecule from which certain molecules could be synthesized and, thus, propagate the individuality of the cell.

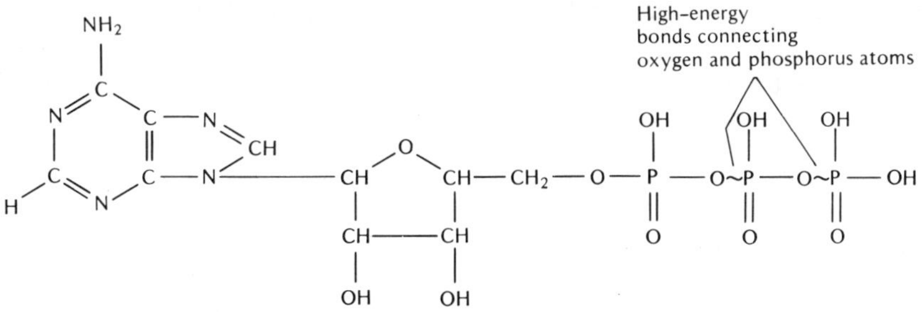

Figure 2–4
Molecular structure of adenosine triphosphate (ATP). The structural subunits found in ATP are found in many other molecules that are essential to the metabolism and structure of living things.

The early energy-storing molecule was probably very much like the molecule which performs this function in many of today's cells — *adenosine triphosphate* (ATP; Fig. 2–4). ATP is composed of molecular fragments which no doubt were present on the early earth. It functions by absorbing energy into its bonds, notably in the two bonds between the three fragments containing phosphorus. This energy can be released as well as absorbed easily without cellular disruption in the presence of the proper enzymes. Such a molecule protects the integrity of the cell from large and uncontrolled bursts of energy. It also functions as an energy storage site and in this way provides a ready supply of energy for the activities of the cell.

At what step did coacervate aggregations develop enzymes and use ATP-like compounds to store the energy of enzymatically broken molecules, which had diffused into the protein aggregate, and use the energy to help maintain a self-perpetuating orderliness? No one knows the answer to this question, but when life began, enzymes and ATP-like molecules most likely were present in the environment. Control and availability of abundant energy was the promising consequence.

The control and availability of abundant energy was not all that was needed to maintain the first life. In addition to the origin of energy-storing molecules, early cells must have developed an ability to propagate the various kinds of molecules which gave them their identity generation after generation. Some coding molecules, such as the gene which we find in contemporary cells, must have come about. A gene is composed of polymers of simple molecules probably present on the early earth. The polymers have the ability to code other molecules needed by the cell, such as proteinaceous enzymes. Hence, the cell must have developed the ability to perpetuate its characteristics through the coming of the gene. In today's cells the genes are highly organized, connected to proteins to form the larger *chromosomes,* the structures enclosed in the cell's *nucleus* (Fig. 2–5).

Thus, out of the matter and energy present in the early environment came life. As we have seen, steps in its development included (1) the synthesis of small molecules out of the original atoms; (2) the synthesis of larger molecules, such as proteins, from the small molecules; (3) the coalescence of the large molecules into membranous structures, reminiscent of today's cells, called coacervates; and (4) the incorporation of enzymes, energy-storing molecules, and propagating molecules (genes) into the membranous coacervates to yield functions similar to those found in contemporary cells. Presumably, from these steps came cells which, once formed, further developed and elaborated into still new levels of organization and function, eventually leading to today's plants and animals.

From environment to cell to organism

The story of the development of cellular life out of the environment has been laboriously pieced together by biologists. Even today some of it

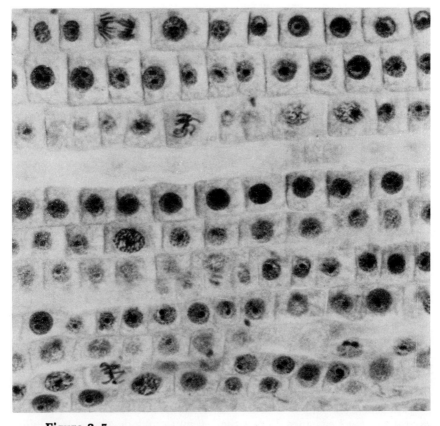

Figure 2–5
Cells in an onion root tip exemplify the cellular nature of most all living organisms, plant and animal. Note the nuclei and cell walls. Some cells are undergoing *mitosis,* cell division. (Turtox, Chicago.)

has not been verified, because so little evidence exists. The story of the evolution of the cell into more complicated structures is even more difficult to construct. Mainly this is because once life appeared on earth, it continued to evolve into many new levels and kinds of organization. Life became self-diversifying. Furthermore, as time went on, the diversification of life accelerated until in most recent times we find an extremely vast and heterogeneous collection of living things inhabiting the earth which obscures the details of life's early transformation.

It is important to remember that the transformation of environment into life was an evolutionary process. This means that any steps leading to life were affected by the environmental conditions in which the events were occurring. Such a close relationship between surroundings and the evolving life-forms can be well-illustrated by examining how the plant and animal kingdoms are thought to have developed, an explanation revolving around a major change in the ancient milieu — the accumulation of oxygen gas in an environment previously *anaerobic,* without oxygen.

Plants and animals presumably evolved far after the development of first primitive cells, possibly as a result of what may have been the first global food shortage. The early cells must have fed by absorbing materials from the watery environment that surrounded them. The abundant food was constantly formed in the seas out of the floating small molecules bathed in sunlight and heat. These early cells, and those today which feed on materials produced in the environment, are called *heterotrophs.* The early heterotrophs fed well at first, but as their number increased and the earth cooled down, the supply of food molecules waned. Early heterotrophs were faced with starvation. They survived that crisis; but how?

Today it is believed that the early heterotrophs did not die, because some of them developed the unique ability to make their own food *inside their cells.* No longer would they depend on the absorption of food molecules formed by chance in their environment. This new synthesis, as any anabolic process, needed energy to occur. Biologists believe that these early heterotrophs used the energy from sunlight, especially the red and blue portions. The sunlight was absorbed in a special kind of molecule, *chlorophyll.* Specifically, biologists think that the cells used sunlight to combine the plentiful quantities of small carbon dixiode (CO_2) molecules available in the primitive atmosphere. The carbon dioxide (CO_2) molecules formed larger carbohydrate molecules bound with converted light energy and emitted oxygen gas as waste. This was the birth of *photosynthesis,* a process in green plants which most living things depend upon today.

The coming of photosynthesis had many consequences, including the initiation of the two main groups of living things. The birth of photosynthesis originated a new breed of organism, one that used sunlight to make its own food within itself. This new kind of cell, called an *autotroph,* marked the beginning of the plant kingdom; the heterotrophs, which retained the ability to break down food and never developed a mechanism to photosynthesize, became the progenitors of the animal kindgdom. Thus, in the early environment, photosynthesis saved early life from starvation and at the same time divided all life from then on into two great groups — the plant and animal kingdoms.

The coming of photosynthesis at the same time permanently changed all life and environment (and thus, once again illustrates the inseparableness of these two ingredients of the natural world). Until photosynthesis occurred, the earth's atmosphere was anaerobic. But as photosynthetic organisms spread over the surface of the moist earth, the abundant waste product oxygen gas (O_2) began to mix into the atmosphere and dissolve in the water. The presence of O_2 in the primitive earth's environment was to have profound changes in the new plant and animal kingdoms.

All the consequences are probably not known, although scientists at least suspect that these events resulted. Oxygen gas probably was toxic to many of the early life-forms which originated and evolved in an anaerobic environment. Its presence killed many early kinds of life and influenced the selection for survival of the O_2-tolerant types. In addition, oxygen probably helped screen out the harsh ultraviolet light streaming from the sun and, thus, prevented this radiation from disrupting the developing metabolic systems and structures in early life-forms. Most likely this was accomplished by the accumulation of *ozone* (O_3) far above the earth, a layer of gas that is still present.

Another major effect of the presence of O_2 was on the process of *cellular respiration,* that carefully controlled catabolic process of all life in which food molecules are disintegrated and the bonding energy freed, some to be lost as heat but much to be stored in the ATP molecules. In the anaerobic environment this process did not involve O_2, of course. However, when O_2 became plentiful, some cells incorporated it into their respiration. The consequence was immense: respiration carried on with O_2 *(aerobic respiration)* released more energy from every food molecule than when oxygen was not involved *(anaerobic respiration).* Thus, more molecules of ATP became available as energy resources than ever before. Such abundant, easily derived, and available energy was a new resource for the developing life-forms. Once again: a change in life produces a change in the environment; a change in the environment effects a change in life.

Although some life-forms did not incorporate this new respiratory method into their metabolism (contemporary yeasts and some bacteria may be derived from this line), others did. The advantages were great, and life accelerated its diversification and migration into new environments.

Other developments, which illustrate well the interaction of life and environment in early times, helped to save life from extinction. For example, early cells developed the ability to divide and, hence, to propagate asexually through the long periods of time. Cells still do this, of course; the process is known as *mitosis*. Presumably, from this ability came *multicellularity*. One explanation for the development of multicellularity is that, for whatever reasons, single cells came together and remained attached. This situation is seen in a contemporary unicellular green alga, *Chlamydomonas*. In times of stress, individual cells aggregate and lose their flagella, which are whiplike cell extensions used in locomotion. When favorable times return, the cells form new flagella and swim away. Another hypothesis proposes that new cells clung together after mitosis instead of separating. However they originated, the advantages of multicellularity probably include greater size, cellular specialization, and the ability of certain cells to perform certain metabolic roles. Such an arrangement could be more efficient for the whole organism than one cell doing all life functions. It also would allow a greater ability to exploit environmental conditions and, thereby, to enhance survival. Out of this development could have come the origin of tissues, organs, and systems, hierarchies of organization found within all the most recent and advanced groups of plants and animals.

Mitosis when it first occurred in early animals and plants probably had at least two consequences: (1) it led to the multicellular condition, with its many survival advantages, and (2) it provided a source of reproduction for the early organisms which aided their distribution. Whether single cells or aggregations of cells, these structures, broken away from parent organisms and cast out into the environment to be distributed by wind or water, helped distribute life around the surface of the globe. Of course, mitosis allowed for no change in the distributed structures, unless, of course, some genes happened to mutate. But rapid change could only take place if greater variation occurred in the offspring of the plants and animals. Eventually, a mechanism evolved that produced variation in offspring, and as a result the origin of new kinds of plants and animals greatly accelerated. The new source of variation was *sexual reproduction*.

A source of change: sex

The coming of sexual reproduction provided an essential ingredient for further adaptation of plants and animals to their environments. That ingredient was genetic change in the organisms. Sexual reproduction is a phenomenon that originated around the genes, those molecules which in even the earliest organisms give cells their identity because of their ability to produce the enzymes and other molecules. The mutation of genes can change the nature of molecules produced and, hence, the individual organism. Sexual reproduction augments this source of variation by providing for *new combinations of genes* in organisms in a population. Specifically, sexual reproduction allows for new combinations by reassorting the genes as they are distributed into new cells, the eggs and sperm, and by recombining the genes in the fertilization of eggs with sperms, that is, in the formation of the *zygote*. These two opportunities for gene reassortment in cells, together with mutation, allow for plentiful variation in a population of any kind of plant or animal. And as an understanding of the synthetic theory makes clear, variation in offspring is an essential ingredient in adaptation of plants and animals to their environments. As environments change through the millenia those organisms in variable populations best able to survive propagated themselves; the least able to survive died. Out of the variation in populations of plants and animals, produced with the aid of sexual reproduction, came the raw material needed for evolutionary processes to work on the early earth.

The significance to evolution of sexual reproduction suggests that this process must have evolved at a very early time in the history of life. How did sex come about? One hypothesis states that two unicellular cells fused to mix their contents, including the genes. Advantages of this state were immediate: two sets of genetic material were better than one, especially if the genes in one of the cells did not provide the organism with characteristics that would allow it to manage well in its environment, or if the environment was rapidly changing. Thus, the mixing of genes gave a better potential for survival than the single sets of genes found in either single cell.

Today, some organisms may or may not reproduce sexually. Those which exist in relatively unchanging environments, such as sponges bathed in the depths of the ocean, an environment that remains relatively constant, often produce asexually. Other, more complex organisms may rely very little on sexual reproduction, a condition that appears primarily to be due to some innate disadvantage, such as the inability or difficulty of sex cells or embryos to mature during very short growing seasons. This is seen, for example, in plants growing during the short summers of the tundra.

Nevertheless, these few exceptions do not negate what is true generally: sexual reproduction is a major source of genetic variation, which leads to diversity and the evolution of populations of organisms. Such variation, the basic condition in life upon which the synthetic theory operates through the millions of years life has been on earth, has led to the elaborate situation that we see in nature today — diverse plant and animal forms interrelated with each other and the physical environment.

Onto land and the development of the kingdoms

Out of the diversity accumulating in the plant and animal kingdoms came organisms structurally and functionally better adapted to live in

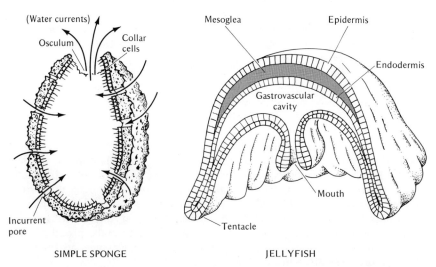

Figure 2–6
A simple sponge and a jellyfish. Note the two layers of cells that essentially make up the body of each.

environments less watery than the seas. These early transition organisms were the first to move into the land environment. At first, they probably stayed near their ancient birthplace, the water. But as the earth continued to dry and large land masses were exposed, successive generations were more able to survive farther and farther from the shores and beaches.

Many of the structural and functional developments that we now see in the most complicated land plants appear to be devices to survive an environment with little moisture. For example, highly differentiated cells that carry food and water throughout the plant body (vascular tissues); stout support cells that help hold the tall plants upright, thus exposing leaves to optimal amounts of sunlight; desiccation-proof reproductive structures, such as thick-walled ovaries which cradle delicate eggs and almost weightless pollen which bear the male sperm, appear to have been evolutionary products of organisms no longer existing in the protective sea but taking their chances on dryer and more hazardous land.

Not far behind the plants moving up on land came the animals. Being heterotrophs, their migration was forced to wait upon the presence of ample food such as that provided by the photosynthesizing plants. Of course, not all plants or animals left the sea. Today, we see descendents of those water-dependent forms, relatively simple-structured creatures which stayed behind — the algae, many protozoans, sponges, the jelly fish, and sea anemones, to mention only a few. But for those that did migrate, major changes in structure and function accompanied the move. For example, whereas many of the sponges and jellyfish were to develop a body plan really no more complex than a double-layered sac of cells, land creatures were soon to develop *bilateral symmetry,* in which their two body sides would be essentially alike, with a distinguishable head as well as a tail end (Fig. 2–6). And for most, two body layers of cells would prove less advantageous than three, and so most animal forms evolved a third body layer, called the *mesoderm,* from which eventually was to develop numerous organs and accessory structures essential to the metabolism of the animal (Fig. 2–7). Many animals also evolved a mouth at the head end for eating and an anus at the rear end for ejecting waste, thus facilitating the operation of bodily functions. Today, these structural adaptations are found in most all recently evolved organisms, such as fish, birds, and mammals.

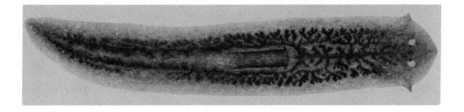

Figure 2–7
A flatworm, an example of an organism derived from three body layers, the middle one (mesoderm) giving rise to systems such as the muscular and excretory systems. (Turtox, Chicago.)

Evolution in retrospect

The details of the evolution of plants and animals is a story too full to tell here. What is important to us, however, are the evolutionary trends seen in the early history of life as it came out of the environment of the young earth. Today, surrounded by the natural world — the product of evolution through the ages — we can look back and see in general what was happening to life as it ascended from the early earth to now. Certainly, one obvious trend was the movement toward complexity. The first life was simple and came out of a lifeless environment, but as time progressed it became increasingly complex in structure and function. Another trend was the diversification of life. Whereas early life was present or manifested in only a few forms, today it occurs in a vast and wide array of

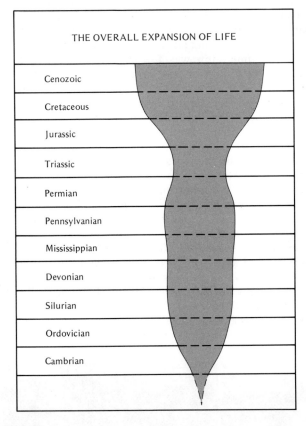

Figure 2-8
Life becomes more diverse the longer it is on earth. The width of the pathway is approximately in proportion to the known diversity of plants and animals at various times in the past. (See Table 1-1 for times in millions of years and predominate forms of organisms.) (From *Life: An Introduction to Biology*, by George Gaylord Simpson and William S. Beck, copyright © 1957, 1965 by Harcourt Brace Jovanovich, Inc., and reproduced with their permission.)

forms (Fig. 2–8). Still another trend was the acceleration of the production of increasingly complex forms as time passed. Finally, as reference to Fig. 1–9 proves, all changes in plant and animal groups since life appeared on earth were related to significant changes in the earth's environment.

Ecosystems: product of the evolution of organisms and environment

The biologist who studies the history of life often is most concerned with the *evolutionary process* that has produced the types of plants and animals in the past or today, or with the *types* of living or fossilized plants and animals themselves. But whether concerned with process or product, his information will always be incomplete, because (1) the biologist was not there to witness the drama of change in previous millenia, and (2) few remains of ancient life-forms have survived the battle of time.

Another characteristic of past life that has not endured for today's biologist is the interrelationships of the numerous kinds of ancient organisms. An ecologist finds this especially vexing, for to him the products of the evolution of life have not just been new and diverse cells or bodies or brains, but the interweaving of the lives of diverse creatures to form a complicated whole always greater than its parts — a natural world, now or ages ago.

What is this "larger" dimension which the ecologist perceives in the natural world? An answer comes in knowing of its parts (hence, the time spent on the synthetic theory of evolution and the spiraling development of life through time), but there are aspects of this overview which the ecologist studies for their own sake. The ecologist focuses on the interrelationships of organisms in their environments, the functional organization and interdependence of the parts. This aspect of the natural world best fits the concept of *system,* as a dictionary definition shows: "an assemblage of objects united by some form of regular interaction or interdependence; an organic organized whole; as, the solar *system;* a new telegraph *system.*" Thus, the ecologist studies the natural world as a system. He calls the organized interaction of plants and animals in nature an *ecosystem.*

Ecosystems have many characteristics, and since the study of them has only recently begun compared to the study of other areas of biology, we have much still to learn. However, one characteristic of ecosystems that has been appreciated for some time is the dynamic balance of its parts, *homeostasis.* This means the functional intermeshing of all parts of the system, both living and nonliving, which results in a life-sustaining relationship for plants and animals with the environment. No Swiss watch, no sophisticated space technology displays more elegantly the intricate balancing of components as well as the ecosystem of a marsh or meadow.

Given the complexity, subtlety, and sophistication of the fine tuning of an ecosystem in homeostasis, we might well wonder what mechanism is required to control the balance of its many elements. Important to the concept of ecosystem homeostasis is self-regulation. Not only do parts of

Figure 2–9
Evidence of the major change in an ecosystem after the introduction of a "foreign" organism — in this case the water hyacinth into a Florida waterway. Top: As St. John's River near Tocoi, Florida, appeared before the introduction of the water hyacinth and after it had been eradicated. Bottom: The river choked with water hyacinths. (Courtesy of C. Zeiger, U.S. Army Corps of Engineers, Jacksonville, Fla.)

a functioning system mesh well, but they regulate each other so that proper functioning is assured. The significance of such self-regulation can be seen in a simple man-made self-regulation system, the temperature-control system in a home. Many home temperature-control systems have at least two elements, a heater and a thermostat. The thermostat turns the heating unit off or on as a response to room temperature. If the heater is on, the room warms up, but eventually the high temperature will trigger the thermostat to turn the heating unit off. When the room cools, the thermostat will turn the heater on again. In this way the room will be more accurately maintained at a constant temperature than if some influence external to the system is episodically introduced.

Biology abounds with examples of self-regulatory thermostatic systems. Such a system can be seen in the interaction in the human body of the female hormones that control ovulation (for details, see p. 222). Another example is the careful control by predators of some animal populations, such as the rise and fall of the deer population on the Kaibab Plateau when mountain lions and wolves were removed (see p. 190). Still another example is the control of plankton in a lake or pond. The plankton in Lake Michigan increase greatly in the early fall and spring when the change in air temperature causes a turnover of the water. Biologists now know that the distribution of oxygen and nutrients accounts for the population increases. But as the air temperature levels off and the water ceases to mix, the blooms fade, because the nutrients are no longer distributed. Thus, homeostatic control in ecosystems, similar to the controls in other biological systems, helps to explain the impressive balance and stability of undisturbed natural areas (Fig. 2–9).

The nature of an ecosystem

An *ecosystem* can be defined as the interaction of the *biotic* (living) and the *abiotic* (nonliving) elements in a particular geographical area. An ecosystem can be a pond or a forest, but it can just as easily be a continent, or the planet, for it is any setting where the abiotic and biotic elements interact.

R. L. Smith in *The Ecology of Man: An Ecosystem Approach* (Harper & Row Publishers, Inc., New York, 1972) defines an ecosystem this way:

> It consists of a physical environment — in the case of the pond, water and the bottom mud and drainage system; in the case of the forest, the atmosphere and the climate, the soil, and hydrological influences. The environment is inhabited by a number of different plants and animals, each of which in turn modifies the climatic, the hydrological, and the nutritional aspects of the environment. Each group of species is made up of individuals, collectively a population, that are held together by some form of social and biological interaction. Populations and individuals within populations do not exist alone but form some kind of an association, not haphazard but orderly and well organized, utilizing and transferring energy and materials. These interacting plants, animals, and environment make up the ecosystem.

What are the elements of an ecosystem? The *abiotic* elements are the water, soil, rocks, and weather conditions. The *biotic* elements are two: the autotrophic *producers*, or food-producing organisms (green plants), and the heterotrophic *consumers*, or feeding organisms. The heterotrophic

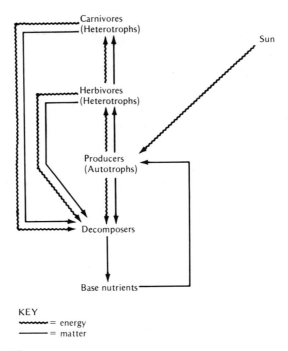

Figure 2–10
Generalized nutritional relationships of organisms in an ecosystem.

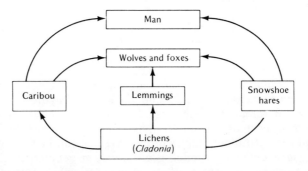

Figure 2–11
Simplified food chain for an Arctic terrestrial ecosystem. (From *Ecology and the Quality of Our Environment*, by C. H. Southwick, © 1971 by Litton Educational Publishing, Inc. Reprinted by permission of Van Nostrand Reinhold Company.)

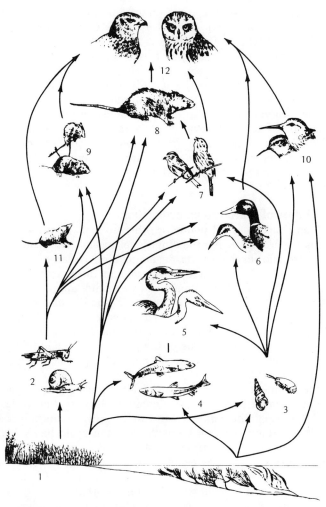

Figure 2–12
Food web in midwinter in a San Francisco Bay area salt marsh. (From *Ecology and Field Biology*, by Robert L. Smith. Copyright © 1966 by Robert Leo Smith. Reprinted by permission of Harper & Row, Publishers, Inc.)

consumers can be divided into the *macroconsumers* ("animals" in the broadest sense) and the *microconsumers (saprotrophs)*, organisms such as fungi and bacteria which help decompose organic material.

All these organisms are related functionally. The autotrophs produce the food, the heterotrophic macroconsumers directly or indirectly are sustained by the autotrophs, and the microconsumers, or decomposers, feed off both (Fig. 2–10). Thus, green plants are ecologically "one up" on all animals, because animals — including man — depend on plants for food. The ecologists refer to these linked relationships of organisms in an ecosystem as a *food chain*. A good example of a food chain is found in the arctic terrestrial ecosystem (Fig. 2–11).

Often the interrelationships of autotrophs and heterotrophs become more complicated than as seen in the simple food chain. This is because in some areas the heterotrophs usually eat many kinds of autotrophs; the consumers, or herbivores, are devoured by several second-level consumers, or carnivores, which in turn may be eaten by numerous *third-level consumers* or even *fourth-level consumers;* and some consumers, known as multilevel consumers, feed from several levels. These intricate food relationships, known as a *food web,* can be seen more clearly in the diagram of the salt marsh ecosystem (Fig. 2–12).

The flow of matter in an ecosystem

When consumers eat plants or other animals in any ecosystem, matter moves from one trophic (feeding) level to another. When an organism from any trophic level dies, its decomposition releases the atoms which make it up and sets them free once again in the environment. Thus, the movement of matter can be thought of as cyclic, although some elements may spend long periods of time stored in the bodies of organisms or in inorganic compounds in the abiotic environment.

This cycling of elements through an ecosystem can be illustrated with many examples, but carbon, one of the prime elements in living organisms, illustrates it well. Figure 2–13 shows carbon's cycled trip from organisms to atmosphere and back again.

Other elements exhibit similar cycles. For example, nitrogen is incorporated by bacteria into significant compounds, essential to the growth of plants. After remaining combined for awhile, nitrogen is released as nitrogen gas into the atmosphere. From there it can again be cycled into compounds essential to organisms (Fig. 2–14).

Phosphorus is another example of a recycling element harbored by organisms for long periods of time. It forms the basis of bone and teeth in animals, is abundant in solid waste, and is incorporated into the significant energy-carrying molecule adenosine triphosphate. Phosphorus also combines with elements in rocks. Its cycling in nature is illustrated in Fig. 2–15.

The flow of matter through the ecosystem from organism to environment can be illustrated easily with many kinds of materials. Each element makes a unique and characteristic trip, but they all illustrate one of the basic truths of ecosystems: the biotic and abiotic ingredients are intimately related even to the point of exchanging the matter that makes them up.

The flow of energy in an ecosystem

Studies show that the flow of energy and matter in any ecosystem differs significantly: matter is recycled but energy is not, because much is constantly lost as energy usable to living things. The energy is dissipated

53 Ecosystems: product of the evolution of organisms and environment

as heat when the bonds of molecules formed by autotrophs with the sun's energy are broken. The amount of energy lost to living systems from broken molecules is great. Studies show that when autotrophs are ingested and the molecules that compose them are broken down by the heterotrophs, only about 10 percent of all energy ingested by the heterotroph is retained; 90 percent of the energy remains unavailable, mostly because it is lost as heat, although some remains locked in the molecules excreted as waste. If the heterotrophs that devour the producing autotrophs are in turn eaten by other heterotrophs, as is usually the case in the wild, then once again most of the energy in the ingested molecules is lost, about 85 to 90 percent of the energy provided by the first heterotrophs. Only from 10 to 15 percent is utilized. Thus, constant input of new energy via photosynthesis is essential or the ecosystem cannot continue.

As a consequence of this large energy loss, every time molecules are passed from autotrophs to the first heterotrophs and then up the feeding levels of the food web, organisms in a given area in the high trophic levels are always *less plentiful* than in low trophic levels and contain a small mass of living material, or *biomass*. The organisms in the high trophic

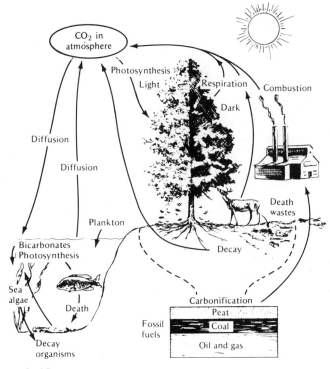

Figure 2–13
Cycling of carbon in the ecosystem. (From *Ecology and Field Biology*, by Robert L. Smith. Copyright © 1966 by Robert Leo Smith. Reprinted by permission of Harper & Row, Publishers, Inc.)

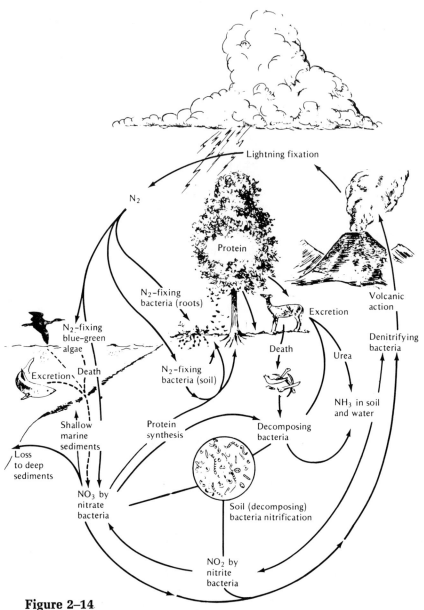

Figure 2–14
Cycling of nitrogen in the ecosystem. (From *Ecology and Field Biology*, by Robert L. Smith. Copyright © 1966 by Robert Leo Smith. Reprinted by permission of Harper & Row, Publishers, Inc.)

levels also contain very little usable energy compared to those composing the low levels. From these generalizations an ecologist can construct "pyramids" which represent the relationship of the trophic levels in terms of usable energy, numbers of organisms, and biomass. The pyramids aid in comprehending the complex relationships of plants and animals in the ecosystem, especially their interdependence (Fig. 2–16).

Several basic assumptions underlie the construction of pyramids:

1. Plants and animals relate in ecosystems in a way that can be represented with trophic levels.
2. The trophic levels that usually exist in nature are the autotrophs, or producers, and various levels of heterotrophs, including the decomposers.
3. Some energy is lost as heat when energy and matter change states.

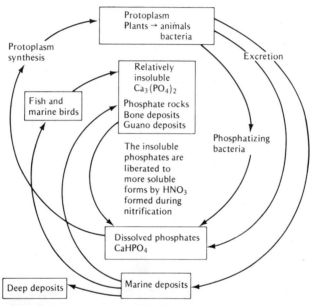

Figure 2–15
Cycling of phosphorus in the ecosystem. (From *Ecology and the Quality of Our Environment*, by C. H. Southwick, © 1971 by Litton Educational Publishing, Inc. Reprinted by permission of Van Nostrand Reinhold Company.)

Ecosystems and man

Stability, one of the principal characteristics of ecosystems, results from the *complex interaction* of diverse plants and animals. If one of the many diverse species dies out for some reason, the relationships of the numerous other species are not affected greatly, because many interdependent species still exist with which to interact. But what happens when a force foreign to the system intervenes to reduce the great diversity and, thus, to reorder the elements in a relationship that is not homeostatic? We are asking here: Has not man influenced ecosystems profoundly? Has he not developed agricultural lands lacking diversity and stability

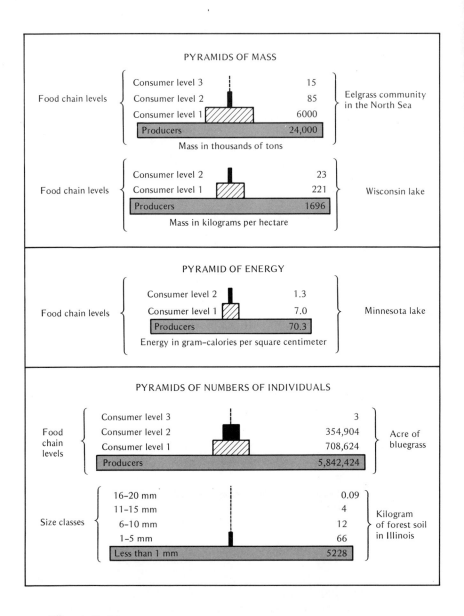

Figure 2-16
The pyramids of mass, energy, and numbers of organisms in a community illustrate the relationships of consumers and producers. (Decomposers are omitted in these pyramids.) The producers are plants; the consumers are animals. Animals in consumer level 3 feed on animals in consumer level 2, which in turn feed on animals in consumer level 1. Animals in consumer level 1 are herbivores and feed on the plants (producers). (From *Life: An Introduction to Biology*, by George Gaylord Simpson and William S. Beck, copyright © 1957, 1965 by Harcourt Brace Jovanovich, Inc., and reproduced with their permission.)

and, thus, destroyed or altered stable ecosystems? (See Fig. 2–17.) Of course, ever since the agricultural revolution man has been attacking natural ecosystems. Testimony of his activity to date are the billions of acres of farmland that support single crops, such as rice, corn, or oats.

But what price does man pay for eradicating diversity in the creation of such single-purpose farms? The lack of self-regulation that results demands constant monitoring, adjusting, and manipulation to induce continued production of these artificial areas. Man must apply abundant fertilizers, because no natural process of fertilization is allowed to occur; he must cultivate extensively, because the crops cannot grow on uncultivated ground; he must spray herbicides and pesticides, because pests constantly threaten to destroy the crops. (The magnitude of such intervention can be appreciated better when it is realized that, for example, approximately 2.5 pounds of active ingredients of fungicides are needed every year to control diseases on each of approximately 1.5 million acres of potatoes in the United States.)

Man has created unstable ecosystems to feed himself more reliably, but occasionally his efforts backfire. Extensive plantings of one crop encourage great losses from disease. Panic occurred in 1970 when a new species of fungus attacked corn. In that year the United States had 46 million acres (about 80 percent of the total corn crop of the country) planted with a variety that was susceptible to this parasite. The threat of a national corn disaster was great. Since then, corn with different genetic makeups is being grown, but the threat still persists. Similar situ-

Figure 2–17
Aerial photograph which shows the extent of a citrus grove in Florida, a man-made monoculture that lacks diversity and stability. (Courtesy of State of Florida, Department of Citrus.)

ations in the past with other crops, notably oats and wheat, illustrate the kind of threat to man's existence that can result when man maintains unregulated, and consequently unstable, ecosystems.

Knowledge of the factors operating in an ecosystem should give man more insight into their management. The creation of a crop with a common gene base on which man depends may be totally unwise in the future, especially in view of the feeding problems already facing the world. Greater genetic variability would mean less susceptibility of the crop to a single pathogen. The stability inherent in diversity possibly should be a consideration in harvesting our forests. Such a harvesting policy would speak against the common logging practice of cutting all timber in an area, and would provide for a practice of taking only mature trees, leaving the young. And knowledge of the significance of maintaining the optimal size of a population in an ecosystem should make man worry greatly about the effect of his own population increases on the global ecosystem.

In the past, man has disrupted natural ecosystems with little anticipation of consequences beyond the satisfying of his immediate desires. Disaster has often resulted. Any future careless manipulation of ecosystems may bring man to a point of no return. Although his history on earth in general has been one of altering the natural ecosystems in the direction of allowing him to live better (as the next few chapters help to show), ecosystem alteration today appears at least to be an erosion of the quality of man's life. If so, modern man's responsibility must be to manipulate ecosystems knowledgeably and responsibly. To this point, R. L. Smith, an ecologist, says in *The Ecology of Man: An Ecosystem Approach:*

> *Few of us ever consider ourselves part of a vast ecosystem. Yet each day we are utilizing both primary and secondary production and step directly into a food chain. Our wastes, our garbage, the addition of pesticides and chemicals to air and water and soil put us directly, for ill or for good, into the mineral cycle. Our pollution problems, our food problems, our population problems are all ecological ones. If we are to survive, we must gain far better knowledge of how ecosystems function and apply those principles to the systems, cultivated or natural, that we exploit and upon which we depend.*

Summary

In general, the intent in this chapter was to show the intimate relationship that exists between the environment and the various kinds of life on earth. This concept was emphasized by pointing out that originally only environment existed and that out of the early environment life arose. As time went on, life was threatened with extinction many times, and probably many forms did die out. Yet many kinds of life did persist, because they evolved new structural and behavioral adaptations to the often-changing environmental conditions. Eventually from this arose a great abundance of diverse life-forms, plant and animal, in a variety of environments. The collection of diverse organisms in a particular environment is known as an ecosystem. An ecosystem is characterized by a complex interaction of its parts. Sadly, this complexity has not always been understood or appreciated by man. Hence, ecosystems have been unknowingly, and sometimes designedly, tampered with to the extent that today the very existence of many organisms and environments is threatened. Since man himself is an animal, and as such is an ingredient in ecosystems, even his existence may be in jeopardy. To assure ourselves of a future existence with quality, we need to understand the rules by which nature plays.

Supplementary readings

Calvin, M. 1969. *Chemical Evolution: Molecular Evolution towards the Origin of Living Systems on the Earth and Elsewhere.* Oxford University Press, New York. 278 pp.

Colbert, E. H. 1969. *Evolution of the Vertebrates: A History of the Backbone Animals through Time,* 2nd ed. John Wiley & Sons, Inc., New York. 535 pp.

Delevoryas, T. 1966. *Plant Diversification.* Holt, Rinehart and Winston, Inc., New York. 145 pp.

Emlen, J. M. 1973. *Ecology: An Evolutionary Approach.* Addison-Wesley Publishing Company, Inc., Reading, Mass. 493 pp.

Goin, C. J., and O. B. Goin. 1974. *Journey onto Land.* Macmillan Publishing Co., Inc., New York.

Grobstein, C. 1974. *The Strategy of Life,* 2nd ed. W. H. Freeman and Company, San Francisco. 174 pp.

Murdoch, W. W. 1971. Ecological systems. Pages 1–28 in W. W. Murdoch, ed., *Environment: Resources, Pollution, and Society.* Sinauer Associates, Inc., Stamford, Conn. 440 pp.

Odum, E. P. 1969. "The Strategy of Ecosystem Development." *Science,* vol. 164, pp. 262–270, reprinted in R. L. Smith, *The Ecology of Man: An Ecosystem Approach.* Harper & Row, Publishers, Inc., New Work, 1972.

Odum, E. P. 1971. *Fundamentals of Ecology,* 3rd ed. W. B. Saunders Company, Philadelphia. 574 pp.

Oparin, A. I. 1953. *The Origin of Life,* 2nd ed. Dover Publications, Inc., New York. 270 pp.

Riley, H. P. 1970. *Evolutionary Ecology.* Dickenson Publishing Co., Inc., Belmont, Calif. 113 pp.

Romer, A. S. 1972. *The Procession of Life.* Doubleday & Company, Inc., Garden City, N.Y. 384 pp.

Scientific American. 1970. *"The Biosphere."* W. H. Freeman and Company, San Francisco. 134 pp.

Scientific American. 1971. *"Man and the Ecosphere."* W. H. Freeman and Company, San Francisco. 307 pp.

Smith, H. W. 1961. *From Fish to Philosopher.* Doubleday & Company, Inc., Garden City, N.Y. 293 pp.

Smith, R. L. 1974. *Ecology and Field Biology,* 2nd ed. Harper & Row, Publishers, Inc., New York. 850 pp.

Whittaker, R. H. 1970. *Communities and Ecosystems.* Macmillan Publishing Co., Inc., New York. 162 pp.

Chapter 3

And God said, Let us make man in our image, after our likeness: and let them have dominion over the fish of the sea, and over the fowl of the air, and over the cattle, and over the earth, and over every creeping thing that creepeth upon the earth.

So God created man in his own image, in the image of God created he him; male and female created he them.

And God blessed them, and God said unto them, Be fruitful, and multiply, and replenish the earth, and subdue it; and have dominion over the fish of the sea, and over the fowl of the air, and over every living thing that moveth upon the earth.

Genesis 1:26–28

The arrival of man

Through the millenia following the creation of living substance billions of years ago, life differentiated into myriad plant and animal types, a result of the forces of evolution working incessantly through the ages in the interplay of organisms and environment. By 70 million years ago, on a limited part of the earth, presumably today's Africa, certain animal forms and environmental conditions converged to produce the ancestors of the creature that now is spread over the earth — man. The developments that led to his dominion are the subject of this chapter.

We shall look briefly at the details of the progenitors of modern man. Visible throughout the chapter should be how the mechanism of evolution worked with organisms and environment to produce the human population today. We shall also consider the discovery of these ape-men, and eventually man-apes, and the kind of lives we think they lived. When you finish reading this chapter, you will be familiar with your own biological history, and thus, you should have gained a greater sense of who you are, both as an individual and as a member of the human tribe.

Ancestors in the family tree

Man has been evolving for millions of years, and like all creatures, he is a product of changes in life and environment. What characterizes the early environment in which his ancestors evolved? His beginning has been traced to about 70 million years ago, in the Paleocene epoch (Table 1–1), when much of the earth was warm and humid. For 30 to 40 million years vast tropical forests reached great distances north and south of the equator. Numerous species of animals came and went in this environment, but the lush vegetation and the millions of years allowed some animal lines to persist and slowly evolve into many diverse and complicated forms.

Numerous among those varied inhabitants were the *prosimians*, the ancestors of modern monkeys, apes, and man. Some of the early prosimians resembled contemporary tree shrews (Fig. 3–1). As big as cats, long-tailed and eaters of grubs and buds, they had sharp-visioned eyes. They also had claws for holding objects. Judging from the contemporary tree shrews and tarsiers (Fig. 3–2), these claws were probably borne on fingers that moved only in unison. Most important, they had a flicker of intelligence not present in other mammals of the time.

Few fossils have been found of the descendants of these prosimians. Probably, this is because their lush habitat did not provide conditions for preserving bones. Then, late in the Oligocene epoch, much of the habitat changed: forests dwindled into fragments, and grasslands developed. With the drier environment came more fossils, today a testimony of increased speciation of this ancient prosimian during that time.

One dry area that still persists is the Sahara desert. A small part, known as the *Fayum Depression,* is located on the desert's eastern edge and has proved to be the one major site of fossils of the animals that dwelled in this area when the environment began to change about 40 million years ago. Fossils found there during this time suggest that from the shrewlike progenitor came two major groups of animals. One group included the monkeys, probably best represented by a fossil monkey

Figure 3-1
The common tree shrew in Asia may be considered a transition between insectivore ancestors and the primates. (San Diego Zoo photo.)

called *Oligopithecus*. The other group included the apes, one of special significance known as *Propliopithecus*. The discovery of *Propliopithecus* and *Oligopithecus* is immensely important because it establishes that both monkey types and ape types had begun to evolve very early, about 40 million years ago.

Differentiation of the monkeys and apes presumably continued in the next 15 millions years or so, leading to the beginning of the Miocene epoch 25 million years ago. Many diverse fossils of each type have now been found. Apparently, the monkeys stayed arboreal, because we find most of their modern descendants in trees, but apes explored new territory, taking with them evolutionary developments gleaned from their ancestral past in the trees, such as stereoscopic vision and dextrous manipulative fingers with nails. Most important, they took with them a highly developed brain. Among new territories for these life-forms to explore was the grass- and brush-covered *savanna*, where new food opportunities awaited. The apes were probably slight of build and, hence, difficult for their new predators to catch. For some reason, probably as an adaptation to ground life, they developed the ability to walk on two legs. Thus, by this time the apes and monkeys were greatly different.

Was the two-legged variety, the ape type, an ancient "man"? No anthropologist would claim that, for man was not to appear for many more millenia, not until about 12 million years ago. This ape line did evolve more apes: in Africa and Europe a gibbonlike ape (*Pliopithecus*), 20 million years ago; and a variously sized chimpanzee type, called *Proconsul*, about

64 Chapter 3: The arrival of man

Figure 3–2
The face with large eyes, the short muzzle, and larger brain suggest that the tarsier is more closely related to the anthropoids than is the tree shrew. (San Diego Zoo photo.)

the same time (Fig. 3–3). Probably, *Proconsul* evolved into the modern chimpanzees and gorillas. Like modern apes, its teeth are U-shaped, with the rows of molars parallel, its back molars are as far apart as those near the front, and the roof of the mouth is not arched (Fig. 3–4).

But another fossil may be related to man. Found in India in the 1930s, about the same time that *Proconsul* was discovered in Africa, it was named *Ramapithecus*. Specimens have since been found in other places, such as China, Germany, and Spain. *Ramapithecus* has an important tooth and jaw structure. Its teeth curve in an ever-widening arc and its palate is high, which suggests that it may be the fossil creature which first shares characteristics with contemporary man (Fig. 3–4). Little else is definitely known of *Ramapithecus*, except that he was bipedal, which freed the arms for carrying, fighting and bluffing, feeding, and using tools. (Although no evidence supports this, perhaps the tools were used in its encounters with others or in securing food.)

From 12 million years to approximately a few million years ago little is known of any ancient form leading to man, because few fossils have been found (Fig. 3–5). But from a few million years ago, new forms of creatures evolved which indicate further evolutionary development toward modern *Homo sapiens*. Next we shall consider these developments and how they were discovered.

Figure 3–3
Fossil ape, *Proconsul*, from Kenya, Africa. (From *Life: An Introduction to Biology*, by George Gaylord Simpson and William S. Beck, copyright © 1957, 1965 by Harcourt Brace Jovanovich, Inc., and reproduced with their permission.)

The Australopithecines
Fossils of the earliest type of man known were found in 1924. A crew blasting in a limestone quarry near Taung, South Africa, found fossilized remains of ancient animals which they sent to Raymond Dart, professor of anatomy at the University of Witwatersrand, Johannesburg, South Africa. Earlier a rare baboon skull had been found in the quarry, so Dart had arranged for all subsequently "interesting" fossils to be sent to him at his Johannesburg home. When he opened the box, he hoped that the rock pieces might contain another "find." Luck was with him, for near the top of the debris lay embedded in stone a piece of skull apparently bigger than a baboon's skull and even bigger than a chimp's skull.

Dart picked and chipped away the stone embedding the skull, occasionally using his wife's sharp knitting needles. After 73 days he uncovered the face of the buried skull. Its slanting forehead, snoutish face, and large eye orbits resembled those of fossilized and contemporary apes. But its larger cranial capacity, lack of a supraorbital ridge (a heavy bone above the eyes), and unique set of teeth suggested that it was something other than an ape (Figs. 3–6 and 3–13).

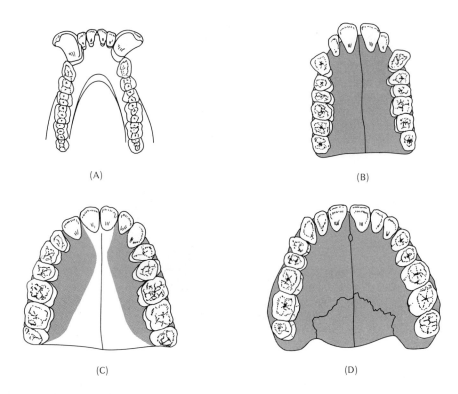

Figure 3-4
Comparison of the teeth of (A) a monkey, (B) an ape, (C) *Ramapithecus*, and (D) modern man. The curved arc of *Ramapithecus* and modern man contrast with the U-shaped arc of the monkey and ape. (Reprinted with permission of Macmillan Publishing Co., Inc., from *Man and the Environment*, by A. S. Boughey. Copyright © 1971 by Arthur S. Boughey.)

Dart decided that he had found the "missing link" between man and ape. He called it *Australopithecus africanus* (South African ape-man). It soon became popularly known as the *Taung baby*, because its milk teeth were intact and the permanent molars were just beginning to erupt. The Taung baby was estimated to be 6 years old when it died.

News of the Taung baby spread rapidly. Soon skepticism developed as to the authenticity of Dart's discovery. In fact, *Nature*, a prestigious British scientific journal, refused to publish Dart's description of it. Nevertheless, at least one colleague was convinced of the significance of the find. Robert Broom, a Scottish physician sent Dart an inquiring note, then appeared in Dart's laboratory to establish the details for himself. After one week of intense investigation he wrote an article expressing his convictions of the significance of Dart's discovery. The article was published in *Nature* in 1925.

Ancestors in the family tree

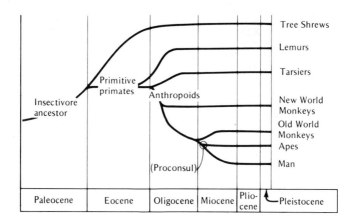

Figure 3-5
The evolutionary paths of the primates during the last 75 million years. (From *Bioscience*, by R. Platt and G. Reid, © 1967 by Litton Educational Publishing, Inc. Reprinted by permission of Van Nostrand Reinhold Company.)

An immature specimen of ancient ape-man was an inestimably valuable discovery. How much more valuable would be the discovery of an adult Australopithecine. Broom secured an appointment as curator of vertebrate paleontology and physical anthropology at the Transvaal Museum in Pretoria. Now he could persist in such a search himself. In 1936 in Sterkfontein, an old mining town near Pretoria, Broom discovered remnants of the first adult Australopithecine.

Broom persisted in his search. Then in 1938 an upper jaw with one molar was presented to him. The specimen was discovered in nearby Kromdraai by a local schoolboy. From a cache of ancient relics the boy produced a well-preserved piece of lower jaw and from his pocket, as Broom said, "four of the most beautiful teeth ever found in the world's history." Reconstruction of the pieces produced an exciting discovery: a skull resembling *Australopithecus* but with a flatter face, heavier jaw, and longer, more human teeth. Broom labeled the specimen *Paranthropus robustus* ("robust near-man"). *Paranthropus* probably stood 5 feet tall — a foot taller than *Australopithecus* — and weighed more. Digging resumed in the area after World War II. More fossilized human remains were found, including many teeth and a nearly perfect pelvis. In addition, peculiar chipped stones were discovered, very likely some of the first tools.

Broom and Dart were not the only anthropoligists who had discovered ancient pebble tools while searching for early man in Africa. Since 1931 Louis and Mary Leakey had also been digging in the ancient rocks. The Leakeys, however, were working in Tanganyika's Oldovai Gorge, a miniature Grand Canyon approximately 25 miles wide and 300 feet deep

68   Chapter 3:   The arrival of man

Figure 3–6
Comparison of the reconstructured faces of *Australopithecus*, Java man (*Homo erectus*), Neanderthal man (*Homo neanderthalensis*), and Cro-Magnon man (*Homo sapiens*). (From *Life: An Introduction to Biology*, by George Gaylord Simpson and William S. Beck, copyright ©1957, 1965 by Harcourt Brace Jovanovich, Inc., and reproduced with their permission.)

(Fig. 3–7). Although the Leakeys had been searching longer, they faced the frustration of not finding what they were after, fossilized human remains. Unfortunately, all that they uncovered were abundant pebble tools. Hopefully, they felt that some day, if they persisted, they would uncover the bones that went with the tools.

Then on July 17, 1959, 28 years after they initiated digging at Oldovai, their efforts paid off. Louis remained at the base camp that day because of a fever. Mary went alone to the site. A recent rockslide had

Ancestors in the family tree

Figure 3-7
Louis S. B. Leakey, Mary Leakey, and son, Phillip, search for remains and tools of creatures who lived 2 million years ago in Olduvai Gorge in Tanzania. The umbrella protects from the sun; the dogs guard against snakes and wild animals. (Courtesy of National Geographic Society. Photo by R. F. Sisson.)

exposed previously buried sections of the canyon. She glimpsed a fragment of skull. Searching higher up the cliff she also found large and shiny brown-black teeth that she recognized as more advanced than monkey or ape. Marking the site, Mary Leakey drove hurriedly back to camp, shouting as she neared, "I've got him! I've got him!" Leakey leaped from his sick bed, fever forgotten. He and his wife raced back to the site.

Leakey records his feelings on seeing the shiny molars. "I turned to look at Mary, and we almost cried with sheer joy, each seized by that terrific emotion that comes rarely in life. After all our hoping and hardship and sacrifice at last we had reached our goal . . . we had discovered the world's earliest known human."

A few months later the Leakeys displayed a reconstructed skull of their find, built from the teeth and nearly 400 bone fragments later found at the site. They named the ancient ape-man *Zinjanthropus boisei* (from "zinz," Arabic for east Africa, and Charles Boise, the financial supporter of Leakey's searches).

How did *Australopithecus, Paranthropus,* and *Zinjanthropus* compare? The *Australopithecus* was slender and weighed under 100 pounds. Probably his jaw was slightly thrust forward, or snouted, to accommodate the well-developed canine teeth and molars. He probably walked upright a

Figure 3-8
Australopithecines possibly used old bones from prey, stones, and sticks as hunting tools. (From Z. Burian, *Prehistoric Man,* Artia, Prague, 1961.)

good part of the time. *Paranthropus* and *Zinjanthropus* were apparently larger — standing approximately 5 feet tall. They presumably weighed about 140 pounds. Possibly the males and females differed slightly in size, but both were large boned and robust. A ridge on top of the skull to which muscles could attach suggests that they had massive jaws containing well-developed molars.

They all shared some characteristics. In general, they were ape-like, snoutish like a chimpanzee, with slanting forehead. Their cranial capacity was between 450 and 600 cubic centimeters (cm^3), somewhere between the chimp and the gorilla. The *occipital foramen,* the hole in the base of the skull through which the spinal cord enters the skull and joins the brain, was located more forward in each of the three than in the apes. This suggests that the head was held more vertically than in the apes. Remains of leg, foot, and pelvis bones suggest strongly that they all walked upright. The teeth, which included well-developed molars, canines, and incisors, suggest an omnivorous eating habit (Fig. 3-8).

The great similarity in the ape-man fossils has provoked some anthropologists to simplify the taxonomy of the group. Overwhelming similarities between *Zinjanthropus* and *Paranthropus* have led some scientists to classify both as *Paranthropus*. A few anthropologists have gone further. Since all remains appear very similar, they group all forms under

Australopithecus, the slighter form known as *A. africanus* and the more robust form known as *A. robustus.* This simple taxonomy is gaining popularity.

Frequently, the question is asked: Where do the Australopithecines fit into the ancestry of modern man? The age, characteristics, and distribution of the remains of these early ape-men bones allow for fairly valid

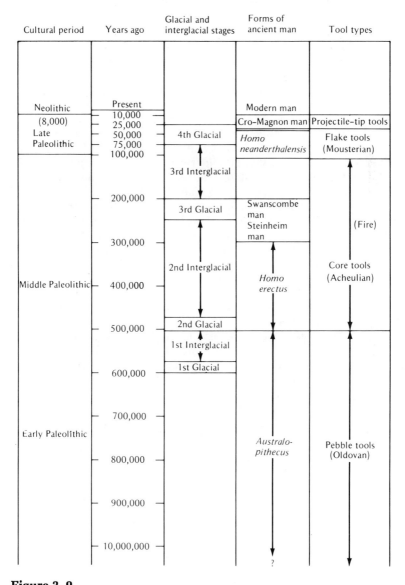

Figure 3–9
Some significant events relating to man's evolution from approximately 1 million years ago.

speculation. Apparently the Australopithecines were the first of the early manlike hominids, although they certainly appeared as much ape as man. These ancient forms have been dated fairly reliably. For example, the Leakeys' *Zinjanthropus* was found between two layers of ancient lava flow. Measurement of the slow decay into argon of the radioactive potassium contained in this volcanic material reveals the age of the stratum containing the relics to be between 1 and 2 million years. Other Australopithecine fossils have been similarly dated, or found with bones of animals known to have lived at specific ancient times, or discovered in geological formations known to have occurred at specific times. Thus, the ape-men reliably are believed to have lived between approximately 500,000 years ago to possibly over 2 million years ago (Fig. 3–9).

The dating of these fossils suggests that possibly only *A. africanus* was a direct early ancestor of modern man, that *A. robustus* was another type, who died out as *A. africanus* evolved. Fossils of *A. robustus* suggest that he lived unchanged for approximately 1 million years, or up until about 500,000 years ago. *A. africanus* apparently lived at the same time, but his fossils suggest that he underwent modification with time. Why do anthropologists believe that *A. robustus* died out? Mainly because no fossils of this species more recent than 500,000 years ago can be found, whereas fossils of *A. africanus* are more recent. Another bit of evidence may throw light on the fate of *A. robustus*. All the tools found with the ancient Australopithecines appear to accompany *A. africanus*. Some anthropologists believe this suggests that *A. robustus* did not have tools, hence was disadvantaged in his environment, and eventually became extinct.

Why it took 1.5 million years for *A. robustus* to become extinct is difficult to explain. One reason may be that *A. robustus* and *A. africanus* were not living in the same environment and, hence, did not compete directly. The teeth of *A. robustus* indicate that he was primarily an herbivore. Scientists believe that the area he frequented was part plain and part forest at that time, so he probably stayed in the areas most covered with vegetation, the forest and the bushy parklike country between forest and plains. *A. africanus,* on the other hand, probably ranged farther. His teeth suggest that he was an omnivore. He may well have eaten fruits and berries in the wet seasons and maybe roots in the drier times. The numerous tool artifacts found with his remains indicate that he was probably also a hunter, first of small animals and later, as he grew larger, bolder, and smarter, of larger creatures, such as antelopes and horses. Thus, each of the two species of ancient man occupied his own place in the environment.

Presumably, *A. africanus* survived longer because he possessed tools that allowed him to dig, hunt, spread out into new areas, and perhaps even skirmish with *A. robustus*. Having tools may have provided another opportunity for *Australopithecus* — to think about their application. Hence, in time, natural selection could have worked to preserve those ancient apemen who used tools in new ways to increase their survival chances. Feedback occurred therefore, which encouraged an increase in mental ability. Man, the tool user, was becoming man, the thinker. If *Australopithecus* was becoming more intelligent, was he also developing speech? No one knows for sure, but he most likely was able to signal his fellows, and like the modern apes, may have had a call system to express certain feelings. At least, as Richard Carrington says, there is nothing in the jaw structure to prevent the development of speech in Australopithecines, so it may be that these early toolmakers conversed with each other about the weather, hunting conditions, and other topics that we humans are concerned about today.

Homo erectus

The Australopithecines disappeared from Africa and Asia about 500,000 years ago (Fig. 3–8). At that time, another manlike creature began to walk the earth. This man was first found by a young Dutch scientist, E. B. DuBois. Fascinated by recent discoveries of manlike fossils, he asked his government for money to support a digging expedition to Java. DuBois believed remnants of fossil man would most likely be found there, because a close relative of man, the orangutan, still survived in that area. DuBois was denied support, so he joined the army as a surgeon and requested assignment to Sumatra, conveniently near Java.

After a persistent search in Java which uncovered a rich cache of fossil animals, in 1890 DuBois found the jawbone of an early manlike fossil. The following year he found a molar and a thick piece of cranium, and the next year he uncovered a thigh bone. Finally, in 1894 he published his conclusion: the discovery of an ancient man, perhaps the "missing link" between ape and man. DuBois dubbed it *Pithecanthropus erectus* (erect ape-man). It was commonly known as "Java man."

DuBois's hesitancy to announce his discovery can be appreciated more when it is realized that only a few decades earlier Charles Darwin had published his earthshaking *The Origin of Species*. The implication to man of his hypothesis concerning the evolution of species alarmed many, especially religionists. A later book of Darwin's, *The Descent of Man*, examined that suggestion more closely. The public was in no mood to be forced to consider further DuBois's proposal that the relics he had found in Java might well be evidence that the evolutionists had a point after all.

DuBois's published opinion was promptly denounced. He defended his position patiently at first. Then with increasing annoyance he withdrew the remains from public view, locked them in a strong box, and permitted no viewers for the next 25 years. Only in 1920 was he persuaded to allow the remains to be studied.

What did DuBois's Java man look like? He was probably about 5 feet tall, and like the Australopithecines, he stood erect. His skull was thick and he bore a distinctly low-browed look because of a bridge of bone that ran across his forehead just above his eyes. His head, wider than high, was held upright. His mouth held teeth like ours. His chin and forehead receded. His skull had a cranial capacity nearly twice as large as that of *Australopithecus*, nearly 900 cm^3 (Figs. 3–6 and 3–13).

A few years later the announcement of the discovery of new fossil men closely resembling *Pithecanthropus erectus* was better received. These discoveries were made at Choukoutien, China, near Peking in 1927 by Davidson Black. This site was chosen because of the age-old reputation of the area as a source of "dragon bones," ancient animal fossils ground up for use as medicine. Furthermore, Black had a hunch that ancient man probably lived in China. At first, in 1927, the uncovered human remains were sparse. Only three teeth were found. Persistent searches in 1928 and 1929 yielded little more.

More than 1,000 boxes of animals were collected, but only a few more human teeth, small pieces of human bone, and a nearly entire skullcap were found. Nevertheless, from this small evidence Black named a new ancient man, *Sinanthropus pekinensis* (Chinese man of Peking), Peking man.

Later, more relics were found. In the 1930s Black discovered another skull. This was enough evidence for Black's successor, Franz Weidenreich, who was continuing the search and systematically studying the remnants found. He confirmed Black's hunch: Peking man was indeed an early human. Most convincing evidence was his cranial capacity of about 1,000 cm^3, a brain size very near that of modern man.

This skull with a large brain capacity and artifacts found in the Choukoutien caves provide some idea of how Peking man related to his environment. He had tools. He also had fire. Some tools were found among charred pieces of bone and wood over hard-baked clay hearths. Bones in the caves indicate that he probably cooked meat over his fires and that he was a hunter. He may have enjoyed venison most, or perhaps deer were simply very plentiful; in any case, more than three-fourths of the bones are of deer. Bones of the boar, bison, ostrich, and sheep are also strewn about. So are split human bones, which suggests that he may have been a cannibal.

The large animal bones found with the fossils and Peking man's large brain capacity suggest that there had to be some cooperation — most likely verbal communication — among individuals in order to hunt and trap such large animals. Further suggestion of communication is provided by his tools, called "flake tools." It is easy to imagine him verbalizing their intended purpose to others.

Such tools were not found accompanying the bones of DuBois's discovery in Java. Recent searches in Java have produced some artifacts that may be tools, but in general, no idea of Java man's life is known today.

How do Java man and Peking man compare? In 1939, S. H. R. von Koenigswald, after finding a lower jaw and a skull, traveled to Peking to compare his Java man relics with Peking man. Amazingly, the two kinds of ancient men resembled each other so closely that Weidenreich and Von Koenigswald agreed that the fossils probably represented the same species. Java man was renamed *Pithecanthropus erectus,* and Peking man was called *P. pekinensis.* Today, the taxonomy is often further simplified, and the humanness of each is emphasized by calling both *Homo erectus* (Fig. 3–10).

It was fortunate that Weidenreich and Von Koenigswald met when they did. Shortly thereafter, in the fall of 1941, the spread of World War II prompted workers at Peking to remove the remains of Peking man to a

Figure 3-10
Java man, *Homo erectus*, discovered by E. B. DuBois near the river Solo in Java in the shadow of volcanoes. *H. erectus* probably roamed in bands searching for fruit, roots, and possibly small animals. (From Z. Burian, *Prehistoric Man*, Artia, Prague, 1961.)

safer place. The collection was boxed, put aboard a train, and sent toward the coast to meet an American ship. War in the Pacific area began on December 7. The remains of Peking man have not been seen since. Hope still prevails that they have survived and are only waiting rediscovery.

In Java at about the same time, Von Koenigswald distributed the Java man relics to friends so as to preserve them during the invasion of the Japanese. After the war the relics were collected. They are now on display in Europe.

The relationship of *Australopithecus* and *Homo erectus*

Ever since the discovery of *Australopithecus* and the Java and Peking forms of *Homo erectus*, scientists have speculated on their evolutionary relationship. The appearance of *Homo erectus* as *Australopithecus* was dying out is intriguing. The most obvious assumption would be that *Australopithecus* evolved into *Homo erectus*; however, no clear picture exists of this happening. The variation found in fossils may be representative of that which naturally occurred in each species, rather than as evidence for interbreeding or the evolution of *Homo erectus* from *Australopithecus*. Clearly, *Homo erectus* did not appear *de novo* any more than

Australopithecus, but it will probably never be known when particular variations initiated the divergence of the two species.

However, what can be detected in a comparison of *Australopithecus* and *Homo erectus* are different stages of man's cultural evolution. The use and manufacture of tools, and probably speech, were more developed in *Homo erectus* than in *Australopithecus*. Here may have occurred cultural seeds that would later blossom into the elaborate science and technology of contemporary man.

Homo neanderthalis

Before Darwin published *The Origin of Species* in 1859, human fossils which Darwin was surely aware of, but which he did not comment upon in the book, were found in Germany. A nearly complete skeleton of a fossil man was unearthed in the Neander Valley. Other such skeletons had been found probably as early as 1700 — the first discovery of early human remains ever to be recorded — but they were essentially ignored. The ancient man represented by the fossils was labeled *Homo neanderthalis*, after the valley in which it was found (Fig. 3–11). Today, the bones of nearly 100 specimens of this man have been uncovered in such areas as Palestine, Iran, Iraq, southern France, Italy, Yugoslavia, Germany, southern Russia, Czechoslovakia, and Spain.

The Neanderthals mostly have been found in areas of the world where relics of *A. africanus* and *H. erectus* have not been found. This

Figure 3–11
Small communities of Neanderthal man (*Homo neanderthalis*) lived in caves or camped under the sky in fair weather. (From Z. Burian, *Prehistoric Man*, Artia, Prague, 1961.)

suggests that Neanderthal man or his predecessor was more footloose and migrated more widely than either of the earlier two types. Furthermore, he was moving into regions that were destined to become colder than the African forests and plains. The Europe that Neanderthal man migrated into about 110,000 years ago was probably warm, but by 75,000 years ago, the last of four great glaciers covered Europe.

Neanderthal man apparently lived successfully through this cooler time. In fact, the extensive glaciation of Europe may explain the structural diversities found in Neanderthal specimens. Although all appear to be about 5 feet tall and heavily built, two general variations are apparent. One type, called the *conservative* type, has a very heavy cranium, with large brow ridges, receding chin, and sloping forehead. The other type, called the *progressive*, has a less-pronounced slope to the forehead, less-pronounced brow ridges, and a higher cranium. In general, they all have a brain capacity one-third greater than *H. erectus*.

The effect of the glacial environment on this ancient man may be one of the most obvious examples of how environmental conditions have helped evolve species into new forms, again illustrating the intimate relationship between life and environment. For example, one hypothesis, the glacier hypothesis, proposes that the ice sheet separated what was a fairly homogeneous Neanderthal population into two separate groups. Enforced isolation during the glaciers' reign allowed time for each of these separated populations to evolve its own uniquenesses. In western Europe, isolated and without the opportunity for genetic interchange with other groups, the population gave rise to individuals of the conservative type. In eastern Europe and the Middle East, because the glaciers lay to the north or in the Carpathian or Balkan Mountains, ancient Neanderthalers moved about over a wide area and presumably freely interbred, to give rise to the progressive type. Certainly, the geographical location of the sites of discovery of the two types correspond with the glacier hypothesis.

Other hypotheses have been proposed. Because the progressive type appears to have many features of modern man, the suggestions have been made that perhaps the progressive was evolving into modern man. Another hypothesis claims that modern man *migrated* into the populations of the Neanderthals, possibly from the east, and perhaps interbred with the Neanderthals. In any case two types of Neanderthals are known.

In general, both types of Neanderthal man are our stereotypic caveman, with a stooped, long body, short arms and legs, and heavy head. However, within his thick skull developed a brain that theoretically, at least, enabled him to think about yesterday and tomorrow and entertain spiritual thoughts (Figs. 3-6 and 3-13). We have evidence of his spiritual concerns. For example, in one interesting site in France, apparently all skeletons of a family, except one, faced west. Typically, the skeletons seem to have been buried on their side, with the feet drawn up under the buttocks. In eastern Russia a skeleton of a boy was found surrounded with a ring of ibex horns. In Italy a skeleton was found within a ring of white stones.

Other evidence of spiritual concerns, perhaps expressed ritualistically, has been found. In some caves, bear skulls were found arranged in a pattern, and in one cave in Switzerland, seven bear skulls were found in a niche in the floor of a cave. All the skulls faced the entrance to the cave. Similar finds have occurred in France and Germany.

Was the bear a symbol of respect, or hatred? Did the men who placed them there attempt to show fear or respect of the larger and stronger bear? Stone age cultures today in Africa and Australia are known to show a great respect for the animals with which they compete. American Indians typically feel themselves to be a part of the life scheme of the other animals, and name themselves and their clans after animals whose characteristics they admire or fear. Today's Finns and Lapps have epic stories that feature the bravery and the strength of the bear. If only it were possible to imagine ourselves in a world without explanations for birth or death, sunrise or sunset, or the change of the seasons, we perhaps could know the thoughts of those early men. About early man Loren Eisely in *The Immense Journey* (Random House, Inc., New York, 1957), comments:

> *For the first time in four billion years a living creature had contemplated himself, and heard with a sudden, unaccountable loneliness the whisper of the wind in the night reeds. Perhaps he knew, there in the grass by the chill waters, that he had before him an immense journey.*

Neanderthal man presents us with many mysteries. Perhaps the most intriguing is what caused him to die out. His fossils appear to be no younger than 30,000 years — about the time that Cro-Magnon man, a true representative of *Homo sapiens,* appears. Hypotheses for the Neanderthal extinction are numerous. Was he unable to survive the wet, severe cold; was he unable to compete for game with the more adept *H. sapiens;* was he merely eradicated by an enemy; or did he catch a devastating disease? No one knows.

Homo sapiens

Neanderthal man survived during the great glaciations of Europe. One can imagine his responses to these conditions — the use of tent shelters for prolonged hunting trips, the development of warm clothing, and the exploitation of caves as shelters. Successful competition among individuals and groups must have been the rule in the Neanderthal's daily life. However, the coming of a modern man when the last great glacier covered the north meant a new competition from which Neanderthal did not emerge the winner.

Modern man, *Homo sapiens* (Fig. 3–12), probably appeared about 35,000 years ago (Fig. 3–9). At least we find an abundance of remains from about that time forward. Some anthropologists suggest that he may have occurred even earlier in Europe, but evidence is meager. The suggestion comes from the discovery of a nearly total skull at Steinheim (Steinheim man) near Stuttgart in 1933, and skull fragments found at Swanscombe (Swanscombe man) near London in 1935. Other fragments, including an incomplete skullcap and a piece of frontal skull bone (Fontechevade man),

Homo sapiens

Figure 3-12
Cro-Magnon man (*Homo sapiens*) lived in larger communities than *H. neanderthalis*. Hunting, a major preoccupation, was presumably carried on in groups, as this attack on a mammoth illustrates. (From Z. Burian, *Prehistoric Man*, Artia, Prague, 1961.)

in France found with the remains of animals known to have lived before the last glacier suggest that modern man may have been in Europe even earlier, but definite evidence is lacking.

In any case, abundant fossils and artifacts indicate that he was widespread over Europe from 35,000 years ago. His bones tell us he was tall and more slender and graceful than Neanderthal man. His head bulged less in the back, and the top of the head was higher (Fig. 3-13). His forehead lacked the great brow ridges and was very high compared to all the other hominids who had lived. He walked completely upright. It has been said that if early *H. sapiens* were available today, they could be dressed in contemporary clothes and go unnoticed (Fig. 3-6). Their brain capacity was about the same as the Neanderthals, and the motor area was not well-developed, but the *cerebrum*, the region associated with conceptualizing, had reached a unique refinement.

Evidence of such refinement is found on the walls of the caves he frequented, notably in the Dordogne region of France. The Dordogne River and its tributaries have flowed for ages through an area composed mainly of limestone, cutting caves, grottoes, and overhangs. In many of the caves overlooking the valleys, remains of ancient *H. sapiens* have been collected in great quantity. The first were found in 1868 in a rock shelter invaded by road-widening activities. These specimens, named after the shelter, were called *Cro-Magnon* men.

In addition to the shelters above the ground, some occur almost inaccessibly underground. These are the damp, dark recesses in which Cro-Magnon man left evidence of his remarkable ability to conceptualize. Here occur his famous cave paintings, first discovered in 1940. Today, about 70 sites of Cro-Magnon cave drawings are known in France alone. All are dated between 28,000 and 10,000 B.C. The cave paintings are frequently large, 10 to 20 feet long. The subjects are almost always animals, especially mammals, and of these usually about 12 species are of those of the animals most usually hunted. Frequently, the mammals are being clubbed or speared, are trapped or dying. Seldom are people depicted, but occasionally they do appear, disguised in animal skins. All pictures are in profiles executed in bold outlines in blacks, browns, yellows, and reds. These paints apparently were made from natural materials, such as clay, mixed

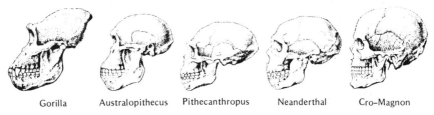

Figure 3–13
A comparison of skulls of an ape and four types of prehistoric men. (From *Life: An Introduction to Biology*, by George Gaylord Simpson and William S. Beck, copyright © 1957, 1965 by Harcourt Brace Jovanovich, Inc., and reproduced with their permission.)

Figure 3–14
In this example of cave art, notice that some figures are superimposed on others.

with charcoal and animal fat. Reaction with the limestone wall helped preserve the brilliance of the colors (Fig. 3–14).

What was the purpose of the art? Was it drawn merely for esthetic purposes? Evidence suggests that it was not. It is speculated that the art may well have been part of magical or religious rituals practiced before a hunt to ensure success. Since Cro-Magnon's life depended on hunting, such an interpretation may be valid. Several characteristics of the art support this interpretation. The subjects depicted are mainly game animals being attacked or dying. At several sites several paintings are superimposed on each other, suggesting that a picture drawn at a particular site produced good hunting, so the site was reused before subsequent hunts. Occasionally, a picture has been found in which only the head has been redrawn. This suggests that the site may have been effective, but time was important, so only the head of the animal to be hunted was drawn — onto the body of a different beast. The 50 or so depictions of animal-clad people found in the caves may be illustrations of *shamans* (priests) invoking magical power or luck to the hunters. No one can be certain of the meaning of the art, but such suggestions appear reasonable.

The tie that Cro-Magnon man had with hunting and game also may be suggested in other paintings, those of animals shown mating and those of pregnant mammals which apparently emphasize fertility. The fact that Cro-Magnon lived in part during the last glacial era, when securing food was difficult, may well account for his concern with the abundance of game animals.

Whatever the purpose of his art, Cro-Magnon man had the ability to conceptualize to a degree no other ancient human could. The able brain, indicated mainly by his art but also by his sophisticated tools, probably accounts for his widespread distribution over the earth; where he came from may forever be a mystery. Where he went is not. From him came the population distributed over the present world. As he migrated he took with him the seeds of modern culture: technology from the Australopithecines, religion from the Neanderthals, and his own contribution, art.

Summary

The major themes in this chapter are that man, like all other organisms, has evolved through millions of years, constantly changing, from prosimian forms to the upright, intelligent *Homo sapiens* that walk the earth today; that the stages through which man evolved show many of the same trends visible in the evolution of other organisms; and that his evolution was always a product of the interaction between himself and the environment in which he existed.

The stages of man's evolution probably began before any recognizable manlike creature existed on earth, about 70 million years ago, presumably in today's Africa. For 30 to 40 million years tropical forests slowly began to evolve the prosimian ancestors of modern monkeys, apes, and man. In time, as the climate changed, the monkey and ape lines diverged, the latter giving rise to forms such as *Proconsul* and *Ramapithecus*. And then about 12 million years ago the first manlike apes, the Australopithecenes, appeared. What environmental conditions produced them and why they eventually became extinct are questions that could be asked about *Australopithecus* and those manlike creatures that followed, *Homo erectus,* appearing about 500,000 years ago, and *Homo neanderthalis,* appearing about 100,000 years ago. It is known that *Australopithecus* invented tools, that Neanderthal man had religious concerns, and that the most modern form of man which followed, *Homo sapiens,* brought to the walls of his caves colorful art.

Thus, man has evolved through many phases of development: early forms died out and later forms incorporated the advantageous characteristics of the first forms, a trend consistent with the evolution of all creatures. By developing a life-sustaining relationship with his environment through the millenia, man has survived to the present day. His long tenure on earth should give us hope for the ability of modern man to put himself right with his environment and, thus, assure himself of a future.

Supplementary readings

Ardrey, R. 1968. *African Genesis: A Personal Investigation into the Animal Origins and Nature of Man.*
Atheneum Publishers, New York. 380 pp.

Boughey, A. S. 1971. *Man and the Environment: An Introduction to Human Ecology and Evolution.*
Macmillan Publishing Co., Inc., New York. 472 pp.

Carrington, R. 1956. *The Story of Our Earth.*
Harper & Row, Publishers, Inc., New York. 240 pp.

Coon, C. S. 1971. *The Story of Man: From the First Human to Primitive Culture and Beyond,* 3rd ed., rev.
Alfred A. Knopf, Inc., New York. 430 pp.

Eiseley, L. 1957. *The Immense Journey.*
Random House, Inc., New York. 210 pp.

Howell, F. C., and the editors of *Life.* 1965. *Early Man.*
Time-Life Books, New York. 200 pp.

Koenigswald, S. H. R. von. 1962, *The Evolution of Man.*
University of Michigan Press, Ann Arbor, Mich. 148 pp.

Korn, N., and F. K. Thompson, eds. 1967. *Human Evolution: Readings in Physical Anthropology,* 2nd ed.
Holt, Rinehart and Winston, Inc., New York. 466 pp.

Lasker, G. W. 1961. *The Evolution of Man.*
Holt, Rinehart and Winston, Inc., New York. 239 pp.

Pfeiffer, J. E. 1969. *The Emergence of Man.*
Harper & Row, Publishers, Inc., New York. 477 pp.

Supplementary readings

Scientific American. 1967. *Human Variation and Origins: An Introduction to Human Biology and Evolution.* W. H. Freeman and Company, San Francisco. 297 pp.

Young, L., ed. 1970. *Evolution of Man.* Oxford University Press, New York. 648 pp.

Chapter 4

It is a truism that a given environment can act as a selective agent, and thus govern evolutionary changes, only if the animal elects to stay in it long enough to reproduce. In general, an animal occupies a given site and continues to function there because forced to do so by external forces. Commonly also, the animal reaches a new environment accidentally in the course of exploration and elects to remain in it. Such choice implies a preadaptation which can be either genetically determined or the result of prior individual experience. In any case an important aspect of the preadaptive state is that the natural selection exerted by the environment is preceded by some kind of choice, not necessarily conscious, by the animal. Whatever the precise mechanisms of this ill-defined situation, common sense indicates that animals and men do not behave as passive objects when they become established in a given environment.

Eventually a change of environment leads to a change in habits, which in turn modifies certain characteristics of the organism. Even when repeated for several generations, such modifications are not truly inheritable, but they may nevertheless foster evolutionary changes. The reason is that continued residence in a particular environment tends to favor the selection of mutants adapted to it. Eventually, such mutations are incorporated into the genetic structure of the species involved.

So Human an Animal
René Dubos

The distribution of life

86　Chapter 4:　The distribution of life

Many biologists never outgrow the urge to collect plants and animals; these scientists usually become *taxonomists,* collectors and labelers of the diverse animal and plant forms that are found distributed around the earth. This enterprise can be wholly descriptive, as was common in the last century, when extensive inventories were taken of the flora and fauna inhabiting nearly every area of the world. Today, taxonomists not only ask, "What is it?" but also ask, "What events in the past, both to the organisms as species and to the environment, account for the presence of certain plants or animals in a particular region?" or, "Why do animals in some places show characteristics not found in animals occurring in other areas?" or, "Why do animals in widely separate geographical areas resemble each other?" Thus, biologists today are concerned with the relationship between types or species of organisms *and the environments in which they are found.* This science is known as *biogeography.*

Biogeographers know that answers to their questions must be sought in the past, primarily in fossils and geological formations, signs of previous eras telling of the distribution and characteristics of ancient life, and in the forces that operate which shaped the environment and the organisms in it.

Evidence from fossils and geology indicates that several trends operate in the evolution of life. Through the ages, life-forms appear to become more plentiful and diverse; environmental conditions that sustain one form of life are not effective in precisely the same manner to sustain any other form of life; a complex interaction occurs among organisms as they struggle to survive; any change in organism or environment affects the other; most forms of life seem to exist during only a short period of geological time; and — most significant in this chapter — the *distribution* of organisms around the globe is a result of the evolution of a sustaining relationship between plants or animals and environment, a relationship so delicate that its change may mean the extinction of a species from that area. (Testimony to this balanced and precarious relationship is the claim made by some scientists that nearly nine-tenths of all the animal species that ever existed on earth are now extinct!)

These evolutionary trends, which have produced the natural world we know today, are the subject of this chapter. When you finish reading it, you should have some insight into how evolutionary forces working through the past, now, and into the future affect organism and environment and account for the structure and the distribution of life forms, including man, as we know them today.

Thus, the distribution of life as an evolutionary product is the one generalization which ties together chapter topics that may at first glance appear diverse and unrelated. These topics include the following: (1) the evidence and hypotheses that trace the factors affecting man's distribution around the earth; (2) the biogeography of two kinds of islands (because the forces affecting the distribution of organisms on islands are easy to see and are most likely the same as those which may explain man's development into races and distribution around the globe); (3) the major land divisions of the world, or *realms,* and the reasons for the distribution of the characteristic animals in each; and (4) the *biomes,* environmental subunits of realms, which contain unique and fascinating plants and animals, all products of evolution working through the ages.

The distribution of man

Man's environmental manipulation through the years, ever increasing as his brain became more sophisticated, today has resulted in a very crafty animal with seemingly great survival potential. Nevertheless, it is true that today certain of his manipulations of the environment seem to be out of hand and may be more life-destroying than life-enhancing (a topic considered later in more detail). What will be the result? Has man gone too far? Has he changed his environment too much and into such a state that he can no longer hope to survive the consequences?

Man's ability to change the environment in the past suggests that he can continue to alter it. Clearly, the theme of this chapter suggests that with adequate basic background knowledge of how organisms relate to their environments in a life-sustaining way, man can right any environmental wrongs that he has brought about and, thus, assure himself, as a species, of a future on earth.

If we know how man obtained his needs from the environment in the past, we may gain insight into how he could continue to do so. Thus, in this chapter we will look at how ancient man, with the aid of tools, handled the environment so as to survive better, increase his numbers, and become distributed around the globe. Always underlying the discussion should be a major understanding: man is what he is today because of his interactions with the environments he inhabited; and the environments in which he lived changed as a result of his presence. Thus, man (like all other organisms) and environment change through the years, each affecting the other.

The effect of tools on man's environment

Man's original environment is thought to have been Africa, because no definite fossil evidence of *Australopithecus* has been found elsewhere. According to many scientists, his range was limited because he had no elaborate tools with which to change the neighboring and different environments so that he could survive in them. Only as he evolved and his tools changed did he have the ability to move into previously inhospitable lands and leave the continent that gave him birth.

The simplicity of the Australopithecine's first tools, called *pebble tools*, explains why they gave him only limited influence over his environment. A pebble tool is merely a stone, ranging in size from a table-tennis ball to a billiard ball, with one end chipped to a sharpened point. This hewn stone, a "chopper," was probably used to hack through rough meat and sinews. The chips are very sharp and so were probably used to scrape and to cut. Pebble tools represent a cultural development known as the *Oldowan industry*, a name derived from Olduvai, the site in east Africa in which Australopithecine fossils have been found (Fig. 4–1).

More advanced tools, most of them known as the *Acheulian industry*,

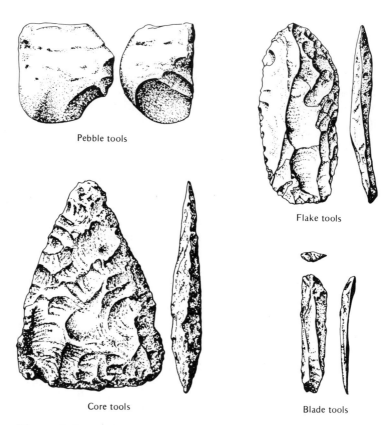

Figure 4-1
The earliest tools of prehistoric man were probably pebble tools. Next, flake tools and core tools evolved, followed by blade tools. With each new type of tool, man could push into new territories.

after a town in France, may have allowed man to move out of Africa. It may never be ascertained who invented the Acheulian tool industry, but in any case, by 500,000 years ago, when refined *Homo erectus* populated Africa and northerly regions outside that continent, he used Acheulian tools (Fig. 3–8). These tools were constructed from flint, quartz, or quartzite, common rocks in western Europe. They were flat and sharp-edged, the product of a new technique — chipping away at a stone over its entire surface, not just at one point as was done in the Oldowan culture. The tools are known as flake tools. The hand axe, which was usually pear-shaped and pointed, and the cleaver, which looked a great deal like our modern axe head, came into being at this time. The most recent tools of this era were presumably hammered out with wood or bone, not stone, because the scars are more regular and the cutting edges straighter (Fig. 4–1).

H. erectus spread north through the Middle East into both Europe and Asia. Without question, his tools probably helped him survive his trip northward in weather conditions known to be severe. In fact, a period of glaciation was known to cover Europe down to 50°N latitude in Europe and 40°N latitude in the East from approximately 500 million years ago until about 475,000 years ago, the same time that H. erectus was in those areas. H. erectus survived the cold probably because of his use of fire (which is well-documented), the continued use of pebble tools, and the development of the Acheulian tool industry, which allowed him to easily gather food and skins for warmth. Once again, changes in the way man behaved effected changes in his environment.

Approximately 110,000 years ago *Homo neanderthalis* arrived, and with him came a new tool culture to accompany the pebble and flake tools. Most of the new tools are known as the *Mousterian industry,* named after the southern French town LeMoustier. Tools of this period are diverse. They include many types of borers, axes, choppers, scrapers, knives, chisels, and planes made of stone and some constructed of bone and antler (Fig. 4–1). The bladed tools were sharp chips of stone carefully flaked from a larger rock. Perhaps in the process a core tool was made as well. Balls of limestone found in one cave suggest that even the bola was made. Without question this collection of tools gave Neanderthal man increased advantage over his environment. Whether it be glacier cold, as it was toward the end of his reign, or mildly warm, as at the time of his appearance on earth, this primitive man had the tools to secure food, prepare hides, and generally adapt himself to the conditions.

Even so, Neanderthal man died out, and for years anthropologists have wondered why. Some believe it was Neanderthal's inability to compete with a new tool culture brought into Europe and Africa by new men migrating from the east. The invader was Cro-Magnon man, a true *Homo sapiens* (Fig. 3–9). His new tool culture included the *projectile-tip tool,* used in arrows and spears. This tool greatly changed man's relationship with his environment. It gave man an advantage in that the hunter or warrior remained more distant from the prey or enemy than ever before. His chances of survival were greater; in fact, eventually Cro-Magnon man became distributed around the world and presumably was the direct predecessor of many modern people. Many anthropologists believe that this successful invasion of strange environments and the use of diverse and effective tools are very closely related.

The development of races

Scientists think Cro-Magnon man migrated into various environments which, in time, subdivided him into races. In Europe he became the Caucosoids; in Africa, the Negroes; and in the east, the Mongoloids. Most likely all three races were derived from a common stock from the east, possibly an ancient *H. erectus.* The specific conditions that tended to encourage *raciation,* or diversification of man into races, are almost totally unknown, but scientists have made some assumptions, or *hypotheses,* which may explain race development in view of geographical and meteorological conditions existent long ago. Whatever the explanations for the origin of the races, very likely they lie in two developments: the evolution of effective tools, which in turn allowed migration; and the evolution of physical characteristics of significant survival value.

Among the hypotheses proposed, one that has gained some support claims that *latitude* (how far north people lived) caused variations in skin pigmentation. In northern latitudes less sunlight strikes the earth; consequently, the natural synthesis of vitamin D, a process that occurs in the skin when exposed to sunlight, does not occur efficiently in dark-skinned people because too much light is screened by the dark pigment. However, lack of vitamin D results in serious diseases, one of which is *rickets;* therefore, because light-skinned people would live more healthfully than dark-skinned people in the more northerly regions, in time the population in the northern latitudes would be composed of those people who survive, the light-skinned ones. According to this hypothesis, dark skin would tend to develop in more exposed areas on earth by the same evolutionary processes. Individuals in the sunnier regions whose skin filters out some of the sun's intensity have better survival chances and, thus, are selected out. As a result, in time a dark-skinned race would evolve in the brighter and hotter regions of the earth. This hypothesis has some credibility because it is consistent with Darwin's ideas concerning the origin of new kinds of plants and animals, and some research supports it. One researcher calculated the *average* daily amount of sun shining on the faces of European infants exposed to fresh air and the sun. He showed that such exposure produces an adequate amount of vitamin D to meet the children's needs. Calculations indicate that similar exposures of the faces of black children did not yield enough vitamin D to prevent serious diseases. Consistent with this is the high incidence of rickets among Negro children in the Northern Hemisphere. Here again is an example of changes in a living organism brought about, in part, by environmental influences.

Perhaps the temperature influenced the development of other racial characteristics. For example, some scientists speculate that the facial characteristics of the Mongoloid race were influenced by a very cold period. Assuming that ancient Cro-Magnon appeared during an intensely cold and glaciated age and maintained populations in the northernmost regions even in the postglacial stage, such reasoning could be valid. For example, the slitlike eye is due to an exterior fold of skin accentuated by fatty insulating tissue and may be an adaptation to intense light reflecting from the snow. The fat, padded cheeks and flat, broad face, the small nostrils, and the reduction of the cold-sensitive sinuses may also be cold-adaptive features. North-moving Mongoloids are thought to have come into North America when land connected the two continents (during the period when the last glacier covered the area), and it is these features, along with projectile-tip tools, that may account for his survival during the migration.

Man's continued migration into North and South America may be due to his resistance to cold and to his tools. Soon after migrating into Alaska, the emigrants distributed themselves rapidly and eventually gave rise to the North and South American Indians, including the Eskimos and the Aleuts. Interestingly, concomitant with the initial entry of the Mon-

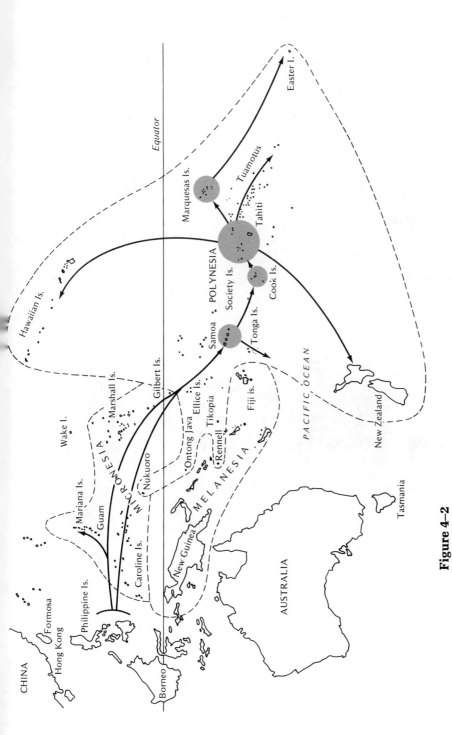

Figure 4-2
The paths of man's migration which settled the Pacific Islands. [From *The Study of Man*, 2nd ed., revised, by Carleton S. Coon. Copyright © 1954, 1961, 1962 by Carleton S. Coon. Reprinted by permission of Alfred A. Knopf, Inc.]

92   Chapter 4:   The distribution of life

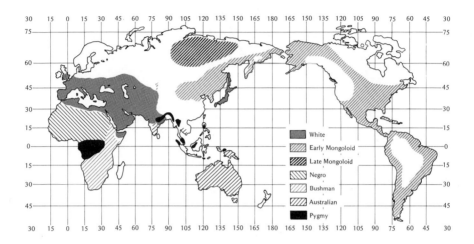

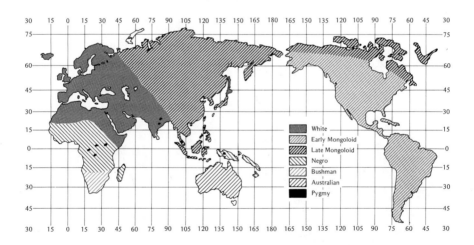

Figure 4–3
The distribution of the races of man (A) in 8000 B.C., at the time of the Agriculture Revolution and (B) in 1000 A.D. (From W. W. Howells, "The Distribution of Man." Copyright © 1960 by Scientific American, Inc. All rights reserved.)

goloids, with their projectile-tip weapons, into the North American continent came the extinction of approximately 40 percent of the continent's large animals. No one is certain of the cause of this mass extinction, now called the *Pleistocene overkill,* although many suppose that it was arrow- and spear-carrying man. Once again, as man increased his ability to manage his environment, he became more widely distributed.

Man's success in changing the environment to his liking probably accounts for his spread over the rest of the world, the final major part being the island regions of the Pacific. Some of the islands in this area have been very recently settled, such as Hawaii about 1,500 years ago, and New Zealand about 100 years ago. The Polynesians, who appear to have been the first human inhabitants in the Pacific region, most likely came from southern Asia or islands near Asia, although it is not known just what routes they followed. Most likely they are derived from a mixture of Mongoloid and Caucasian stock, perhaps at an extremely early time (Fig. 4–2) in the East.

Man eventually became spread out over the surface of the entire earth (Fig. 4–3). The development of effective tools allowed him to manage his environment and probably indicates profound changes occurring within himself, notably the development of an increasingly sophisticated brain. However, wider distribution, whether explained by tools or brains, is only another way to point out that man had developed an effective life-enhancing relationship with his environment, a relationship that characterizes the success of any living thing in any environment. Thus, a better insight into the kind of relationship with the environment that works *for* man, not against him, may help correct any environmental crises that we are now experiencing and help assure a future for us as a species.

The biogeography of the Galápagos Islands

Insight into the kind of relationship with environment that helps sustain life, including man, can be gained by examining the evolution of other organisms in their habitats, whether deserts or forests, the Arctic or Africa, islands or continents. With the hope of gaining such insight, we shall examine life in one of these environments in more detail. We shall look at life on islands because there the environment is geographically limited and only a few species of plants and animals exist. We shall compare two kinds of islands, the Galápagos Islands (oceanic islands) and the British Isles (continental islands).

The geography and kinds of life found on the Galápagos Islands are unique. In fact, Charles Darwin, when in his early twenties, visited there and gathered much information regarding the evolution of plants and animals which later was expressed in a book, *The Origin of Species.* About the Galápagos Islands Darwin wrote: "The country was compared to what we might imagine the cultivated parts of the Infernal regions to be." Yet this archipelago 600 miles west of Ecuador later proved to have plants and animals not seen anywhere else on earth. Included in the list of organisms found on the islands were low, thorny bushes and prickly pear cacti and occasional spots of forest with ferns and orchids which gave shelter to two kinds of mammals (the bat and rice rat), one kind of tortoise, four kinds of lizards, one kind of snake, one kind of flightless cormorant, one species of penguin, and twenty-six kinds of land birds, most notably a group of finches now known as Darwin's finches (Fig. 4–4).

The discovery of these birds had profound influence on Darwin's thinking about evolution and the origin of species to be expressed later in his book. He realized that the finches showed adaptations to the unique

Figure 4–4
Darwin's finches from the Galápagos Islands include the ground dwellers (1–7), which seek food on the ground or in low shrubs, and those which primarily eat insects (8–13). (From Biological Sciences Curriculum Study, *Biological Science: Molecules to Man*. Houghton Mifflin Company, Boston, 1963.)

environmental conditions on their respective island habitats. Yet, although the species appeared native to the Galápagos Islands, he noted that they also remotely resembled forms on the mainland of South America.

Darwin explained the appearances of the finches this way: the ancestral finches migrated or were blown from South America's mainland to one of the then new volcanically formed uninhabited Galápagos Islands. Through the years, as the finches became plentiful, individuals strayed or, more probably, were blown to adjacent islands. There, isolated from others, these aliens evolved in time into the birds Darwin saw.

Specific characteristics of the finches illustrate the outcome of the evolutionary process involving genetic change and isolation. Most islands were inhabited by the three primary types of ground finches — large, medium, and small — which fed respectively on large, medium, and small seeds. On the two southern islands, where only the small and medium species occurred, the medium-sized species was notably larger than on the other islands. On another island only the medium and large species occurred, but the medium finch tended to be smaller than elsewhere. Finally, on still another island, only the small and large species were found, and the small species there was larger than anywhere else. Such variations in size suggested that the birds tended to evolve structures and habits that reduced competition among themselves for food and other necessities.

Very likely, the forces operating to produce new species of finches on the Galápagos Islands also produce new species elsewhere on earth. However, the uninterrupted evolutionary process is not so easily observed or identified on continental land masses as on the Galápagos. Nevertheless, these processes probably help explain the diversity of all organisms in the biosphere and the movement of organisms, including man, into new environments. To gain further insight into how these forces explain the distribution of organisms, let us now look at another type of island.

The biogeography of the British Isles

The Galápagos and islands such as the Azores, Bermuda, Sandwich Islands, and the Polynesian islands, never connected to a continent, are known as *oceanic* because of their isolated origin. The characteristics of the animals on oceanic islands are predictable and can be noted on the Galápagos: the species are few; some vertebrates, especially carnivorous amphibians and freshwater fish, are least plentiful, mainly because no mechanism exists for them to migrate there easily; other vertebrates, such as bats, rats, pigs, and small arboreal mammals, are comparatively abundant, probably because they can be transported on floating materials; birds are comparatively common, but for some reason they are often dull or white and frequently evolve into wingless forms (such as the now extinct *dodo* from Mauritius); the fauna are often large (such as the Galápagos tortoise); and insects are usually wingless, probably because they are derived from wingless forms carried to the islands on floating debris.

In contrast to the fauna on oceanic islands, the characteristics of animals on islands once connected to continental masses, known as *continental islands,* are very different. Generally these forms, although unique, clearly resemble the species existing on the continents nearby, because that is where the animals originated. Often the evolution of these fauna has differed greatly from that of life on oceanic islands. Animals on continental islands have undergone periods of isolated evolution, alternating with times of interchange with continental animal populations, as the islands have been alternately attached and separated from the mainland.

This situation is well-illustrated by the continental islands that compose the British Isles. These islands have a complex geological and biological history. During the Paleozoic and Mesozoic eras many geologists believe they were connected to a great land mass which extended from North America to Greenland and Northern Europe. Probably at the end of the Miocene epoch (approximately 10 to 15 million years ago) the isles lost connection with the North American continent. Consequently, unique animal forms, no longer able to migrate, began to evolve. During this period, saber-toothed tigers, tapirs, and even mastodons shared company with abundant lizards and amphibians.

Then during the next years (the Pleistocene era), the climate of the area vacillated between cool and warm periods, and the animal populations changed accordingly. The especially sturdy forms, such as the mastodons, otters, red deer, and rodents, apparently endured the cold times, but the less resistant died out or migrated into the continent. Then, as the climate warmed, back came the saber-toothed tigers, lions, and Old World monkeys. Subsequent cold temperatures and ice killed most of these forms, but in time warm weather brought in reindeer, musk-ox, voles, horses, pigs, hippopotamuses, and early Stone Age Man; a subsequent cold period killed or drove many of them out.

When the last ice age passed (about 9,000 years ago) many new kinds of organisms migrated into the thumb of land destined to be the British Isles. Here they were trapped when the Straits of Dover sliced the isles from the mainland 7,000 to 8,000 years ago. The thousand years or so between the last glacier and the formation of the straits allowed colonization, but because the emigrated organisms have had only a short time to evolve since then, not many unique species have developed. Today, some meager evidence of the species evolution is detected — for example, the almost identical British and European red squirrels occur, but the British species has developed a lighter tail. Consequently, the British Isles have almost no unique forms. All told, in fact, only about 50 species of mammals (compared to approximately 100 species in Germany and 70 in Scandinavia), 6 species of amphibians (compared to 12 species in France and the Low Countries), and very few reptiles exist there today.

The kinds of animals found in Ireland further illustrate the effect of past geological and meteorological effects on the formation of species. Ireland has fauna similar to Great Britain, but less of it. This is explained as follows: Ireland became separated from Great Britain after the British Isles became isolated from the continent. Ireland was far north, so it contained few animals. As a result, only a few species were trapped there. In fact, today only one-half of the English species of animals are found in Ireland, and none of these show much differentiation. The eight mammal families include small animals, such as rabbits, squirrels, and foxes. Only three of the eight British species of reptiles and amphibians are found.

(Incidently, Ireland does have four animal species not found in England — an earthworm, a spotted slug, a moth, and a wood louse. These creatures *are* found in southwest Europe today, and it is supposed that a small patch of Ireland was not glaciated sometime in its past, and these species were spared.)

Thus, an examination of the history of islands and the kinds of life they support once again clearly illustrates the intimate relationship between life and environment in general. A change in one sooner or later effects a change in the other; in the life-form it can be their appearance, but, as this chapter tries to show, it can also be their distribution.

Realms of the earth

The environmental influences that affected the distribution and evolution of animals on islands, whether oceanic or continental, can also be detected in operation in the past — as well as now — on the large continents. The story is less easily detected there, however, because the land masses are larger, the number of animals greater, and the geological influences more numerous. Nevertheless, today a survey of the total globe reveals that geological forces have affected the distribution of kinds of animals into six discernible collections of animals distributed in separated geographical areas, called *realms*. The six *realms* are labeled the *palearctic,* the *nearctic,* the *neotropical,* the *Ethiopian,* the *Oriental,* and the *Australian* (Fig. 4–5).

Each realm is separated from another by a natural barrier, in most cases large expanses of ocean, but sometimes by mountain ranges, such as the Himalayas. The boundaries of the realms were drawn as early as 1857, based on these natural barriers and the types of birds found in each. In 1876 the boundaries were adjusted a little to accommodate other kinds of animals.

In the next section we shall look in detail at some of the kinds of creatures composing each realm. Such a review has limitations that should be kept in mind: the numerous and immense geological forces in the past which isolate today's realms are not fully known, and so not fully described. Thus, the role of environment in bringing about the distribution of organisms in each realm is slighted. Nevertheless, foremost in mind should be the realization that the kinds of animals in any realm is a result of life and environment operating together in the past.

The palearctic realm

The far-reaching peaks of the Himalaya Mountains and sterile Sahara desert on the south and the stretching seas on all other sides bound the *palearctic realm.* This realm includes all of Europe and the USSR, east to the Pacific coastline. The palearctic realm does not contain great animal diversity. For example, only 28 species of land mammals, not counting

98   Chapter 4:   The distribution of life

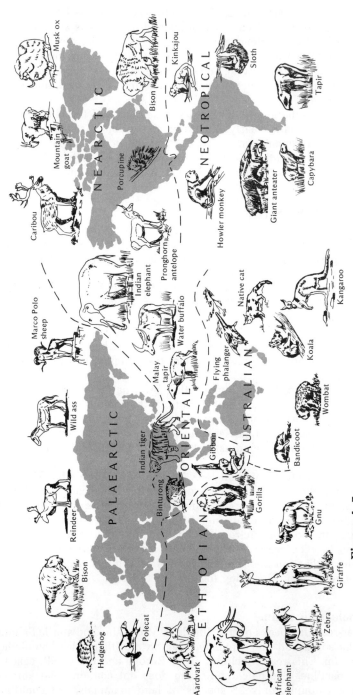

Figure 4-5.
The realms of the world, showing representative mammals. (From *Life: An Introduction to Biology*, by George Gaylord Simpson and William S. Beck, copyright © 1957, 1965 by Harcourt Brace Jovanovich, Inc., and reproduced with their permission.)

bats, inhabit the area, a far smaller number than in other realms. Of the animal families which are there, many are found elsewhere. Only representatives of two families, both rodents, are unique to the realm.

The nearctic realm

The *nearctic realm* shares much with the palearctic. Reaching from the middle of Mexico northward, and from the Aleutian Islands off the coast of Alaska westward, to Greenland eastward, it has weather and vegetational conditions which are similar to the palearctic. Conifers, deciduous forest, tundra, and grassland occur in both. Neither the palearctic nor the nearctic contains many animal families, although both have representatives of four families, which include the wolf, the hare, the moose, the caribou, the wolverine, and the bison. Four other families are shared with the realm to the south, the neotropical, but notably, most families of animals found in neighboring regions are not found in the nearctic. Only a few families are *endemic,* or native, to the nearctic; these include species such as the pronghorn antelope, pocket mouse, and pocket gopher.

The neotropical realm

The *neotropical realm* comprises all of South America, a large part of Mexico, and the West Indies. It is surrounded completely by sea except on the north, where it is connected via Central America to the nearctic realm. This realm has diverse environments, including broad-leaved evergreen forests in the Amazon Valley, grassy plains, deserts, mountain faults, and plateaus. Much of the region is tropical, and unique and plentiful groups of animals are found. For example, no other realm boasts of so many unique mammal families. (Sixteen are endemic, three of these in the order Edentata which include the anteaters, the armadillos, and the sloths.) Other families are represented by such familiar animals as monkeys, opossums, porcupines, guinea pigs, and chinchillas. Bats and birds are also diverse and plentiful.

The Ethiopian realm

Africa south of the Sahara desert and the southern section of Arabia compose the *Ethiopian realm.* It is completely surrounded by sea, except on the north, where it is connected to the palearctic realm. The geography of the realm resembles the neotropical, but it does not stretch into the southern temperate zone. This realm contains the most widely distributed and diverse fauna of any — 38 families of mammals, not counting bats. Some, such as coneys and wild horses, are shared with the palearctic, and Old World monkeys, apes, elephants, and rhinoceroses are shared with the Oriental. The Ethiopian realm is second only to the neotropical in number of endemic families. Unique organisms include giraffes, hippopotamuses, aardvarks, and numerous rodents. Birds are plentiful and many are shared with the Oriental realm.

The Oriental realm

India, Indochina, southern China, Malaya, and the westernmost islands of the Malay archipelago make up the *Oriental realm*. Oceans bound the realm on all sides, except for the Himalaya Mountains on the north. Dense forests cover much of the area. The fauna of the Oriental region resembles that of the Ethiopian in many ways, although it does not have as many unique families. A total of 30 mammalian families, not counting bats, inhabit the area. Included are porcupines, hyenas, and pigs (shared with the palearctic and Ethiopian regions), and monkeys, apes, elephants, and rhinoceroses (shared with the Ethiopian realm exclusively). Birds and reptiles are plentiful.

The Australian realm

The *Australian realm* includes Australia, Tasmania, New Guinea, and some of the islands of the Malay archipelago. (New Zealand and the Pacific Islands are not included because they are considered oceanic islands.) All the land masses are islands and, consequently, are separated from all other realms by extensive ocean barriers (except in the Malay archipelago). North Australia and New Guinea are tropical and contain rain forests; most of central Australia is hot and dry; southern Australia is in the temperate zone. Only nine mammal families are found, not counting bats, eight of which are unique to the realm. Other animals have been introduced by man, such as rabbits, rats, and mice. In general, the realm has few placental mammals, none of them carnivores, but an abundant variety of marsupials, examples of which include the kangaroo, the wallaby, bandicoot, and wombat. Other animals belong to the egg-laying mammal families and include the platypus and the spiny anteater.

The effect of environmental changes on realms

Many environmental changes have helped to form today's realms. In general, the changes have created or destroyed the realms' boundaries or have altered the interior of the realms. In the last 10 million years, the boundaries of realms have remained fairly constant and definite.

The only boundary which is considered somewhat indefinite is that between the Australian and Oriental realms in the Malay archipelago. Here a string of islands tie the realms together. Once an arbitrary boundary was drawn between the Philippines and the Moluccas in the north, between the Celebes and Borneo in the southwest, and between Lombok and Bali southward; but later another line, more sensibly drawn, separated those islands that supported a majority of Oriental fauna from those with a majority of Australian animals. No mountain range or wide stretch of sea separates the realms at this point so that a recognized transitional zone exists between the two. The area is called *Wallacea* and contains islands with a continuum of Australian fauna reaching toward the Oriental realm and a continuum of Oriental fauna stretching toward Australia.

Definite boundaries surrounding most realms today were not always present, however. In fact, past history of the land masses indicates that boundaries have come and gone repeatedly. Some of today's land masses, such as Central America, were submerged in the past, and the presently shallowly submerged land bridge between Alaska and Siberia was previously exposed, continuous land. These changes in the elevation of the land, caused either by the temporary banking of water in ice as seen in the glaciated Pleistocene 1 million or so years ago or to primary geological changes deep within the land itself, have had a profound effect in determining the kinds of faunal collections found in the realms today. So have other environmental changes, such as temperature, moisture, and vegetation.

What are some significant geological events in the earth's ancient past that have led to the distribution of organisms in the realms? Some events have been fairly reliably traced; others are poorly known. The least information is available about the very early geological occurrences. (For example, one hypothesized very early event that would probably affect our understanding of the distribution of life today is the *continental drift hypothesis*. This hypothesis claims that the continental land masses were once close together, probably contiguous, and that geological forces pulled them apart to form separate continents (Fig. 4–6). Starting with the Mesozoic era, geological conditions are fairly well-known. This era, covering about 120 million years, during which the first significant bird and mammal forms evolved, saw the reptilian dinosaurs at their zenith and marks the general emergence of the continents as identified today. The Mesozoic probably included several trends: a change from arid and warm climates to wet and cool climates on the continents; a general sinking of the continents; and extensive mountain building, producing, for example, the Rocky and Andes Mountains.

However, the most significant geological changes which explain the present distribution of animals in the realms today are not from the Mesozoic but from the Cenozoic, especially up to the most recent epoch, known as the *Pleistocene,* which began about 1 million years ago. During the older eras of the Cenozoic (the Tertiary period), the interiors of the continents underwent profound changes, although their basic shapes and sizes probably stayed the same. Mountain ranges were born or remade, and scattered volcanoes sputtered into existence. The Alps and the Himalayas, effective realm boundaries today, rose out of southern Europe and Asia, an area so low as to be covered with the giant inland Tethys Sea; and in low and partially sea-covered North America, the Appalachians appeared, and the already elevated Rocky Mountains were thrust higher. The Andes Mountains continued to rise, and active volcanos pocked Central America. Only Australia remained calm throughout the period. In general, the climate, at least north of the tropical regions, began to cool, a trend continuing today, and the temperate forests, which existed far into the Arctic regions, and the tropical vegetation, which grew in the present temperate areas, began to recede. Such major geological and geographical changes must have greatly influenced the distribution of organisms and, thus, encouraged the formation of today's realms.

Have the times of these significant events been exactly pinpointed? Yes, a great deal of reliable evidence telling when the events occurred

Chapter 4: The distribution of life

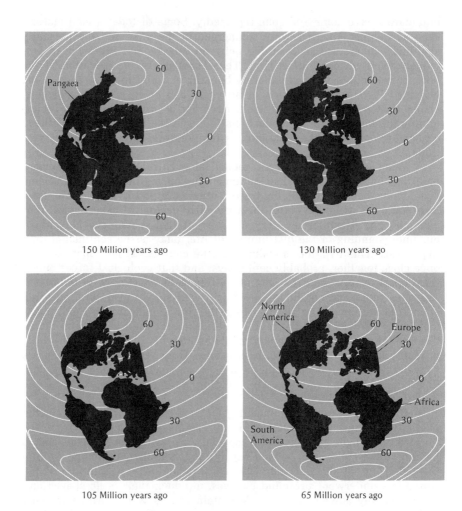

Figure 4–6
The continental drift hypothesis suggests that the continents may have migrated in the last 150 million years.

has come from the dating of fossils and known sequences of geological formations. Africa, for example, appears to have been connected to Eurasia in the early Tertiary, about 70 million years ago, but separated before that, most likely by the Tethys Sea. Connections between North America and Eurasia were apparently made and broken repeatedly across the Bering Strait, probably at least during the Paleocene (70 million years ago), the middle Eocene (50 million), the middle to late Oligocene (35 to 25 million years), and briefly in the first half of the Pliocene (10 million). A bridge between the nearctic realm and the neotropical realm most likely was absent through most of the Tertiary period up until the Pliocene era, 10 million years ago. Then apparently, the connection became complete and remained that way until today. Previously, islands between the two realms may have been the only existing connections. Australia appears to have been isolated from all other areas throughout the Tertiary. Some evidence exists that perhaps the islands in the Malay archipelago formed a minor connection to the Oriental realm, but that otherwise Australia was isolated.

The coming and going of realm barriers greatly influenced the migration of animals from one realm to another. G. G. Simpson, a long-time student of zoogeography, has classified the routes over which animals can move when the barriers are down. Depending on the rate of migration, he

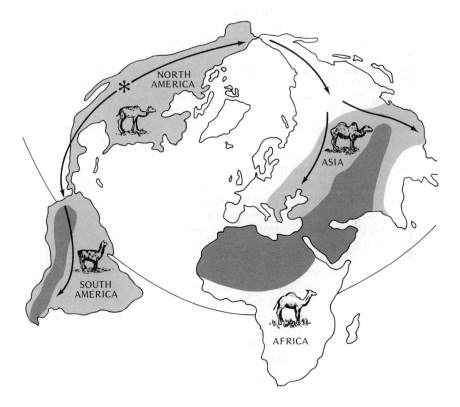

Figure 4–7
The distribution of the camel family during the Pleistocene. The dark areas represent areas inhabited by contemporary members of the group. (From N. D. Newell, "Crises in the History of Life." Copyright © 1963 by Scientific American, Inc. All rights reserved. Adapted from *Life: An Introduction to Biology,* by George Gaylord Simpson and William S. Beck, copyright © 1957, 1965 by Harcourt Brace Jovanovich, Inc., and reproduced with their permission.)

calls these routes *corridors, filters,* and *sweepstake routes.* A *corridor* is a path that allows for the easy migration of most species in neighboring realms. Examples of corridor routes include that path from western Europe into China. As a result of the easy access to adjacent realms, the fauna in these realms are similar. Dissimilar animals in different realms indicate little migration in recent times.

A *filter* is a route through which some species migrate easily, but others do not. Frequently, filters are narrow connections between realms, such as Central America, today's connection between the neotropical and the nearctic realms.

Finally, the route over which the migration of organisms is most unlikely is known as a *sweepstakes route.* In this route the barriers are such that they are seldom crossed. Land organisms on islands such as the Galápagos, or on Krakatoa after its violent eruption, are examples of survivors migrating over a sweepstakes route, undergoing *sweepstakes dispersal.* Sweepstakes dispersal probably explains the settlement of Hawaii.

However the migration of animals occurs, the consequences of faunal interchange are signficant to the animal populations in the realm that has been entered. The biological needs of a newcomer to an area may clash with the needs of an indigent species, and one or the other may die out; or the transplanted species may fill places in the environment not previously filled and eventually interbreed with indigent fauna, giving rise in time to a more diverse realm population.

Today's fauna in North and South America is an example of the results of such activities long ago. In fact, according to G. G. Simpson, historically three influexes of animals across the Central American bridge are thought to have occurred which have helped to enrich the fauna of each continent, especially that of South America. The oldest "stratum" of mammals in South America came during the early Paleocene, when the two continents were connected. Included in this group are the ground sloths, anteaters, and armadillos. Immigrations in the late Eocene to Oligocene came about by sweepstakes dispersal on the island chain which then represented what is now Central America. Braving the partially submerged area were the New World monkeys, among others. From the Pliocene epoch until recently came the deer, camels (Fig. 4–7), tapirs, horses, cats, raccoons, bears, dogs, mice, squirrels, rabbits, and shrews. Eventually, during this last migration, 15 or possibly 16 families invaded South America. (Incidentally, 7 families are also thought to have migrated into North America from South America during this time.)

Examination of South America's fauna today indicates differences among the three strata. Animals of stratum one, the oldest, have diversified the most from related organisms in other realms. Animals from stratum three have diversified least, probably because they have not had adequate time to do so. All together, however, the mixture of fauna in South America is rich and varied, even though many of the forms once there are now extinct.

The development of realms — with unique environmental conditions and unique forms of life — illustrates once again the intimate relationship between life and environment. Out of migration alternating with periods of isolation when barriers change and realms are sealed off, and the ever-working evolutionary forces, come new species of living things. Today the realms are filled with these evolutionary products and, in fact, help give them their identity.

The distribution of organisms within a realm: Biomes

Close examination of any realm reveals that it contains varied environments, or habitats. The factors that produce the habitats are obvious: amount of rainfall, amount of sunlight, mean annual temperature, exposure to wind, type of soil, and the like. Clearly, different habitats encourage diversification of animal and plant species. Hence, the varied organisms within a realm are due not only to the history of the area in relation to other realms (as suggested later in the description of the distribution of the horse), but also to the environments found within the realm. (Again, as we have seen many times before, life and environment are intimately tied; a full understanding of life as we find it today depends on an appreciation of this relationship.)

Generally, six kinds of environment occur within the realms. These six environments, known as *biomes,* are the *grassland,* the *tundra,* the *desert,* the *temperate deciduous forest,* the *coniferous forest,* or *taiga,* and the *tropical rain forest.* All are named according to the predominant form of vegetation, although it should be appreciated that "biome" refers to the kinds of animals as well as the kinds of plants that exist there. (Of course, the plant ecologist would recognize primarily the flora; probably, he would label the six divisions "major plant formations.") The association between the distribution of animals and the distribution of plants is due to the dependence that all animals have on the green plants and the environmental conditions which indirectly affect a diversification of animals as well as plants. For example, areas with 12 to 30 inches of rainfall per year support abundant grass, but as a consequence, such areas also support numerous grazing animals.

Some generalizations can be made regarding biomes. For example, biomes near the equator — no matter in which realm they occur — support organisms that tolerate extreme heat and light. Those biomes farther north support life more tolerant to cold and less light. Such a progression of biomes from the equator northward can also be seen when progressing up a mountainside, because the environmental transitions are similar (Fig. 4–8).

Let us look at the details of the major biomes, those subunits of realms containing unique plants and animals determined by the prevailing environmental conditions. Keep in mind that life and environment interact to produce the characteristics of the organisms in any biome.

The temperate deciduous forest

The temperate deciduous forest biome (Fig. 4–9) occurs in temperate zones. Cold winters and warm summers with evenly distributed and reasonably abundant rainfall (about 40 inches per year) usually support

106   Chapter 4:   The distribution of life

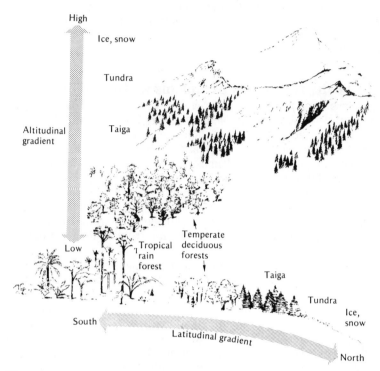

Figure 4–8
The correspondence between the altitudinal and latitudinal biomes in eastern North America. (From *Biological Science,* 2nd ed., revised, by William T. Keeton, illustrated by Paula DiSanto Bensadoun. By permission of W. W. Norton & Company, Inc. Copyright © 1967, 1972 by W. W. Norton & Company, Inc.)

deciduous trees, such as oak, hickory, elm, poplar, and birch. Still found are remnants of a vast deciduous forest biome once stretching across the eastern United States, the British Isles, central Europe, China, and southeastern Siberia. South America also has deciduous forests.

Well-known herbivores that inhabit the temperate deciduous forests are deer in North America, Eurasia, and South America and wild pigs in Eurasia. Predators include cats, wolves, foxes, and martens.

The taiga

The taiga, or coniferous forest (Fig. 4–10) occurs just north of the temperate deciduous forest biome in North America, Europe, and Asia. The growing season is short, from 3 to 6 months. The forests are mostly densely coniferous, with spruce and fir especially, but pines and hemlock, also. Birch, aspen, and alder shrubs are plentiful.

Figure 4-9
The large trees are virgin white oak trees in an eastern deciduous forest. (Courtesy of U.S. Forest Service. Photo by L. J. Prater.)

The soils in the taiga are usually acidic and low in minerals, a condition that results from a great quantity of water seeping through the soil with no significant upward movement of water. Consequently, essential materials, such as calcium, potassium, and nitrogen, are leached away.

Animals found in the taiga include the moose in North America (elk in Eurasia), black bears, wolves, martens, lynxes, and squirrels.

Grassland

Grassland biomes (Fig. 4-11) occur around the world and bear various names: steppe, prairie, plain, campo, pampa, and llano. Probably no other biome occupies so much area of so many realms. One major characteristic ties them all together — the plentiful production of grass. Consequently, they all support great quantities and varieties of herbivores and their attendant carnivores. The vast herds of bison on the North American grasslands are an excellent recent example, but the flourishing and rapidly evolving varieties of horses in past eras is another example. That the horse seems to have developed in the North American continent is not at all surprising in view of the quantity of grass available to sustain it. (See the insert, "The Biogeography of the Horse.")

Figure 4-10
The taiga forest sometimes needs protection from insect pests and lumber companies. (Courtesy of U.S. Forest Service. Photo by J. L. Averell.)

The animals found in this biome in different realms are very diverse, although ecologically they may occupy a similar place in the biological scheme. For example, kangaroos and wallabies are in no structural way related to antelopes, yet they play a similar ecological role in their grassland habitats in Australia and Africa, respectively.

Despite the diversity of fauna found on the various grasslands, grassland biomes show no basic differences no matter where found. Beside grass, the one common characteristic is the low, intermittent water supply. Rainfall may total only 12 to 20 inches per year. If rainfall is more abundant — some grasslands have as much as 40 inches per year — it is usually concentrated within a very brief and intensive wet season. Frequently, the porosity of the soil in the grasslands facilitates runoff, and hence, plant roots do not have a constant and abundant water supply.

The soils of the grasslands are usually extraordinarily fertile. This is because through the years the decaying plant materials have been easily incorporated into the ground. The soil is often slightly alkaline because of the constant reentry of materials such as calcium and potassium by means of the upward movement of the water that results from evaporation. These soils, black earth or brown earth, are today some of the best croplands in the world.

Figure 4-11
Section of grassland interrupted with trees from the edge of a forest belt, naturally eroded by a silt-carrying river. (Courtesy of U.S. Forest Service. Photo by B. Alt.)

The biogeography of the horse
An example of how environment affects distribution

Vegetation and climatic changes in early realms, and alterations in the bridges that bound them, account in large part for the contemporary distribution of the animals in the world. A well-researched example of an animal whose distribution was so affected is the horse.

The first horse may well have originated in western North America (although some researchers feel that central Asia is more likely). It spread rapidly during the Eocene across the nearctic, palearctic, and neotropical realms and differentiated into species of several sizes. From 10 to 20 inches high at the shoulder, with arched and flexible back and long, stout tail, this horse is characterized best by its four front and three hind toes on each foot, each toe terminating in a tiny hoof (Fig. 4-12). The unique teeth suggest a very particular herbivore diet, probably composed mainly of succulent leaves, soft seeds, and fruits. Hence, this early animal, *Eohippus*, was a browser, not a grazer.

Soon after the appearance and spread of the early species of Eohippus, the connection between Eurasia and North America is believed to have

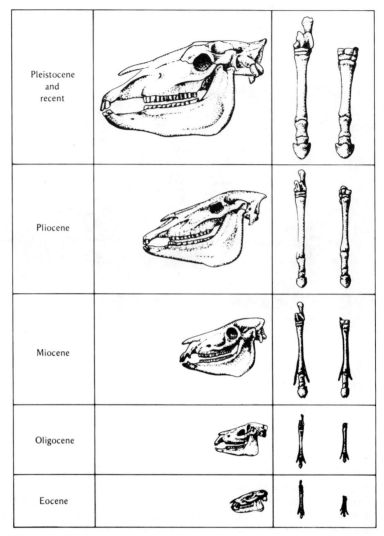

Figure 4–12
Comparison of the skulls, hind feet, and forefeet of ancient and modern horses of successive periods illustrates the transformations that led from a fox-sized ancestor approximately 60 million years ago to the contemporary horse. (From J. F. Crow, "Ionizing Radiation and Evolution." Copyright © 1959 by Scientific American, Inc. All rights reserved.)

disappeared, and migration between the two continents stopped. Horses on each continent continued to develop in continental isolation.

As a result, the European Eohippus diversified into several specialized species which became extinct at the end of the Eocene era, perhaps because their increased specialization disallowed adequate adaptation.

Some researchers believe that factors which contributed to their extinction include increased aridity, which reduced and altered the vegetation, and the influx of other herbivores, such as the rhinoceros and deer, which competed for food.

However, the North American Eohippus did not become extinct; it evolved into an advanced form, called *Epihippus*. The new horse remained small. The forefeet kept four toes and the hind feet three, but the tiny vestiges of two other toes on the hind feet were lost completely. The teeth tended to become more molarlike, an adaptation leading to an effective grazing habit.

At the beginning of the Oligocene, Europe and North America were reconnected, but Epihippus did not migrate into Europe. Epihippus stayed on the North American continent and during the Oligocene evolved into *Mesohippus,* an animal that resembled the horses of today. Mesohippus stood close to 24 inches high or higher, on long slender legs well adapted for running, and had its more-intelligent brain encased in a much larger skull. Each foot bore three toes (not four as in Eohippus), with a pad between and behind them. The teeth were very molarlike and well-structured for grinding and crushing, a development essential for grazing (Fig. 4–12).

At the close of the Oligocene, the history of the horse in North America becomes increasingly complex. Numerous forms developed in a manner seen earlier in Europe. By the early Miocene period, land connections between the two continents were reestablished (after a brief separation in the late Oligocene), and migration began. The numerous and advanced forms invaded Europe and Asia, but by the early Pliocene all the migrating forms died out. One reason may be because they retained a browsing habit primarily, whereas at least two forms that did not migrate developed a dependence on grazing. These were the North American forms (*Parahippus* and *Merychippus*), destined to be part of the line that would lead to the modern horse.

The dentition seen in these horses, one appropriate for grazing, seems to have developed coincidentally with the spread of grasses around the world. This emphasizes once again the influence of environment on evolution in-organisms. Fossilized grass seeds in rock have been found in the same period as fossilized horses with high teeth. These teeth provided durable grinding surfaces. Besides new developments for grazing, these fossils show other adaptive characteristics. They were larger than their predecessors, showed a reduction in side toes, and carried their weight on one central hoofed toe instead of on a pad. (Fig. 4–12).

By the end of the Miocene other advanced characteristics are seen. The form showing them is called *Pliohippus;* most notably, it has lost its side toes and walks on a central hoofed toe. *So significant a development* was the one-hoof characteristic that it would be retained in contemporary horses along with large size (some Pliohippus were as large as today's ponies) and grinding teeth well adapted for grazing (Fig. 4–12).

During the Pliocene, Pliohippus underwent slight adjustments in its physical structure. Included in these changes was a straightening of the molar teeth and a complicating of the tooth surface with folds and pits. Pliohippus was evolving rapidly during the Pliocene into our modern horse, Equus.

Continental changes were occurring that would also influence Equus' distribution. Land bridges between North America and Eurasia and between North America and South America were reestablished for the first time since before Eohippus appeared. Communication was possible into all realms except Australia. Consequently, the very large grass-eating Equus was presented by the end of the Pliocene epoch with the unique opportunity of walking on its one-toed feet into any part of the world except Australia.

Fossilized remains of Equus from the Pleistocene are found everywhere. Most peculiar, however, is the rapid extinction of the group during this epoch from many areas, including North America, soon after its wide expansion. We know that in northern climates they did survive the Ice Age. Also, adequate forage grass seems to have been present, and no evidence exists of a new enemy evolving. Some authorities propose that their extinction could have been due to an epidemic or to their being killed by man for food. Neither explanation is adequate, however. In fact, no one has explained the horse's extinction. Whatever the cause, it must have been a factor to which the animals could not adapt.

Today, naturally occurring descendants of Pleistocene horses are represented by only a few remnants, all in Eurasia and Africa. One descendant (*Equus caballus*), which later gave rise to the domestic horse, once roamed Europe and central Asia. The Romans knew them, and until the Middle Ages they lived in the more remote areas of central Europe. Horses that roamed eastern Russia, known as *tarpans,* may also have been of this group. The Mongolian horse (Przewalski's horse), once found abundantly in western Mongolia, may also have been a naturally occurring variety of *E. caballus,* but it now appears that this horse is also nearly extinct. Wild horses, such as those found in America's west, are not naturally occurring *E. caballus* but domesticated horses that got free. Hence, all that apparently remains of *E. caballus* is the domesticated horse.

Other Equus species still occur: the Kiang or Asiatic wild ass, once common to China, central Asia, India, and Persia; the true wild ass, ancestor of our domesticated donkey, in northern and northeastern Africa; and the zebras of which there are several species, in Africa.

The desert
The grassland biome is characterized by little moisture throughout the year, but its condition seems very wet compared to the desert biome. Deserts (Fig. 4–13) have as little as 10 inches of precipitation per year,

and frequently, this falls during a very limited time. Furthermore, the temperature fluctuations are extreme; the summers produce temperatures of 35 to 40°C or more, and winters and nights, even in summer, may be extremely cold.

As a consequence, the flora and fauna that inhabit the desert must be very well-adapted to endure these harsh conditions. Most of the plants are small and conditioned to spurts of growth and reproductive activity when the rains occur. Some plants complete their reproductive cycles in a few days. Small leaves that reduce transpiration, tough outer coverings, long roots, and cells that store substantial amounts of water are all common structural adaptations for life in the desert.

Animals, too, are well-adapted. Many are very fleet, often eluding predators with a jumping or zigzag motion, an obvious survival mechanism in an area with so little cover. The burrowing habit, drab coloration, and fringed toes for traction are among common characteristics of desert dwellers.

Deserts frequently intergrade with other biomes, such as grasslands. Such conditions are usually produced by geographical or meteorological conditions that produce a quantity of moisture above that found in the desert proper.

Figure 4–13
Not all deserts are hot or grow cacti. Here is a view of desert country in Nevada. (Courtesy of U.S. Forest Service. Photo by L. J. Prater.)

The tundra

The tundra (Fig. 4–14) is the bleakest biome. It lies above the latitude 57°N. It contains a spongy and rough soil surface, the result of alternate thawing and freezing. Beneath the surface lies the permafrost, ground

Figure 4–14
Alpine tundra, showing sparse vegetation and patches of mountain snow. (Courtesy of U.S. Forest Service. Photo by H. E. Schwan.)

perpetually frozen. The temperature is very low most of the year. So are precipitation and evaporation.

Plants that grow in the area have only the soil above the permafrost for their roots. Consequently, all are shallow rooted and small. Grasses and sedges are common, especially on the marshy areas, as are acid-loving plants such as blueberries; lichens, too, are abundant.

In spite of the inhospitable conditions there, some animals do thrive. For example, large mammals occur, such as the musk-ox, the reindeer or caribou, the arctic wolf, the arctic fox, the arctic hare, and lemmings. Birds are abundant during the brief summer but are scarce in the colder months, when the area is in darkness most of the time. The color of many birds and animals that inhabit the area in the colder months is white, an obvious adaptation to the snowy environs. Sometimes even the usually nonmigrating fauna is driven out by the conditions. Snowy owls, for example, sometimes migrate to more southern regions when the winters are especially harsh.

Other adaptations to the environment include a tendency for animals to develop into short or stocky varieties, compared to representatives of their groups found in warmer climates. This tendency, presumably an adaptation to conserve heat, is known as *Bergmann's rule*. Perhaps for the same reason, animals in colder climates also appear to frequently produce shorter appendages, such as ears and tails, than similar representatives in warmer climates. This tendency is known as *Allen's rule*. Some experiments appear to support both rules. For example, mice reared at 31 to

35.5°C have longer tails than those of the same strain produced in temperatures of 15.5 to 20°C, but the mice reared in the colder temperatures have heavier bodies. The domestic chicken, when raised in both cold and warm temperatures, seems to respond similarly. Chickens kept at 6°C during the third and fourth months of their existence were shorter, heavier, and produced shorter toes and tails than chickens of the same flock raised at 21 to 24.5°C.

The tropical rain forest

Rain forests (see the frontispiece, Chapter 1) can occur in the temperate zone (the one on North America's Pacific northeast coast is one example), but they are most impressive in the subtropics or tropics. Rain forests commonly occur in most of northern South America and Central America, southern Asia, central Africa, the East Indies, and in northeastern Australia. A continuous water supply and a long growing season make the tropical rain forests the lushest and most varied of the biomes. Not uncommonly the tropical rain forest will be composed of over 100 species of trees — one forest had over 500! — and characteristically, the individual specimens of any particular species will be greatly scattered, sometimes up to miles apart. Rain forests are notable for their vertical stratification, that is, the layers of varied vegetation. The canopy formed by the spreading leafy tops of trees, which is 20 to 40 feet high, is so effective in intercepting sunlight that the forest floor is usually extraordinarily shady. As a consequence, few plants can survive close to the ground even though high humidity and even temperatures are found there, although liana vines may twist around the trees for support.

Contrary to popular view, the tropical rain forest is not teeming with animals. In fact, most evidence of any animal life at all is at dusk, when the birds and mammals break the usual stillness. During the day only the occasional cry of monkeys or parrots is heard. Vertebrates found in the various rain forests include the musk deer and pigs in the Old World and the rodents and peccaries in South America. Cats are common in most rain forests, invertebrates are not uncommon, and insects are abundant.

The varied distribution patterns of animals as seen in biomes are another testimony to the unceasing interaction of life and environment which characterizes the living world. Part of this world is man, whose history of distribution is a story of increasing adaptation to the environment — of increasing manipulation of the environmental conditions to enhance his chances for survival.

Today, man's distribution is worldwide. His adaptation to the diverse and often harsh environments appears to be successful. Yet the current environmental problems seem to be symptoms of maladjustment. What can man do to maintain a life-sustaining interaction with his world? In the next chapter we shall look at the ways in which animals have behaviorly responded to their environments, so that we may gain insight into man's potential to *do* something about the crisis in his relationship with his environment.

Summary

The main theme of this chapter states that the distribution of plants and animals around the earth is a result of the interaction of the organisms with their environment through long periods of time. This distribution, a study known as "biogeography," is a result of evolution. Insight into how interaction of organisms with their environments has influenced their distribution can be gained (1) from a study of the spread of early man into diverse parts of the globe , (2) the forces affecting the distribution of plants and animals on oceanic islands such as the Galápagos Islands and the continental islands such as the British Isles, and (3) an examination of the major land divisions of the world, or realms, and the characteristic animals which exist in each, and (4) a study of biomes, environmental subunits of realms, and the unique plants and animals found in each. Hopefully, out of the study of biogeography will come better understanding of the forces that enable man to inhabit the environments in which he is now found and the delicate relations that exist between him as a species and his surroundings.

Supplementary readings

Allen, D. L. 1967. *The Life of Prairies and Plains.* McGraw-Hill Book Company, New York. (Published in cooperation with the World Book Encyclopedia). 232 pp.

Braun, E. L. 1950. *Deciduous Forests of Eastern North America.* McGraw-Hill Book Company, New York. 596 pp.

Buxton, P. A. 1923. *Animal Life in Deserts: A Study of the Fauna in Relation to the Environment.* Edward Arnold, Ltd., London. 176 pp.

Cain, S. A. 1944. *The Foundations of Plant Geography.* Harper & Row, Publishers, Inc., New York. 556 pp.

Cloudsley-Thompson, J. L. 1954. *Biology of Deserts* (Proceedings of a Symposium on the Biology of Hot and Cold Deserts). The Institute of Biology, London. 223 pp.

Darlington, P. J. 1957. *Zoogeography: The Geographical Distribution of Animals.* John Wiley & Sons, Inc., New York. 675 pp.

Dice, L. R. 1952. *Natural Communities.* University of Michigan Press, Ann Arbor, Mich. 547 pp.

Elton, C. S. 1958. *The Ecology of Invasions by Animals and Plants.* Methuen & Co., Ltd., London. 181 pp.

Freuchen, P., and F. Salomonsen. 1958. *The Arctic Year.* G. P. Putnam's Sons, New York. 438 pp.

George, W. B. 1966. *Animal Geography.* William Heinemann, Ltd., London. 142 pp.

Krutch, J. W. 1968. *The Desert Year.* The Viking Press, Inc., New York. 270 pp.

Leopold, A. S. 1961. *The Desert.* Time-Life Books, New York. 192 pp.

Loomis, F. B. 1926. *The Evolution of the Horse.* Marshall Jones Company, Francestown, N.H. 233 pp.

Milne, L., and M. Milne. 1962. *The Mountains.* Time-Life Books, New York. 192 pp.

Oosting, H. J. 1956. *The Study of Plant Communities,* 2nd ed. W. H. Freeman and Company, San Francisco. 440 pp.

Simpson, G. G. 1951. *Horses: The Story of the Horse Family in the Modern World and through Sixty Million Years of History.* Oxford University Press, New York. 247 pp.

Stonehouse, B. 1971. *Animals of the Arctic: The Ecology of the Far North.* Holt, Rinehart and Winston, Inc., New York. 172 pp.

Weaver, J. E., and F. W. Albertson. 1956. *Grasslands of the Great Plains.* Johnsen, Lincoln, Nebr. 395 pp.

Chapter 5

Granted that modern biology accepts the primary importance of natural selection in evolution, the question remains: "What is it that has survival value?"

Social Behavior from Fish to Man
William Etkin

Behavior and survival

Chapter 5: Behavior and survival

In the first four chapters we examined the close relationship of life and environment which through millenia constantly has produced new and often innovative plant and animal forms. Specifically, we looked at the evolutionary process which accounts for change (Chapter 1), and how this process explains the development on earth of life in general (Chapter 2) and man in particular (Chapter 3). In Chapter 4 we considered another evolutionary product of the interfacing of living things with their environment — the distribution of life, including man, around the world. In this chapter we shall look at one more product of the relationship of organisms and their environments through time: the *behavior* of animals.

Behavior, as well as structure, can determine the survival of a species. The ameba's simple twisting in response to probes of light and man's ability to adjust to immediate exigencies and to foresee future threats are examples of a wide range of behavior which accounts for the survival of these life-forms. But to emphasize the significance of the evolution of appropriate behavior for the survival of an organism is not to deemphasize the significance of the structures which permit that behavior. In fact, the behavior of any organism is a result of the possession of the appropriate structures to allow the behavior to occur. Thus, as the behavioral responses of organisms become complicated, the structures that account for those behaviors also become complicated. A single-celled ameba evidences simple movements; man, with an elaborate central nervous system, including a mysteriously functioning brain, participates in complex social interactions ranging from peaceful communal living to aggressive war.

In this chapter we shall be concerned in detail with evolved animal behavior as it explains survival and the structural developments that have evolved to allow for this behavior. We shall begin by looking at the possible origin of some of man's behavior, especially his social behavior (all the while realizing that it is with a complex cerebrum and nervous system which allows this complicated behavior to occur). Then we will look at the behavior of simple animals and the simple structures which accompany their responses. Finally, we shall look at some of the organisms which behave not in a sophisticated manner, as the human being, or simply, as the ameba, but on the continuum between these two extremes. The view of behavior as an evolutionary product and a survival mechanism will, thus, be viewed in an evolutionary sense fully consistent with one of the major trends evidenced in any evolutionary study of life: that organisms seem to develop structurally, and in this case, behaviorly, from the simple to the complex.

Out of such a view, spanning the entire animal kingdom but citing specific representatives, you will gain additional insight into the nature of the evolutionary process and how closely the process binds animal life to its environments. Hopefully, out of such insight will come suggestions for the formation of a behavioral strategy that will help us, as another animal species, to relate more reasonably with our environments and, thus, ensure ourselves of a future.

Man's social origins

Man's past can be studied in many ways. In Chapter 3 we viewed it as *physical anthropologists;* in this chapter we shall approach it as *cultural anthropologists.* Both approaches are equally valid, although the goals are different. Physical anthropologists are striving to know how man looked, but cultural anthropologists are concerned with how he behaved. In some ways the physical anthropologist has an easier task, because he can produce substantive evidence in the form of fossilized bones and skulls. These bits of evidence (even a single molar!) can lead to valid generalized notions of the appearance of ancient man.

The cultural anthropologist has a more difficult time. Because he is concerned with the behavior of ancient man the evidence that he seeks is not as easily produced as a bone or tooth, legacies that tell physical details — how browed the face of ancient man, how long his bones, how shuffling his gait, how stooped his shoulders. Instead, the cultural anthropologist delights over a broken tool or pottery piece, which at best only hints at what ancient man did. Fragments tell only fragmented stories. Yet, despite the difficulty in proving how early man behaved or how that behavior came about, it did happen, nevertheless. Thus, even from very few facts the cultural anthropologists can hypothesize about what may have occurred. Some hypotheses appear sound as evidence accumulates; others fall quickly into disrepute. But because abundant definite evidence is lacking and because man's behavior certainly began some time and in some way, a cultural anthropologist's subjective opinion is not without value.

Let us look at some of the explanations for man's early behavior offered by the cultural anthropologists, recognizing that they may be hypotheses which may or may not bear up in the future. Even a brief view of some of the tentative explanations for man's social behavior will help drive home the point that behavior is an evolutionary product and that — even for man — its beginning probably had survival advantages.

Many cultural anthropologists believe that many aspects of the complicated human society in which we live had their origins at a very early time in man's history, probably during the hunting culture which followed the earlier life style of food gathering. These anthropologists point out that many of the human characteristics that would lead to survival while on the hunt or while participating in a hunting party, such as the ability to relate events quickly and to remember accurately, are among the characteristics that we value today in our most complicated societal activities. Perhaps out of these early hunting activities these traits — selected in a manner thoroughly consistent with Darwin's theory of natural selection expounded millenia later — eventually led to the most notable aspect of being human, the possession of considerable intelligence.

Very likely one of the first results of the development of superior mental ability from the hunting activity was the development of a system of communication, language. It is not hard to imagine situations developing wherein prehistoric man, gathered together in his hunting party, needed rapid and efficient relay of information. Perhaps from no more than vocal inflections similar to what is still heard among apes in packs, language developed. People disagree as to the forces leading to its development, and many factors have been cited. For example, it may be that the differentiating roles of the sexes helped greatly. In a hunting culture, the roles probably often caused the males to be out of hearing or sight of the females. Consequently, the kinds of cries or signs adequate for hunting closely together, such as seen in ape packs today, were probably inadequate for early man. In time, great selection pressure would encourage the male returning from the hunt or the female returning from foraging to develop refined verbal symbols to represent the events that confronted each of them while they were separated.

Another development toward man's becoming "social" was the differentiation of the male and female roles. Hunting most likely turned out to be a male occupation, probably because the females were homebound, childbearing and childcaring. Babies carried on a hunt could be endangered or might cry and frighten prey; other siblings might need protection when the hunting father was absent. Thus, hunting probably selected for strong, fast-moving males and for females who tended to manage the home front well. Probably, this was augmented by the development of secondary sex characteristics in the female such as breasts for childfeeding and a large pelvis for easy births.

No doubt this evolution was further augmented by the loss of *estrus*, the brief and marked period of sex proclivity during which the egg is ready for fertilization. The female, instead, was developing a year-round sexual receptivity, an advantage in that year-round sexual receptivity eliminated neglect of offspring. (Today, it is observed in some animal species that during estrus sexual activity becomes the prime and often total concern of the female, to the point where she may neglect to eat or becomes oblivious to others in the group who are not sexually active males.) Thus, human offspring still needing maternal care were not neglected.

Whatever the reason, year-round sexual receptivity freed sex from having only a procreative function and led to another social development, pair bonding. The ability to decide when and with whom to meet sex needs probably encouraged the development of a more permanent union. A mate sexually available most any time, who possessed pleasing characteristics, probably proved too desirable to abandon after one mating. This, along with the ability of a hunting male to provide for only one female and offspring, probably encouraged pair bonding. Marriage is most likely the contemporary result.

According to some cultural anthropologists, the evolution of the establishment of the pair bond, or marriage, probably led to other developments, for example, the elaboration of the complex social interaction between the sexes. Today, the procedures leading to the gratification of the sex needs of males and females and the establishment of a temporary or permanent pair bonding are so integrated into the overall social structure that it is often difficult to discern where the boundary lies between sexual activity and nonsexual social behavior.

The consequences of pair bonding were many but possibly included other social behaviors, such as the development of the taboo on incest. The social unacceptability of incest may result from the disruptive influence of the competition within the family that would accompany indiscriminate sexual activities among family members. Such a prohibition may have originated, together with the taboo against premarital sex found in some societies, with the prolongation into adolescence of the dependency period of offspring. Adolescent children able to produce offspring, yet still dependent on parents for training and acculturation, possibly were prevented from producing offspring through these taboos. (Prohibition of incest may also have evolved because of the observation that inbreeding among any species often produces an expression of recessive characteristics, many of which have little or no survival value.)

One of the survival advantages of marriage was the beginning of human beings to accept responsibility for each other. Such acceptance of responsibility, probably initially to share food, was the real beginning of man's acculturation as a species. The first marriages were statements that males and females have responsibility for each other and for their children. Further cultural development probably occurred when individuals in a family learned to care for others outside that group, thus recognizing "kinship" with grandparents or uncles and aunts. Today, many of our societal institutions, such as religious beliefs or laws, are the result of man's learning to be concerned for his fellow beings, a cognizance of the "brotherhood" of man.

So modern man evolved, according to some cultural anthropologists, not only physically, but also behaviorly. Today, the truth of the above hypotheses will perhaps never be known. Yet something of the sort described must have happened, for most likely much behavior developed because of its survival advantages. To what extent basic instincts, for example, aggression, now govern his life and to what degree modern man has the ability to choose his behavior are controversies. Perhaps the day will arrive when the human organism understands himself more fully, and the problems of education or government, as examples, will be greatly simplified. Until that day, perhaps the best man can do is to hypothesize his behavior in view of its evolutionary importance and to look for clues for the significance and origin of his behavior in the activities of other organisms.

Responses of simple animals

The claim that much of human behavior may have persisted because it provides (or did provide) survival advantages is difficult to demonstrate. In fact, many of the hypotheses concerning the origin of man's behavior may forever remain unproved. Yet an investigation of the behavior of other animals suggests clearly that many activities characteristic of animals often structurally simpler than man rather convincingly show survival advantages. From a study of a selected few examples, a fair inference appears to be that much of man's behavior — no matter how complicated it appears now — probably at one time, if not now, helped to keep our species alive.

In the remainder of the chapter we shall look at animals that are in many ways simpler than man. We shall look at their behavior and the

124 Chapter 5: Behavior and survival

structures which allow that behavior. Two impressions should come clear: (1) as the behavior becomes increasingly sophisticated, so do the structures that allow for the behavior; and (2) the behavior of animals, whether ameba or man, is an evolutionary product, a result of the interaction through time of life-forms and their environments.

Some of the simplest behavior is seen in the ameba and other one-celled *protozoans*. In general, protozoans have an ability to move from a situation that is not life enhancing into one that is. Amebae, for example, respond very negatively to a strong light and will change their line of movement toward a darker area. On the other hand, they respond positively to food particles and warm temperatures. The amebae move by mere extensions of protoplasm from the cell body proper. These extensions, the *pseudopods*, "flow" the animal in a new direction (Fig. 5–1).

The *paramecium* is a cell similar to the ameba, except that it has a more constant shape and tiny *cilia*, or protoplasmic extensions, which move in unison to propel it through water. Like the ameba, the para-

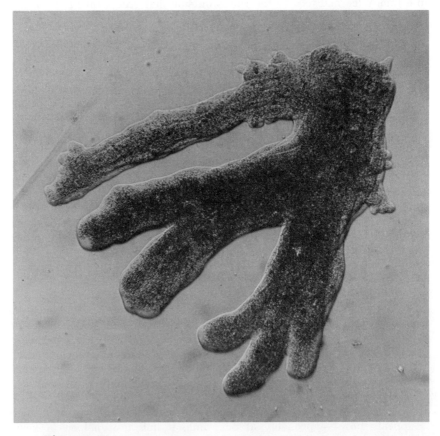

Figure 5–1
The ameba, a common one-celled organism, moves by extending lobes of protoplasm, the pseudopods. (Turtox, Chicago.)

mecium is attracted to some stimuli and repelled by others. When confronted head on with a bubble of carbon dioxide, for example, it will back up, move approximately 30 degrees to the side, and try again. This it will do until it has moved around the barrier. Such behavior is perhaps the crudest sort of *trial-and-error* behavior known, although all animals through man evidence it (Fig. 5-2).

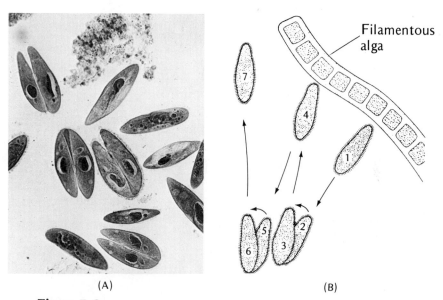

(A) (B)

Figure 5-2
Outer covering of cilia, which are rapidly vibrating extensions of protoplasm, propels the paramecium. Left: Group of paramecia, some conjugating, others feeding. Note the obvious nuclei and cilia. (Turtox, Chicago.) Right: Path of a paramecium as it avoids an obstacle.

Simple behavior patterns accompany simple body plans. The behavior and structure of sponges is consistent with this maxim. Sponges are not much more than many cells aggregated, some of which are differentiated. One type of differentiated cell appears ameboid and simple, yet it has a comparatively refined ability to respond to stimuli. Apparently, the sponge has evolved a condition in which the burden for responding to the environment resides in only a few cells.

A coelenterate has a more elaborate structure, a nervous system composed of nerve cells (Fig. 5-3) that relays impulses fairly well. In addition, some cells, such as the highly structured *nematocyst* (Fig. 5-4), show a very high level of response.

The various kinds of worms are generally considered more recently evolved than protozoa or sponges. They also show more sophisticated behavior and structure. For example, some *flatworms* also have a net of nerves, but others, considered more advanced, have evolved several long nerve cords in addition to the nerve net (Fig. 5-5). Near the anterior

126   Chapter 5:   Behavior and survival

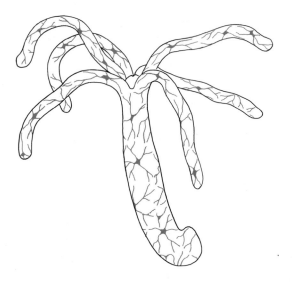

Figure 5-3
A hydra, showing the nerve net.

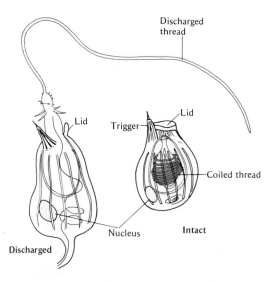

Figure 5-4
Nematocysts are specialized cells that are helpful in securing food found on the tentacles of the hydra. Right: Nematocyst with coiled stinging thread ready for discharge. Left: The thread discharged and ready for the entanglement of prey.

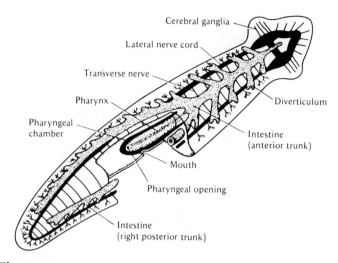

Figure 5-5
Ladder-type nervous system in the flatworm planaria. Note the cerebral ganglion.

end of these animals is located two swellings in the cords, structures called *ganglia*. This more elaborate nervous apparatus allows more complicated behavior. For example, flatworms move through the water with a rippling swimming motion (which can be disoriented by the removal of the ganglia) and a looping crawl (inhibited by removal of the ganglia). Flatworms also respond obviously to stimuli. They will turn away from strong light, some chemicals, and contact with some objects. They may even learn. Worms, swimming in a dish with a roughened bottom, can learn not to cross that area if the dish is vibrated each time they near it.

Earthworms have structures reminiscent of the flatworms, although the ganglia appear to have greater coordination than in flatworms. Their removal affects the tone of the muscles and general sensitivity, although it does not affect locomotion. Learning is demonstrated by placing worms in a **Y**- or **T**-shaped tube. As the worms approach the junction in the tube, an electric shock discourages them from taking one route. Eventually, some worms will learn not to travel into the wired tube arm even when no electricity is turned on.

The behavior of insects suggests that this group probably has an efficient nervous apparatus, too. The refined coordination of their bodies and sophisticated response to stimuli is possible because of the presence in many of a ganglion, or concentration of nerve tissue, in each of their body segments, and an especially large one in the head. The ganglia in the body are very functional and help explain why some insects stay alive even when their heads are gone. For example, the male praying mantis can still copulate with the female even after she chews his head off, a usual courtship activity for mantis. His body ganglia apparently carry on even when he has "lost his head."

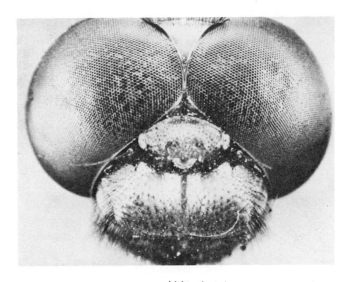
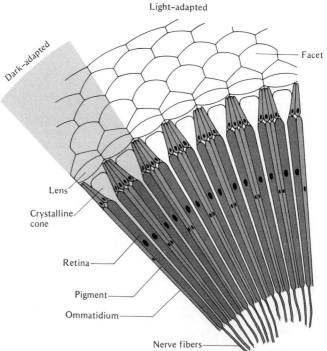

Figure 5–6
Top: The compound eye of the dragonfly is composed of nearly 30,000 units known as ommatidia. Bottom: A cross section through a compound eye reveals the structure of the ommatidia. (From R. Buchsbaum, *Animals without Backbones,* University of Chicago Press, Chicago, 1948.)

The compound eye is an excellent example of the sophisticated sense organs that insects have developed (Fig. 5-6). The eyes are domed and, hence, are in a position to receive numerous stimuli. Each eye is composed of many — sometimes thousands — crystal cones at the points of which are nerve connections to the brain. Some insects also have unusual ability to detect odors and tastes because of receptors in their antennae and around their mouth and feet.

The behavior of bees

Many experiments can be cited to illustrate the sophisticated behavior of insects, but the pioneer work of Karl von Frisch and coworkers with communication among bees is perhaps most outstanding. How do bees that locate nectar-bearing flowers communicate this important information to bees back in the hives? Von Frisch, a German zoologist, was intrigued with this and other questions regarding the behavior of bees. As a result of his curiosity and careful investigations, he discovered some very interesting behavior patterns in these insects, particularly relating to their ability to communicate.

Von Frisch found that bees require a long time, even days, to discover a dish of scented sugar water within 50 meters of the hive, but within a few hours after its discovery by one bee, tens, even hundreds of bees from the same hive will frequent the dish. Observations prove that returning bees communicate their find by giving fellow bees bits of scented sugar water and dancing a peculiar circle dance for 30 to 40 seconds (Fig. 5-7). Via the dance, the bee apparently relays the approximate location of the dish of sweetened water, because other scented dishes placed in the immediate vicinity are also soon visited. Similarly, if scout

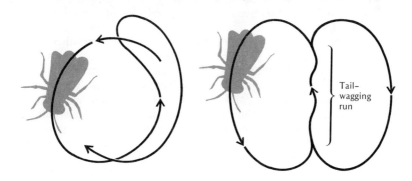

Figure 5-7
The circle dance of the honeybee relays information about food located near the hive.

Figure 5-8
The waggle dance of the honeybee relays information about food located more than 100 meters from the hive.

bees find an especially sweet flower, they mark the flower with a scent produced in a special gland and then return to the hive to alert other bees by transferring some of the flower's nectar.

If the source of sweetness is more than 50 meters distant, the bees dance differently (Fig. 5–8). The bee will move a short distance, all the while wagging its body. Then it will turn to the right, make a loop, and repeat the wagging segment of the dance. It will then turn to the left, make a loop, and again make a wagging motion with its abdomen. Usually, the bee takes 1.5 seconds to complete the dance, although it can require up to 7.5 seconds if the food is 6,000 meters away. The frequency of vibrations given off by the vibrating flight muscles of the dancing bee help tell other bees how far to fly.

The bees usually dance on the vertical combs in the hive, but they also dance on horizontal surfaces. In either case, the bee can successfully communicate the direction from the hive of the sweetened water or nectar.

If the dance is performed on a horizontal surface, the straight-line portion of the dance points toward the food. If the dance is performed on a vertical surface, the straight-line waggling dance is performed at an angle to an imaginary vertical line. This angle is equal to the angle between the sun and the food source. If the food is located by flying toward the sun, the straight-line part of the dance is performed by heading straight up. These signals are correctly interpreted by the other bees, and they will then fly out of the hive in the proper direction, normally no more than 3 degrees off course.

How does the bee "know" in which plane to dance the straight-line waggle? The clue comes from the blue sky. If the hive doorway is covered or if the sky clouds over, the bee cannot correctly relay the information concerning food. Light rays from an area of blue sky, even a small one, are essential for communication. Research now indicates that the bees can actually detect the plane of the light waves coming from the sun. Through the ages the bees have evolved sensory mechanisms to interpret this information.

Other information is communicated from one bee to another in ways not yet thoroughly understood. The distance of the food from the hive is recorded on the flight from the hive to the food apparently through the amount of sugar the bee needs to consume to "fuel" its flight. For example, a strong tailwind speeds the bee along, requiring the bee to expend less energy to get to the food. As a result, the bee communicates erroneous information, that the food is nearer the hive than it really is. Also, the scout bees "record" the position of the sun on the outward flight. By the time they return to the hive and communicate the position of food, the sun has probably moved. No matter, because the bee in some unknown way compensates for this lapse of time before it communicates the position of the food to the other bees. (See the insert, "How Cells Relay Information.")

Thus, bees have evolved very sophisticated behavior patterns around the problem of food getting. Accompanying the evolution of this behavior has been an evolution of very complicated physical structures which allow the behavior to occur. The bee's behavior and nervous apparatus are one more illustration of what is generally true in the animal kingdom, from ameba, to bee, to man: behavior and behavioral structures have evolved together, both the result of the interactions between life-forms and their environments.

How cells relay information

The extraordinary communication among bees has intrigued biologists for many years. Today, it is known that no structures other than elongated nerve cells, sometimes arranged into specialized sense organs, are present to account for the behavior of bees.

It is the *responsivity,* or *irritability,* of certain cells which must be looked to for a basic explanation of the ability of an organism to respond to environmental stimuli, that is, to behave. In the lower organisms, such as the sponge, single cells are especially adept at responding to stimuli. In the higher organisms, which show sophisticated behavior, groups of nerve cells, or *neurons,* account for the animal's behavior.

These neuron associations are often chains of nerve cells. An end of the neuron chain lies in the area where the stimulus is picked up; the other end lies in the region where the effect of the stimulus occurs. Commonly, branch chains of neurons come off the main chain, and hence, the stimulus can be communicated simultaneously to more than one point in the body of the organism.

How a stimulus is transmitted along the series of neurons to the point of action is now fairly well-known. This was a major issue in biology for some time, particularly since it was realized very early that individual nerves in a nerve chain do not touch. Rather, the ends of the extensions of the elongated nerve cells merely lie very near. Today, it is known that as the membrane of the nerve cell changes permeability, the nerve impulse moves along the nerve. Consequently, many positively charged sodium ions which are held outside the unexcited nerve cell move into it, and fewer of the positively charged potassium ions held inside the unexcited nerve cell move outward. As a result, the nerve impulse can be identified as a wave of electronegativity moving rapidly along the nerve-cell chain. When the nerve impulse reaches the junction of the two nerve ends, called a *synapse,* it appears to leap the gap. This is accomplished by the secretion of certain chemicals such as *acetycholine* from the end of nerve cells carrying the impulse. When acetycholine moves across the gap and touches the end of the nearby nerve end, it changes the permeability of its membrane, and the nerve impulse continues. Acetycholine is then destroyed by another secreted chemical, *cholinesterase.* If cholinesterase is affected so that it cannot operate effectively, for example, with a pesticide, the nerves are constantly stimulated, the muscles constantly contract, and the animal suffers *tetany,* or constant muscular action.

The route of the nerve impulse may involve the brain in man and other vertebrates, an organ that is composed primarily of the bodies and extensions of many nerve cells. However, in these higher organisms, many of the nerve chains seldom send messages to the thinking part of the brain. Rather, the stimuli and their responses are managed solely by nerve chains which make up part of the nerve cord in the backbone and throughout the body. A diagram of a simple nerve chain, part of which is housed in the spinal cord and which could effect behavior in the human being without involving the brain, is shown in Fig. 5–9. This simple neuron chain is called a *reflex arc.* In invertebrate animals such as the bee, the ganglia would house nerve cells and, thus, function analogously to the spinal cord in vertebrates.

The coming of the brain

The bee is an example of how many invertebrates have successfully coped with their environment, not with a complicated brain, but with numerous ganglia as refined sensory organs. Other notable examples of invertebrates could be cited — such as the octopus, an invertebrate that

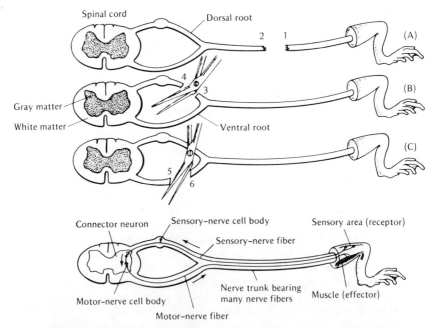

Figure 5–9
Simple reflex arc composed of sensory, connector, and motor nerves, and a receptor and effector. (From *Biology: Its Principles and Implications,* 2nd ed., by Garrett Hardin. W. H. Freeman and Company, San Francisco. Copyright © 1966.)

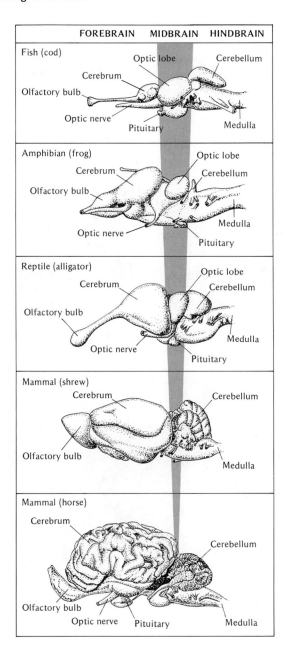

Figure 5–10
Comparison of brains of various vertebrate. Note the relative reduction in the size of the midbrain and the increase in the size of the forebrain in the brains of animals arrayed from simple to complex. (From *Biological Science*, 2nd ed., revised, by William T. Keeton. Illustrated by Paula DiSanto Bensadoun. By permission of W. W. Norton & Company, Inc. Copyright © 1967, 1972 by W. W. Norton & Company, Inc.)

134 Chapter 5: Behavior and survival

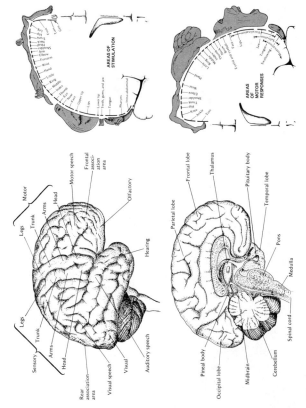

Figure 5-11
Left: Areas of the cerebrum which relate to various areas of the body. (From Lorus and Margery Milne, *Animal Life*, Prentice-Hall, Inc., Englewood Cliffs, NJ., 1959). Right: Relative size of the areas of the cerebrum responding to the stimulation and motor responses of various body parts. (The parts of the body of a figurine are drawn approximately proportional to the size of the area of the cerebrum devoted to each.) (Reprinted with permission of Macmillan Publishing Co., Inc., from *The Cerebral Cortex of Man*, by W. Penfield and T. Rasmussen. Copyright 1950 by The Macmillan Company.)

even learns fairly easily — but space doesn't permit a cataloging of those invertebrates "that made good."

Instead, let us consider the *vertebrates,* animals with backbones. The vertebrates, including recently evolved man, long ago took a different path of structural development. Consequently, their adaptation to the environment has been more sophisticated. Vertebrates evolved the brain, an elaboration of the nervous system structure which allows for nervous activity and complex behavior impossible for any invertebrate, even the insects.

The basic plan of the brain in all vertebrates is the same. Brains appear to be composed essentially of three main parts, each with a cavity filled with liquid and continuous with the hollow liquid-filled spinal cord. These three parts are generally known as the *hind brain,* the *middle brain,* and the *forebrain.* It is in each of these parts where the trends in complexity can be detected as the animals become more sophisticated.

In general, the hind brain shows these evolutionary trends. The ventral portion becomes increasingly specialized as a control center for visceral functions, such as respiration and rate of heart beat, and the dorsal part, or *cerebellum,* concerned with the coordination of muscle action and maintaining balance, becomes more developed. The midbrain is large in the lower vertebrates, mainly because it is the center of control for visual activity, but in the more recently evolved vertebrates, such as man, this part, known as the *optic lobes* of the brain, becomes much less obvious. The forebrain shows an opposite trend. This is the part of the brain called the *cerebrum,* in which the higher mental activities, such as thought and creativity, occur. As the vertebrates become more evolved, they show enlargement of this area. In the lower vertebrates the cerebrum is mainly a control center for the detection of odors and consists primarily of lobes called *olfactory lobes.* The cerebrum is the center for high levels of mental activity and allows vertebrates to adjust to their environments. Hence, its presence explains the survival of many vertebrates through the ages.

An examination of the cerebrum in contemporary vertebrate groups explains the diverse adaptive abilities of these animals. In fish the cerebrum is mostly olfactory lobes and explains their dependence on an acute sense of smell for survival. Amphibians and reptiles show an elaboration of the cerebrum but in an area involved with odor detection and overlying another section of the cerebrum, experimentally shown to be responsible for other behavior. This interior portion of the cerebrum grew into the thinking portion in the birds, an elaboration of the trend already established in the amphibians and in the reptiles, the group from which the birds are thought to have been derived (Fig. 5–10).

In mammals, including man, the cerebrum developed differently, however, which explains why mammals today show the greatest mental activity of any animals. The cerebrum in mammals appears to have originated evolutionarily as an elaboration of the anterior part of the forebrain. This expansion of brain tissue appears to have developed unhampered by overlying tissue, evolving into a very large area in the more recent groups that show high mental development. It could be reasoned that man eventually evolved because the cerebral tissue developed relatively unhampered by overlying tissues and then expanded to provide for exceptional neural activity. The particular areas in the cerebrum that appear to control man's sophisticated mental abilities have now been plotted and are shown in Fig. 5–11.

Instinct and learning

Man's behavior appears to be a combination of "learned" and innate *instinctive* behavior. The distinction between learned and instinctive behavior is a controversy that has raged for years among biologists, psychologists, and educators. Although a great deal of research has helped to clarify the issue, confusion still remains because some behavior appears instinctive, some learned, and some both.

The problem does not come from unclear definitions of instinct and learning. *Learned behavior* is the adaptive change in an individual's behavior as a result of experience. *Instinctive behavior* is the genetically determined behavior not based on the previous experience of the individual. Rather, the problem comes from the inability of scientists to recognize the category to which various behavior patterns belong.

What characterizes instinctive behavior? Instinctive responses are usually identified in multicellular organisms with well-defined nervous systems and elaborate sense organs. Responses due to simple irritability or responsiveness of cells, as seen most obviously in the ameba's response to bright light, are *not* instinctive. Instinctive behavior appears to be triggered by a stimulus, or "releasing mechanism," to which the organism must consequently respond unless learning has conditioned the response.

Excellent examples of instinctive behavior in invertebrates are found among the insects. For example, male mosquitoes are attracted to a tuning fork vibrating at a pitch similar to that of the wings of the female mosquito. The vibration is the releaser, and the male mosquito has no control over his response. And paper that carries the scent of female mosquitoes induces male mosquitoes to copulate with it.

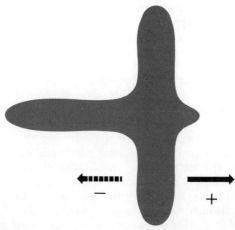

Figure 5–12
Model of a bird that elicited fright reactions in a turkey flock when moved to the left and no response when moved to the right.

Some insect releasors are known to be secreted odoriferous chemicals called *pheromones*. Female silkworms, for example, produce only 0.00000001 gram of perfume, yet they can attract male silkworm moths from distances greater than 2 miles. Ants deposit a chemical from the tip of the abdomen along pathways leading to food which causes other ants to follow.

Instinctive behavior can also be identified in vertebrates, although learned behavior very often also occurs to confuse the issue. For example, a silhouette of a pattern shown in Fig. 5–12 suspended on a wire over a turkey flock provokes no response if it is moved toward +. However, if it is moved toward −, the turkeys become excited and scatter as if threatened. Although proof is lacking that the turkeys instinctively are frightened by the model when it moves in the direction that gives it the appearance of a hawk or similar bird of prey and are unmoved when it appears as a goose or duck, this seems to be a reasonable explanation.

Instinctive behavior can be recognized in the behavior of the black-headed gull, too. These birds, when nesting, always remove broken eggshells from their nests after the young birds hatch. In addition, the adult birds instinctively remove any similar object placed in the nest during breeding season. Studies show that nests that contain only eggs are less often frequented by crows and other gulls that feed on the eggs and young gulls. "Junk" in the nests, as well as broken eggshells, attracts the robber birds. Apparently, removal of shells and junk has definite survival value for the species and, hence, could have evolved into an innate behavior pattern, or instinct.

The nesting oyster catcher offers another example of instinctive be-

Figure 5–13
Seagull striving to incubate an oversized egg. (Time-Life Picture Agency. Photo by Thomas McAvoy. © Time Inc.)

havior in the vertebrates. This bird appears to respond most positively toward a large egg rather than a small egg in its nest. In fact, in an attempt to place an extremely large model egg under itself for proper incubation, it has been observed to neglect its own small egg. The oyster catcher also seems to respond similarly to the number of eggs composing its "clutch." More attention is shown to a clutch of five than to three (Fig. 5-13).

Fish also appear to show instinctive behavior, and in some instances the triggering mechanisms have been identified. For example, the odor of water appears to pilot salmon back to the area in which they were spawned. A chemical may exist in the water which characterizes the spawning area, because if the odor receptors of the fish are blocked, they cannot find their birthplace.

The stickleback fish also shows elaborate instinctive patterns, again in reproductive behavior. In the spring, the stickleback male develops a reddish underside, and the female's body swells. The large belly of the female apparently is the stimulus that causes the male stickleback to swim in a zigzag motion toward her. This, in turn, is the stimulus, along with the red belly, which causes the female to swim toward him with head up. Such a signal "releases" him to swim toward the nest that he has already constructed. The female follows. This stimulates him to thrust his head into the nest and turn on his side. At this point the female enters the nest. The male is stimulated to thrust his snout against her abdomen, and her eggs are released into the nest. The emission of her eggs stimulates the release of his sperm, and the eggs are fertilized. Experiments

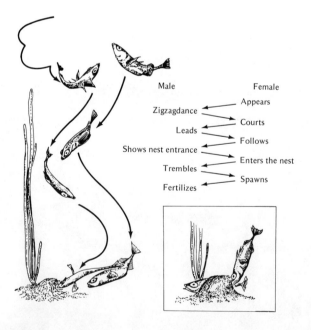

Figure 5-14
Left: Motion behavior of the stickleback fish. Right: Schematic representation of the relations between male and female sticklebacks. (From N. Tinbergen, *The Study of Instinct*, The Clarendon Press, Oxford, 1951.)

show that the sequence of events is an elaborate instinctive behavior pattern, with each event in turn triggering another event (Fig. 5–14).

Often, however, it is not so easy to identify purely instinctive behavior in the vertebrates. Experiments with the chaffinch illustrate how the interaction of instinct and learning leads to complicated behavior. Young chaffinch raised alone usually cannot learn the adult song of the chaffinch, although they can easily learn a simplified and characteristic chaffinch call. If at 6 months of age the young chaffinch hear a recording of an adult chaffinch, the young birds learn the song quickly, even if it is played backward. Clearly, the young chaffinch have the innate ability to sing but need to learn the details from actually hearing the chaffinch song.

An interaction of instinct and learning in determining behavior can also be seen with other organisms. Konrad Lorenz, the famous ethologist, experimented with newly hatched ducks and geese. Immediately after hatching, the young birds tend to *imprint* on the first moving object they see. If the eggs are hatched by an adult bird, the young properly imprint on her. However, if the eggs are hatched artificially, the young can imprint on other animals, such as dogs or cats, or even moving inanimate objects, such as toy trains, especially if they make a noise. They have even imprinted on Lorenz (Fig. 5–15).

One trend generally recognizable in the behavior of invertebrates and vertebrates is the increasing dependence on learning in the more recently derived groups. That is, invertebrates show the most dependence on instinctive response and the least dependence on learning; mammals show the most reliance on learning.

Figure 5–15
K. Lorenz, a pioneer in the study of animal behavior, with an imprinted snow goose. (Courtesy of K. Lorenz. Photo by H. Kacher.)

Learning is characterized by the application of information stored in the nervous system to new situations. Learning can be differentiated into four categories: habituation, trial and error, conditioning, and insight.

Habituation

Habituation is evidenced even by the simplest of animals. For example, an ameba exposed to a bright light will at first tend to move out of the area or contract and extrude any partially eaten food. However, if the ameba is subjected to this experience several times, it will soon ignore the stimulus and go about its business. People are especially effective in eventually disregarding stimuli that do not matter. Any urban dweller knows well, if he stops to think about it, that most of the numerous stimuli impinging on his senses during a day do not provoke a response; in fact, they are not even noticed.

Trial and error

Another type of learning is *trial and error*. A newly hatched chick, for example, will peck randomly at objects that can be differentiated from the background. However, as it matures it becomes more judicious, and soon pecks only at those which are edible. In addition, its pecking accuracy will improve. Most animals appear to evidence some trial-and-error behavior.

Conditioning

Conditioning as a form of learning was brought to light by the famous experiments on dogs of Ivan Pavlov, a Russian physiologist. Recall that Pavlov rang a bell each time he fed his dogs. After a few such events, Pavlov discovered that the dogs had learned to associate the ring of the bell with food; the dogs would salivate when the bell was rung, even when no food was brought. In essence, the dogs had substituted one stimulus for another in evoking a response. For many researchers Pavlov's experiments suggested a general explanation for animal behavior. (See the insert, "A Way to Learn.")

Certainly, conditioned behavior is also easy to detect in many invertebrates. Von Frisch found it in bees, for example. He laid out many squares of paper, each a shade of gray, except for one square of blue. On each of the squares he placed an empty dish, except that on the blue square he placed a dish filled with sugar water. The bees repeatedly came to the dish on the blue square and ignored the rest. In time, Von Frish removed the food, yet the bees still returned to the blue square.

Insight

The final example of learning is *insight*, behavior found most frequently in the most advanced animals and least often in the simpler forms. Insight is the ability to apply to new situations stored information gathered from previous experiences. Many of the daily events most of us experience demand that information gathered previously be applied. The ability to retain information and apply it to a new situation varies in people just as it does with other kinds of higher animals, but whatever the degree of ability it is usually recognized as *intelligence*. Unfortunately, a facility to apply previous experiences in a situation of one type is not necessarily the same in a situation of a different type. Previous conditioning may well determine that insight learning cannot occur easily. Such a situation, for example, may exist in those people who may be extraordinarily adept with words, that is, have a high verbal ability, but little facility with mathematics. The innate talent in a particular area, as well as previous conditioning, affect the ease with which an individual may learn by insight. Clearly, if a person has a high musical talent, he most likely will learn music faster than if he did not. Hence, measurements of intelligence must be carefully made so as not to overlook areas of a person's learning potential.

Of interest is evidence which indicates that human beings are not the only animals that show a high degree of insight learning. Recent observations in the rain forests of Africa by Jane Goodall and her husband suggest that chimpanzees also learn in this fashion. They have been ob-

Figure 5–16
A chimpanzee uses a tool she made by stripping down a blade of grass. With this handmade tool she pokes into a termite mound for insects. (Courtesy of National Geographic Society. Photo by Baron H. van Laurick.)

served, for example, to break off twigs, strip them of their leaves, and use them to probe into termite hives. The termites cling to the probes, the sticks are withdrawn, and the chimpanzee has a meal (Fig. 5–16).

A way to learn
Operant conditioning

Recent studies by the noted psychologist B. F. Skinner suggest that another type of conditioning exists beside that studied by Ivan Pavlov. Skinner calls it *operant conditioning*. Its significance lies in its implications to human behavior, especially learning.

Operant conditioning differs from the Pavlovian type, classical conditioning, in several ways. Recall that classical conditioning involves an unconditioned stimulus, a conditioned stimulus, an unconditioned response, and a conditioned response. In Pavlov's experiments the meat Pavlov fed to his dog was the *unconditioned stimulus*. The dog's salivation was the *unconditioned response*. The bell Pavlov rang at the time he presented the meat was the *conditioned stimulus*. The salivation eventually produced by the bell alone was the *conditioned response*. The simultaneous occurrence of two stimuli, in this case the ringing of the bell and the sight of the meat, is called *reinforcement* and eventually produces a conditioned response.

Operant conditioning differs from classical conditioning in the following ways: operant conditioning is based on the occurrence of operant behavior, that is, behavior which seems to occur without an initial stimulus or which appears spontaneously. The kicking legs of a newborn baby or the meaningless verbalization of a toddler can be

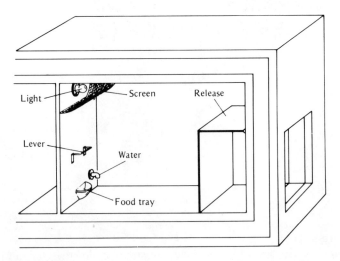

Figure 5–17
The Skinner box with the front cut away to show the interior.

considered examples of spontaneous behavior. If spontaneous behavior is rewarded or reinforced, the organism soon learns selectively to perform that act. For example, the doting parent praising a child for those of his verbalizations that sound like words may be helping the child to speak.

In the laboratory B. F. Skinner illustrated operant conditioning with animals in his "Skinner box" (Fig 5–17). The inside of a Skinner box is plain except for a food dish attached to one wall with a protruding bar above the dish. Above the bar is a small light bulb which an experimenter can light at will. A hungry rat is put into the box. The rat will move restlessly about and in his movements will touch the bar. This causes a food pellet to fall into the food dish. Soon the rat presses the bar presumably *on purpose,* to feed himself. In other words, the rat has learned through reinforcement that pressing the bar will present him with food. It is evidencing operant conditioning.

Some characteristics of operant conditioning may relate to human learning. One of these is the continuation of the conditioned response even when the reinforcing stimulus is gradually removed. For example, a rat will continue to press on the bar even though food is supplied only occasionally as a result of his activity. A pigeon that learned to peck at a spot by receiving a small quantity of grain as the reinforcement continued to peck at the rate of 6,000 or so responses an hour even though food was supplied as a reinforcement on an average of only 12 times an hour.

Another relevant characteristic of operant conditioning is illustrated in experiments that utilize *secondary reinforcements.* A secondary reinforcement in the Skinner-box experiments is a response that accompanies the occurrence of food when the rat pushes the bar. For example, if the experimenter turns on the light every time the rat touches the bar, reinforcement of bar pressing occurs, and eventually most rats will press the bar only in the presence of the light. Experiments with secondary reinforcements have produced interesting results. If the responses of the flashing light and the occurrence of food when the animal presses the bar are eliminated, the rat soon ceases entirely to press the lever. However, if the light is reconnected, the rat soon learns that pressing the bar turns on the light. Soon he will be pressing the bar very frequently, even though food *never* reappears.

Additional, or secondary, reinforcement can lead to even more complicated behavior, *token learning.* An experiment with chimpanzees illustrates this. Through operant conditioning chimpanzees learned to accept poker chips instead of food in a Skinner box. The chimpanzees were then taught to deposit the poker chips in a machine to obtain food. After the animal had been successfully conditioned, it would work as diligently for poker chips as it would for food itself, even hoarding poker chips occasionally before using them in the machine to obtain food.

The implications of operant conditioning on the behavior of human beings is thought by some to be considerable. B. F. Skinner claims that operant conditioning provides information on optimal learning. He maintains that students can learn more effectively if they are rewarded, or reinforced. His ideas on learning have given rise to *programmed learning*, in which a student will answer a question and then be rewarded immediately with the correct answer. Skinner also suggests that the group behavior of human beings is due to a great extent on reinforcers supplied to human beings through social contacts. Much individual behavior, for example, appears to be the result of the social approval that comes from behaving in certain ways. Providing the proper kinds of rewards, or reinforcements, can, therefore, determine what kind of individual and group behavior will occur. The only problem is what behavior is worth rewarding and by whom. B. F. Skinner examines a society based on operant conditioning in his controversial book *Walden II*.

The nature of memory

One of the most intriguing searches in science has been for the cause of memory, the essential component in learning. At one time researchers believed that possibly a particular storage area of the brain was the site of each individual memory, or *engram*. The belief came mainly from the evidence that electrical stimulation of the brain of a person during brain surgery could provoke a clear remembrance of past events and from futile experiments in which parts of the cerebrum of other primates were destroyed in attempts to erase specific memories. No specific loci could be discovered for individual memories; rather, if one-tenth of the brain were destroyed, an indiscriminate one-tenth of the animal's knowledge was lost.

Thus, the question of what a memory is continued to be asked. Then in the 1950s a researcher, James McConnell, made some interesting observations in experiments with the flatworm planaria. He conditioned the flatworms by flashing a bright light on them while they were swimming. He followed it with an electric shock. Soon the shocked worms contracted only when the bright light was shown, even if the shock did not follow. Then he cut in half animals that contracted only in light 23 of 25 times and allowed each half to regenerate the other end, a feat not uncommon to planaria. After regeneration, each full-grown head or tail end showed an undiminished ability to respond to the light stimulus alone. Control animals conditioned and uncut showed a similar response; control planaria unconditioned and cut did not show the response. Thus, proof seemed to exist that the memory deposit, or engram, was not lodged in the planaria's largest ganglia, those in its head end, but rather in the animal's nervous system as a whole.

Later, a biologist experimenting with rats suggested that ribonucleic acid (RNA) may be related to memory. He found large amounts of RNA in nerve cells. He also found that rats taught complicated tasks accumulated more RNA in their neurons than untrained rats.

McConnell, still bothered by the inability to specifically locate memory in his planaria, then performed an experiment using the data concerning RNA from rats. He decided to condition more planaria, feed them to untrained planaria, to determine if the untrained planaria fed RNA from trained planaria would become conditioned without being subjected to the light and shock. His results confirmed his hypothesis, that RNA in some fashion seems to have something to do with memory in animals.

More work is necessary to substantiate this hunch, but injections into elderly people of RNA extracts or chemicals that promote RNA synthesis has improved their failing memories. The coded nature of RNA makes it a likely candidate for the storage of information, just as DNA stores the characteristics of cells. However, nothing at this point is known as to how the RNA molecule might be specifically involved. As a consequence, most investigators receive McConnell's work with some reserve. Their response is similar to the statement in *Biology and the Future of Man* (P. Handler, ed., Oxford University Press, New York, 1970): "These claims are presently regarded with great suspicion and are taken seriously by only a small group of investigators."

Social behavior

Earlier in this chapter we talked about the hypotheses that cultural anthropologists have proposed to explain the origin of man's social behavior. We concluded then, as when discussing the behavior of single organisms, that to have persisted, behavior and structure of organisms very likely has (or had) some survival advantage. With this theme in mind let us now look at the social behavior of animals other than man.

Sooner or later any discussion about the behavior of organisms will focus on social behavior, the interaction of individuals with others, especially of their own kind. In this area is included any behavior patterns that appear to accompany mating, the rearing of the young, and the maintenance of a social structure that involves individuals other than mates or offspring. Once again, attempts to differentiate the aspects of social behavior that seem to stem from instinctive response from those that come from learned responses is extraordinarily difficult.

One of the problems ethologists face in studying any behavior, especially social exchange, is the youthfulness of the science. Although more and more interest is shown concerning the behavior of animals, most systematic and scholarly investigations of this kind are recent (although the awarding in 1973 of the Nobel Prize in physiology and medicine, usually awarded to a molecular or medical scientist, to Konrad Lorenz, Karl von Frisch, and Nikolaas Tinbergen suggests that recognition is coming to the area, and more research activity will soon follow). Consequently, much is yet to be learned before basic and reliable generalizations can be made as to how animals behave socially. Nevertheless, some observations have been made which give insight into the structure of the social activities of animals and testify to the life-giving advantages of behavior, which at a passing glance often appears purposeless.

It is important in the following discussion of the types of behavior to realize that the categories are very arbitrary and that it is difficult to clearly separate behavior patterns. Mating behavior, for example, is very much connected to territoriality in some species, although for ease of discussion both will be considered separately.

Mating behavior

Essential for the propagation of the species is fertilization of the egg and successful rearing of the young. Consequently, some of the most frequently observed and most intense behavior is that which involves mating.

All the behavior patterns concerned with mating that have been observed appear to do one or more of the following: (1) synchronize the reproductive interest of the two sexes, (2) persuade the female to persist in her "nesting" behavior, (3) orientate the two sexes so that interaction can occur, and/or (4) isolate them in a nest or shelter so that fertilization of the eggs can result. An excellent example of a behavior of an animal in which all four factors have been found is the mating ritual of the stickleback fish, already described.

Another example is the mating behavior of the canary, which also illustrates the timing, persuasion, orientation, and isolation factors seen in the mating behavior of the stickleback fish. Apparently, the length of springtime days stimulates the production of hormones in both the male and female birds. As a consequence, the male is stimulated to sing and display, activities that further stimulate the production of hormones in his mate. Consequently, the female begins to gather nesting materials as the eggs begin to develop within her ovaries. After the construction of the nest, the female's hormones stimulate some of her breast feathers to be shed, revealing the "brood patch," a denuded, extraordinarily sensitive area stimulated by contact with the nest. The presence of the male and her own hormones leads to copulation, followed shortly by ovulation, stimulated primarily by the contact of the brood patch with the nest, a contact that also apparently stimulates her constant incubation of the eggs.

The behavior of the stickleback fish and the canary emphasize that mating behavior which appears to be instinctive and built into the animals is really often composed of a series of acts, each triggered by a particular stimulus. In canaries the initial stimulus for reproduction is the amount of sunlight available. In other animals it appears to be hormone secretion, although as more is known, exterior stimuli may be disclosed.

Behavior that apparently synchronizes the two sexes to permit sexual activity can be seen in the elaborate rituals of other birds. For example, male and female Brandt's cormorants, like many other water birds, engage in a great deal of mutual stimulation before mating. The female makes overtures to the male by bending her head and neck backward over her back and raising her tail. The male is then stimulated to do the same, which exposes the brilliant blue pouch on the underside of its neck. Mating ensues, after which the male brings nesting materials to the female. Together they carry it to the nesting site, and nest construction begins.

The showy colors usually characteristic of males are apparently used to advantage in many species of birds. The peacock, noted for its lovely plumage even when not displaying, is breathtaking in its mating ritual — even to people. During the mating prelude the peacock will spread and rustle his large tail fan. Occasionally, he will turn his back toward the female and then, in a burst of color and sound, he will suddenly wheel and spectacularly confront the female. Many other examples of such courting behavior could be given, especially among the birds, but also among fish and mammals.

One of the activities that sometimes precedes mating is the establishment of the pair bond. In some birds this procedure is very elaborate. For example, early in the mating season herring gulls form large aggregations. A female may approach a male and begin to move submissively about him, all the while calling quietly. He often responds most threateningly, a display that may terminate in the regurgitation of food. The female will aggressively seize the food. Eventually, mating occurs, and egg laying ensues. The earlier activities do not cease, however, for during nesting both birds appear to engage in the behavior as a kind of social exchange. These activities apparently guard against the breakup of the pair and neglect of the young.

Care of the young

Much behavior observed between young and adult is yet to be explained. Nevertheless, some experiments have been done that throw light on the seemingly peculiar antics of parents and offspring. Much of the information has been on birds, although data concerning fish and mammals are increasing all the time. The nesting and feeding habits of some birds indicate strongly that much rearing behavior is "released" at the proper time by the occurrence of simple stimuli. For example, young herring gulls newly hatched will peck at the tips of the parents' bills. The adult then regurgitates foods onto the ground and retrieves a small portion to feed to the young. As the young bird pecks at the adult's bill, it obtains some regurgitated food. Experiments demonstrate that the coloration of the adult's bill stimulates the pecking of the young. The adult has a yellow bill with a red patch near the tip of the lower mandible. Models substituted for the adult's bill showed that greatest response was to low, moving, long, thin, downward-pointing, red-spotted models (Fig. 5–18).

Experiments with the "gaping" reactions of young birds show similar results. Nikolaas Tinbergen exhibited a model (Fig. 5–19) to young thrushes and found that the young would gape at the part of the model that seemed most headlike in proportion to the rest. Changing proportions of the model altered the part to which the young would direct their attention. The birds appeared to respond to the *relative* size of the disk parts, not to shape or *absolute* size. An interesting response of parent birds to gaping by the offspring is illustrated by the Old World cuckoo and the New World cowbird. Each lays eggs in a songbird's nest, where they often hatch. The young then push the songbird's eggs or newly hatched young out and take over as the sole young to be doted on by the foster parents. Although the baby songbirds may lie on the nest periphery, they will be ignored by the adults if the young parasite bird is hungry enough to gape. The adult will respond only to the largest mouth; hence, will stuff the foster child and allow its own young to die (Fig. 5–20).

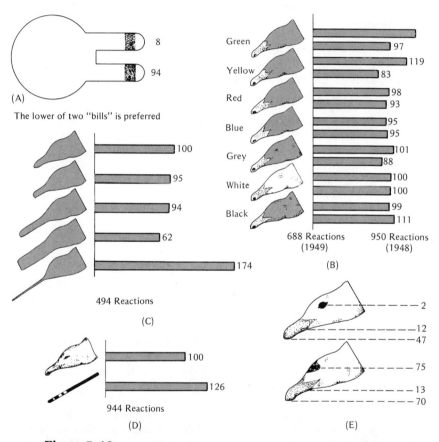

Figure 5–18
Models used to elicit response in young gull chickens. (The frequency of various responses is listed to the right of the models.) (A): The lower of the two simulated bills is preferred. (B): The color of the adult bird's head is significant in eliciting response from young gulls. (C): The long thin bill shape evokes the most response. (D): A thin red rod with three white bands evokes more reaction that a three-dimensional plaster head. (E): The number of responses aimed at the bill tip, the bill base, and the red eye in a normal and misplaced location. (From *The Herring Gull's World*, by Niko Tinbergen. © 1960 Revised edition by Niko Tinbergen. Published by Basic Books, Inc., Publishers, New York.)

Other social behavior

As with mating and rearing of young, interesting behavior patterns can be discerned which appear to bind together many members of the same species, more than just male and female or adults and young. The advantages of group living are many. Protection from predators is important. Birds in flocks are less likely to be attacked by hawks than single flying birds, and the probability of any one being taken is less. Fish in large schools are less vulnerable than individuals. Some members of large groups of mammals and birds, such as flocks of crows, post sentinels which sound alarms when danger threatens. In colder climates animals in large numbers often aggregate and, thus, presumably stay warmer than single members. Evidence also suggests that individuals of some species tend to eat more when they are in groups rather than alone. Single chickens, for example, fed to satisfaction, will tend to resume eating when placed with a hungry flock. Other advantages may also be true, but more work is needed to reveal them.

Animal societies seem to be held together in various ways. Frequently, elaborate behavior determines the roles played by individual members in the group. Chickens evidence such role playing or social dominance in the form of a hierarchy, or "pecking order." One chicken usually dominates the others in a flock. The next most influential hen can dominate all hens except the first. Each hen has some position of influence in the flock. High-ranking hens have first use of nesting boxes and of the feeding and watering troughs. Just what factors help determine these positions is not known, although an inherited tendency for aggression or passivity seems to be part of it.

The advantages of the peck order seem to be a conservation of energy in the flock as a whole. Established roles prevent the aggression seen in establishing the peck order from being a constant. Once the roles are established, the individuals waste little energy confronting other hens.

Another device to establish dominance in a population is "territoriality." Territoriality often prevails as part of the mating ritual. Sometimes pair bonding cannot be established if a territory for breeding and raising young has not been staked out. Generally, it is the male that establishes the territory and wards off other males that attempt to enter it.

Figure 5-19
Models with two heads used to elicit gaping in nestling thrushes. (From N. Tinbergen, *The Study of Instinct*, The Clarendon Press, Oxford, 1951.)

150   Chapter 5:   Behavior and survival

The advantages of establishing a territory are clear. The male with his mate or mates can exist without significant disturbance once its area is determined. Ample food for the young within convenient reach is assured. If the males of a species do not establish a territory, they cannot reproduce. How territorial boundaries are defined is still unanswered. Usually, this is the male's choice, but occasionally the female will also participate. In some animals mutual defense appears to cement the pair bond, to the obvious advantage of rearing of the young.

Variations on this theme are known. In defense of their territory the paired male and female South European green lizard will each fend off only members of its own sex, thus ensuring that the two strongest members of the population will be in proximity to breed. The male stickleback will defend his territory but will be passive if taken out of it. Two male fish have been placed in a test tube, and the two tubes held in each other's territory. Each fish in his own territory is alert and aggres-

Figure 5–20
Adult hooded warbler feeding a young cowbird. (Courtesy of National Audubon Society. Photo by G. R. Austing.)

sive, but each fish in the other's territory is notably unaggressive. The farther from the home territory, the more mild-mannered each becomes.

One prime result of establishing a territory is the elimination of combat or aggressiveness over a prolonged period, a function similar to that provided by the chickens' peck order. In the long run energy is saved within the populations. Nevertheless, to initially establish the territory, as a rule, requires intense fighting. Once established, many signs, such as scent from glands or urine, help indicate territorial boundaries to intruders and turn them away before a confrontation is necessary.

Animals will fight on the rare occasions when confrontation is inevitable. Apparently, the goal is not the death of one or the other, merely establishment of who owns the territory or is stronger. Battling rattle-

Figure 5–21
Wounded soldiers in the Vietnam conflict. (Wide World Photos.)

snakes could kill each other with one nip. Instead, however, they quarrel by entwining their "necks" and just push each other around with their heads.

The defense of territory by rattlesnakes or stickleback fish prompts consideration of man's tendency to territorialism. This subject is controversial, but at least a few writers on animal behavior suggest strongly that we, too, tend to claim an area "as our own." Desmond Morris, author of *The Naked Ape,* cites examples of "marking" a territory – the business executive who immediately sets out his name plate, paperweight, personal pen tray, and photograph of his wife upon moving into his office, and the housewife who decorates her house or apartment with bric-a-brac or ornaments.

Conflict over territorial boundaries may, in fact, be the origin of war. Morris believes war resulted when technology provided for a demonstration of power without opportunity to witness a submission of the antagonists, as happens within other animal species. If only signs of submission on one side or the other could be viewed, the conflict would not be prolonged until one or the other antagonist dies. As Morris points out, antagonism over territory, that is, ample areas for individuals to propagate the species, is a life-sustaining act. Only submission of one animal in any conflict over territorial boundaries is needed. Murder of another member of the species in war can lead to extinction of the species if the conflict is global and the weapons devastating enough, as our hydrogen bomb is (Fig. 5–21).

Man has engaged in more limited conflicts, however, and these along with other exigencies have contained his population size in the past. Yet, despite his battles, famines, and diseases, man's population has grown. In the next chapter you will read of man's increasing dominance and management of his environment, which resulted in a population size so great that in the near future it may threaten his existence as a species as much as any global war.

Summary

In this chapter we have examined (1) some hypotheses that account for the origin of man's social behavior, (2) examples of simple organisms whose behavior and structure seem to have evolved concurrently, and (3) examples of instinctive and/or learned behavior, especially as seen in the social behavior of more complicated organisms. Underlying all of these topics were these generalizations: the behavior of animals is as much an evolutionary product as their structure; animal behavior evolved most likely because it had survival value; man's behavior today is probably as much an evolutionary product as the behavior seen in simpler organisms; insight into the solutions of our environmental difficulties may lie in an examination of the manner in which other organisms relate in life-sustaining ways to their environments.

Supplementary readings

Alland, A., Jr. 1967. *Evolution and Human Behavior*. Doubleday & Company, Inc. (Natural History Press), Garden City, N.Y. 243 pp.

Blum, H. F. 1961. "Does the Melanin Pigment of Human Skin Have Adaptive Value?" *Quarterly Review of Biology*, vol. 36, pp. 50–63.

Cohen, Y. A., 1968. *Man in Adaptation: The Biosocial Background*. Aldine Publishing Company, Chicago. 386 pp.

Comfort, A. 1966. *The Nature of Human Nature*. Harper & Row, Publishers, Inc., New York, 222 pp.

Davis, E. 1966. *Integral Animal Behavior*. Macmillan Publishing Co., Inc., New York. 118 pp.

Dethier, V. G., and E. Stellar. 1964. *Animal Behavior: Its Evolutionary and Neurological Basis*, 2nd ed. Prentice-Hall, Inc., Englewood Cliffs, N.J. 118 pp.

Eibl-Eibesfeldt, I. 1970. *Ethology: The Biology of Behavior*. Holt, Rinehart and Winston, Inc., New York. 530 pp.

Etkin, W. 1967. *Social Behavior from Fish to Man*. University of Chicago Press, Chicago. 205 pp.

Evans, R. I. 1968. *B. F. Skinner: The Man and His Ideas*. E. P. Dutton & Co., Inc., New York. 140 pp.

Harrison, R. J., and W. Montagna. 1973. *Man*, 2nd ed. Appleton-Century-Crofts, New York. 458 pp.

Klopfer, P. H. 1970. *Behavioral Ecology*. Dickenson Publishing Co., Inc., Belmont, Calif. 229 pp.

Klopfer, P. H. 1973. *Behavioral Aspects of Ecology*, 2nd ed. Prentice-Hall, Inc., Englewood Cliffs, N.J. 200 pp.

Lerner, I. M. 1968. *Heredity, Evolution, and Society*. W. H. Freeman and Company, San Francisco. 307 pp.

Lewis, J., and B. Towers. 1969. *Naked Ape or* Homo Sapiens? Humanities Press, Inc., New York. 134 pp.

Morris, D. 1967. *The Naked Ape: A Zoologist's Study of the Human Animal*. McGraw-Hill Book Company, New York. 252 pp.

Pfeiffer, J. E. 1972. *The Emergence of Man*, rev. ed. Harper & Row, Publishers, Inc., New York. 477 pp.

Tavolga, W. N. 1969. *Principles of Animal Behavior*. Harper & Row, Publishers, Inc., New York. 143 pp.

Tinbergen, N., and the editors of *Life*. 1965. *Animal Behavior*. Time-Life Books, New York. 200 pp.

Von Frisch, K. 1971. *Bees: Their Vision, Chemical Senses, and Language*, rev. ed. Cornell University Press, Ithaca, N.Y. 157 pp.

Chapter 6

*In the darkness with a great bundle of grief
   the people march.
In the night, and overhead a shovel of stars
   for keeps, the people march:
       "Where to? What next?"*

The People, Yes
Carl Sandburg

The growth of the human population

The behavior and structure of contemporary man are products of his evolution through millenia. Today, man's wide distribution and abundance are evidence of how well these attributes have helped him maintain a vital relationship with his environment. Yet, despite his apparent success as a species, he now faces a crisis.

The crisis? Near the year 2000, when a college student, 18 years old in 1970, is approximately 50 years old, nearly twice as many human beings may inhabit the earth. The 1972 population of 3.782 billion may increase to 7+ billion. Even as early as 1980, nearly 1 *billion more* people may crowd our planet.

The impact of such a "population explosion" can barely be imagined. We shall consider in detail later some predictions concerning the human situation; but, for an example, think of the traffic problem that exists now in a metropolis such as New York City and try to consider the congestion with twice as many inhabitants. Indeed, man's population increase, of fearsome size and rapidity, may be a crisis more threatening than annihilation with the hydrogen bomb.

But because early man was not as able to relate successfully with his environment as is modern man, he was not always so abundant. During his early tenure on earth he often lost in the battle for survival. Consequently, his numbers stayed small. His population inexorably increased, however, as he evolved an ability to change his environment. In this chapter we shall look in more detail at the history of the human population growth, an index of man's successful interfacing with his environment, and what it means to a contemporary society.

Population development in the past

Through the ages the growth in the size of the human population is a direct result of man's increasingly successful attempts to relate to his environment. Consider the effect of tools, for example. The food gatherer of eastern and central Europe, Cro-Magnon man, learned to use a variety of very specific tools during the last ice phase. He even made tools to make tools. He perfected the bow and spear. He hunted on open tundras in central Europe and Russia, where large mammals, such as bison, reindeer, wild horses, and mammoths, migrated in summer to feed. As a result, he ate well and endured even in an inhospitable climate. Today, testimony to his success as a tool user in those areas are large *middens,* ancient garbage heaps of animal bones, found along those migratory paths. (Scientists estimate that one midden at Predmost in Moravia contains the bones of 1,000 mammoths.) In central France, too, hunting tools — and temperate climate — favored the survival of the early inhabitants. On limestone steppes they pursued musk-oxen, horses, bison, and other large edible mammals and no doubt fished the rivers with the ancient fishhooks recently uncovered in his shelters, the nearby caves. Tools gave early man ability to modify his environment, to emancipate him from the previously uncompromising environmental conditions, and to allow development of a social organization and culture, including ritual and art. With such successful interfacing with his environment, man's numbers increased.

How big was the human population just after the last ice age, about 9000 B.C.? No definite answer is possible. However, some experts, considering all factors, believe that by the year 6000 B.C. as many people existed on earth as India *adds* to its population every 4 or 5 months today — 5 million people! His population grew as more cultural developments increased his ability to cope with environmental exigencies. Consider the significance of the beginning of agriculture. When the last of the great ice coverings left Europe about 9000 B.C., the climate warmed, marking the end of the Paleolithic era, or Old Stone Age, and initiating the New Stone Age. From then on, instead of collecting wild berries and roots and chasing migrating animal herds, he domesticated them. The Agricultural Revolution was underway.

Let us digress for a moment to look at some details of the Agricultural Revolution so as to demonstrate how a life style more conducive to survival than ever before came out of man's early attempts to farm.

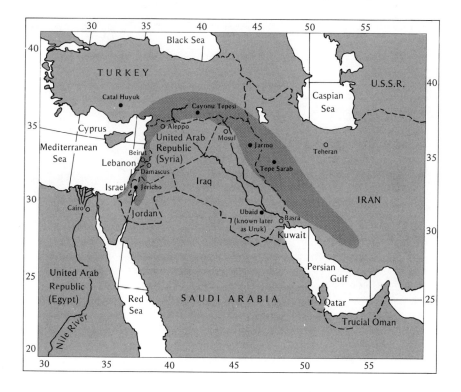

Figure 6–1
Cereal crops are believed to have been first domesticated in the area known as the Fertile Crescent. (Reprinted with permission of Macmillan Publishing Co., Inc., from *Man and the Environment*, by A. S. Boughey. Copyright © 1971 by Arthur S. Boughey.)

158　Chapter 6:　The growth of the human population

Agriculture began early in several places on earth, including Africa, America, and Asia. In Asia the Fertile Crescent region (Fig. 6–1), extending from the alluvial plain of Mesopotamia through Syria to the east coast of the Mediterranean to the Nile Valley, contains some of the oldest remnants of an agricultural civilization. Ancient grains of wild and domesticated wheat have been uncovered in the ruins of Jarmo in Iraqi Kirdistan near the Zagros Mountains, a thriving city in 6000 B.C. Here wheat and probably barley, derivatives of wild grasses similar to some still found growing, were the first cereal grains domesticated. Chromosome studies of wild grasses and tame wheat suggest that common wheat is derived from a grass of central Europe and western Asia, crossed with an earlier wheat type. The lineage of wheat, recently deduced, exemplifies early man's adjustment of his environment so as to live more easily (Fig. 6–2).

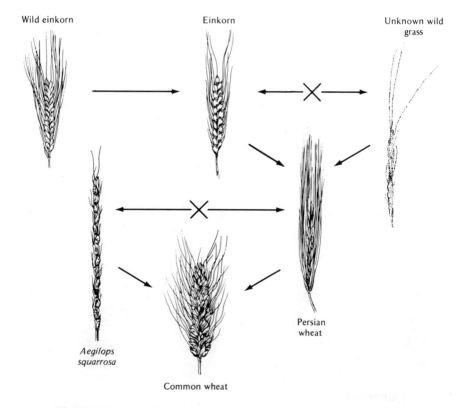

Figure 6–2
Outline of the evolution of common wheat. (From P. Mangelsdorf, "Wheat." Copyright © 1953 by Scientific American, Inc. All rights reserved.)

Down through history man has continued to domesticate plants and, through selective breeding, to improve them. For example, Neolithic man cultivated flax and grapes. More recent domestications include potatoes, first grown by Peruvian Indians in the first century A.D. and shipped to Europe as late as 1540 to 1560 A.D., and sugar beets, first popular in the early nineteenth century. The original plants of each hardly resemble the highly refined forms that exist today.

Neolithic man probably domesticated animals soon after he tamed plants. Grain crops produce food for man and animals. Corralling the tamest wild animals would avoid the long dangerous hunts imperative in Paleolithic times and increased man's chances for survival. After harvest, the crop fields provided grazing for herded animals, whose manure, in turn, may have been used to fertilize next season's crops. Dogs, probably first used to aid in hunting, date from at least 6000 B.C., and domesticated sheep and goats date from 5000 B.C. By 2500 B.C., several breeds of cattle existed, some derived from a humped breed common to India and Mesopotamia in 4000 B.C. similar to today's zebu. Eventually, man used animals' energy as well as their flesh or hide. Cattle pulled ploughs in Iran, India, and Egypt by 3000 B.C. and wheeled vehicles by 2000 B.C. Even earlier, by 3500 B.C., the ass was a beast of burden.

The results of the Agricultural Revolution eased man's relationship with his environment by modifying the harshness of the natural surroundings. As a result, he, too, was changing — culturally, or behaviorly. The significance of such a change was far-reaching. More abundant food freed man to concentrate on activities other than food searching. Trade in foodstuffs demanded money systems and alphabets. Decorated pottery, small colored models from tombs, and painted reliefs found near the Fertile Crescent testify to early man's new freedom to create.

But man had only begun to modify his environment. In addition to artistic creations, prehistoric man created new varieties of animals. Man must have recognized that desirable characteristics in animals can be bred into offspring if the right parents are bred instead of killed for food. Probably, wool is a product of such careful breeding. The wool of wild sheep is merely a fine down between the hairs. The sheep, domesticated about 6000 B.C. for meat, did not produce wool until millenia later. In fact, the Egyptians did not know of it until 3000 B.C.

Thus, out of the Agricultural Revolution, man's successful manipulation of the environment through domestication and selective breeding, came many changes, including new cultural developments and population growth. Previously, the difficulty of obtaining food kept his numbers small and his life short, but cultivation removed such limitations. More food could support more people, and more people, in turn, could grow more food. Children, a nuisance to nomadic food gatherers, now were welcomed to help tend animals and harvest crops. Consequently, man's numbers swelled.

Of course, population increase had momentary setbacks. Despite new affluence, occasional famines still reaped their dead. Great Britain endured 200 famines between 1000 and 1846 A.D., and China reportedly suffered 1,828 periods of famine in the 2,019 years previous to the early twentieth century. Millions of deaths due to starvation have been recorded in this century, too: in 1918–1922 and 1932–1934, 5 to 10 million deaths in Russia; in 1920–1921, 4 million deaths in China; and in 1943, 2 to 4 million deaths in India.

Wars, too, constantly beseiged civilization. No accurate death counts exist for many of the conflicts through history, but certainly, war has been frequent and devastating enough to have affected the size of populations of many countries.

But despite these checks, the population grew. As early as 6000 B.C., small villages developed in areas where food was especially plentiful, such as the Nile Valley. By 3000 B.C., settlements expanded into urban areas, and new lands were settled. Through the Bronze and Iron Ages, and into the Christian era, human life became increasingly free of threats and exigencies. The result at Christ's birth, according to the best estimates, was 200 to 300 million people on the earth.

The population continued to grow by fits and starts as man increasingly modified his environment. By 1000 A.D., western Europe had already begun to exploit energy sources not used before. Besides grinding grain, water power was adapted to other industries, such as sawing wood, and eventually even to forcing the bellows that fed blast furnaces. By the 1100s, wind power was slave, too. Inexpensive energy freed man for creative endeavors, eased sustenance, and allowed Europe's villages to increase populations as well as industries. By 1000 A.D., the world population stood near 275 million and grew steadily until 1348.

Yet some forces persisted to which he fell victim. The plague, for example, occasionally still struck. Between 1348 and 1350, this disease probably killed between 20 and 25 percent of Europeans. Repeated epidemics reduced the population even more, until by 1400 the total European population was only 60 percent of the estimated population before the plague.

But European man endured the plague as he had endured thousands of earlier threats and emerged ever more successful in modifying his environment; consequently, his numbers increased. By the 1500s he burst gloriously into that period of intense intellectual and emotional activity, the Renaissance, in which he cultivated an objectivity that enabled him to fill this period with more discoveries for the time than perhaps any era of his existence so far. The growth of science, technology, art, philosophy, and religion testifies to his ability to more ably interface with his environment and, thus, increase his survival chances. New technology, for example, produced new agricultural techniques. Fertilizers, crop rotation, and extensive soil cultivation all increased crop yields. Historians suppose that an agricultural boom occurred in China as well as in Europe. As a result, the world's population grew to nearly 500 million people by 1650.

By 1850 the world's population reached 1 billion. The doubling of the population between 1650 and 1850 was remarkably fast. More life-enhancing environmental conditions help explain the rapid increase, although all details are not known. Ample food from improved agricultural practices, the industrial revolution, improved sanitation, and the discovery of new lands for expansion are all significant factors to explain Europe's increase. Most significant, however, was the freedom from disease that resulted from improved medical techniques. Today, leprosy, tuberculosis, and polio are only three examples of diseases that once ravaged populations but are now under control in many countries. Certainly, improved medicine accounts for the evolution of Europe from a society with high birth and death rates to a society with low birth and death rates.

Asia's 2.5- to 3-fold increase in population in this period is even more difficult to explain, although political stability and new agricultural practices, both significant adjustments of the environment, were factors.

Incidentally, the rapid increase of the world's population between 1650 and 1850 is considered validated by the actual counts that were taken when England and France counted heads in their Canadian colonies. Other countries followed: Ireland, 1703; Sweden, 1748; Denmark, 1769; United States, 1790; and Great Britain, 1801. Estimates of world population between 1650 and 1850 were compiled from such data and are fairly reliable. The population of some countries, however, such as China, has persistently been questioned. Even recent censuses in that country may be erroneous. The 1953 census may well be inaccurate by 90 million.

How rapidly has the world's population grown recently? From the best estimates it seems that the world population grew about 0.3 percent per year from 1650 to 1750. The rate increased to approximately 0.5 percent between 1750 and 1850 and to about 0.8 percent from 1850 to 1950. This means that the world population increased from 1 billion in 1650, to 2 billion in 1930, to 2.5 billion in 1950. (Figure 6–3 illustrates this rapid growth; Table 6–1 shows the decreasing number of years required to double the population.)

The cause of man's population increase always appears to be linked to his increasing competency to modify his environment; his in-

Table 6–1
World population increase

Year	World Population Total*	Years to Double Population
1650	500 million	200 years
1850	1 billion	
		80 years
1930	2 billion	
1960	3 billion	45 years
1975	4 billion (?)	
1980*	4.5 billion (?)	
1990*	5.8 billion (?)	
2000	7.5 billion (?)	29 years (?)
2004	8.0 billion (?)	
2058	25.0 billion (?)	

*High predictions are used to show the most rapid doubling times.
Source: Data from "Constant Fertility/No Migration," Table A3.5, *World Population Prospects as Assessed in 1963*, United Nations, New York, 1966.

crease in this century is no exception. The attainment of 2 or more billion human beings in this century is due primarily to the elimination of those past conditions that maintained a high death rate. The one serious epidemic of this period, the influenza epidemic of 1918, was mild in comparison to the plague epidemics and other killing diseases of earlier times. Between 1850 and 1930, as the death rate of some nations dropped, populations increased. This phenomenon occurred in the countries in western Europe near the turn of the century, about the same time they were undergoing major industrialization. Many of these countries at the beginning of this century, following the drop in death rate, underwent a depression of the birth rate. (The United States followed this pattern in the 1930s.) Reasons for this drop probably include the expense and impediment to mobility of large families and the replacement of manual labor with machines on farms. The populations of both the United States and western Europe from then on grew slowly, but industrial and economic growth was rapid. Countries that underwent these changes in the recent decades are the most prosperous in the world. They are called *developed countries, D.C.s.*

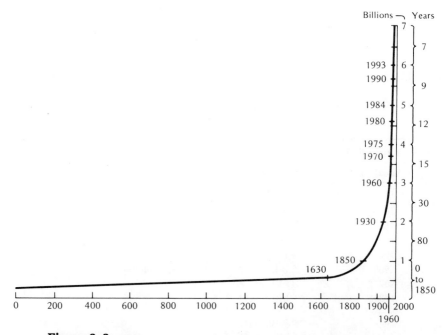

Figure 6–3
World's population growth since Christ's birth. (From Agency for International Development, *Food vs. People,* Washington, D.C., 1968.)

Unfortunately, other countries do not have the history of industrial development's maintaining a pace with population increase. At best, they have only recently imported from the developed countries some products of modern technology and industry. Birth rates and growth rates have not developed gradually. Consequently, they have many people and no well-developed economy to support them. These nations are called *underdeveloped countries, U.D.C.s.*

Population growth: The present and the future

Through the ages man has increased his numbers slowly and inexorably. Clearly, the reasons for his prodigious production into the billions lie in his increasing ability to adapt himself to his environment or his environment to himself. Man's increase through the ages testifies to his biological triumph over other species and his surroundings.

Yet the products of his successful environmental manipulation, billions of people, must have biological needs met in order to survive. What is the prospect for man's survival as a population? Can he continue to successfully manipulate and guarantee himself existence? Will his past success now work against him? A look at the details of the picture of world population now and as projected into the future will help to delineate the prospects for man into the years to come.

The United Nations' estimates

In 1963 the United Nations examined all data concerning current population growth around the world. The analysis included projections of growth rate nationally and globally through the year 2000. Such projections were derived from age-specific birth rates, migration rates, death rates, and reduction of birth rates in U.D.C.s but not major disasters such as war. (See the insert, "Birth Rates, Death Rates, and Growth Rates.") All predictions were considered within a range, with low, high, and middle points. These predictions for the years between 1960 and 2000 are tabulated in Table 6-2.

Table 6-2

The United Nations' predictions of world population (billions)

	1960	1970	1980	1990	2000
Low	2.998	3.545	4.147	4.783	5.449
Medium	2.998	3.593	4.330	5.188	6.130
High	2.998	3.660	4.551	5.690	6.994
Uncontrolled fertility	2.998	3.641	4.519	5.764	7.522

Source: *World Population Prospects as Assessed in 1963*, United Nations, New York, 1966.

In mid-1973 the word population, according to the Population Reference Bureau, was 3.86 billion. Note that this estimate is between the medium and high UN estimates made in 1963. If the medium estimate proves to be the accurate prediction, world population will be 6.130 billion by 2000. If the high prediction comes true, as many scientists believe will occur, the population will reach 6.994 billion in 2000, nearly twice the number of people on the earth in 1970.

The UN considers another prediction in addition to low, medium, or high. It is based on the assumption that the present fertility rate will continue. The estimate for 1970 for the world was 3.641 billion, only slightly higher than obtained; for 2000 A.D., 7.522 billion, a greater prediction than the "high" estimate and more than twice the 1970 population. Which prediction will prove most accurate? Even at the low prediction, and that is unlikely, the increase in the world's population by 2000 A.D. is a frightening specter that should concern all thinking people.

However, you should know that predicting world population growth in the past has been a risky business. The predictions proved too modest or too extravagant. The UN, for example, in 1957 proposed that in 1970 the world population could range from 3.350 billion (low) to 3.500 billion (high). This prediction high of 3.5 billion was attained in 1968.

The world's population in 1973 was 3.860 billion, and its rate of population growth was approximately 2 percent per year. This means that in approximately 35 years, if the present rate continues, the world's population will double, approaching 7.720 billion by 2008 A.D. (Fig. 6-3). A 2-percent rate of increase for several decades obviously will produce a mass of people impossible for the earth to accommodate. One scientist, Harrison Brown, estimates that, by the year 2058, the present rate of increase will yield 25 billion people; by 2100, 50 billion! Brown assumes that energy will be available at a near-zero cost and that the human diet will be greatly altered from what we know now. Another estimate, by Paul Hauser, predicts 20.4 billion in the year 2068. Paul Ehrlich calculates that this rate of increase will yield the impossible quantity of 6,000,000,000,000,000 people per square yard of land and sea in 900 years. Clearly, the carrying capacity of the earth will be reached before then.

One frequently hears in the United States, especially in states of low population density, that it is difficult, even impossible, to believe the predictions of population growth! This is understandable: the increase is an average that represents dissimilar rates of growth in individual regions. This means that the populations of some areas of the world are increasing at a rate greater than 2 percent, although some are increasing at a lower rate.

Birth rates, death rates, and growth rates

The number of births and the number of deaths per year per thousand people determine the growth of a population. These data are known as the *crude birth rate* and the *crude death date*, respectively. These rates are conveniently derived with the formula

$$\text{crude birth (or death) rate} = \frac{\text{total birth (or death)}}{\text{midyear population}} \times 1{,}000$$

Population growth: The present and the future

Because these data are easily derived they are frequently used, yet they do not disclose the specific factors that influence the growth or decline of the population. For example, the reasons or causes of death, although not revealed in the crude death rate, may influence the growth rate for that year or longer and, hence, comment significantly on the state of the population. A cataclysmic event, a hydrogen-bomb holocaust or a devastating earthquake such as struck Peru in 1970 killing 35,000 — obviously pushes the crude death rate higher than usual. So does an increase in deaths due to lung or heart disease during severe smog attacks. Childhood epidemics will escalate the death rate, too.

Significant information is often obtained by analyzing data concerning a limited segment of the population. Breakdowns by sex and/or age are especially important. For example, if the crude birth rate drops, it is worthwhile to determine if fewer women now make up the population as compared to other times. Also valuable would be a determination of the *fertility rate,* that is, the number of babies born per 1,000 women in the 15- to 44-year range. Instead of fewer women in the population depressing the birth rate, perhaps women are producing an average of fewer offspring than at previous times. Analysis of death rates by age or sex are significant, too. Outbreaks of diseases peculiar to particular age brackets, such as childhood, produce data that specifically explain an apparently increasing crude death rate.

Subtraction of the crude death rate from the crude birth rate produces an index of the increase in population for that year. This increase in population is called the *growth rate.* Whereas the crude birth and death rates are numbers of individuals per 1,000, the growth rate represents the increase of individuals per 100 population. Hence, it is a percentage. It is derived this way:

$$\text{\% population increase per year} = \frac{\text{number of births per thousand} - \text{number of deaths per thousand}}{10}$$

For example, assume 45 births per 1,000 people in a population and 20 deaths per 1,000 people. Then,

$$\text{\% population increase per year} = \frac{45 - 20}{10} = 2.5\%$$

You should appreciate the significance of the rate of growth of a population over a span of years. Consider the elementary business arithmetic of compounding interest. This analogy explains clearly the problems facing rapidly growing countries and, consequently, the world.

Assume that 2 percent interest is charged per year on a loan of $1,000, taken for 8 years. Then:

166 Chapter 6: The growth of the human population

Year	Starting Balance ($)	Interest at 2% ($)	New Balance ($)
1	00.00	20.00	1,020.00
2	1,020.00	20.40	1,040.40
3	1,040.40	28.80	1,061.20
4	1,061.20	21.22	1,082.42
5	1,082.42	21.65	1,104.07
6	1,104.07	22.08	1,126.15
7	1,126.15	22.52	1,148.67
8	1,148.67	22.97	1,171.64

Now consider the rate of increase over the original $1,000 at not only a 2-percent interest rate but at greater and lesser rates, too:

% Interest on Original $1,000	Approx. Years Needed to Double Original $1,000 (= $2,000)
0.5	140
1.0	70
2.0	35
3.0	24
4.0	17

Calculating compound interest on a loan, as shown above, is an excellent analogy to appreciate the rapid growth of the world's population. Merely substitute the quantity of people in any country or the world, and according to their rate of increase per year, you can obtain an impression of the rapid growth of their populations.
You should especially note the time needed to double the population.

D.C.s compared to U.D.C.s

The countries reproducing at a rate less than 2 percent per year are generally considered to be economically well off. They are the developed countries (D.C.s) described earlier. According to the United Nations, these nations usually include Europe, the USSR, United States, Canada, Japan, temperate South America, and New Zealand and Australia. Other characteristics of these countries include, besides a low rate of population increase, low infant mortality, low death and birth rates, economic independence with high per capita incomes, a well-fed and well-educated populace, and only approximately one-third of the population in an economically dependent group, that is, under 15 years of age or over 64 years old.

The underdeveloped countries, which maintain a birth rate greater than 2 percent, are generally regarded to be economically struggling. They include east Asia (less China and Japan), most of Africa, and most of Latin America except temperate South America. Characteristics of these countries include a relatively high birth and infant mortality rate but declining or low death rate, poor diets, low income and literacy levels, and approximately two-fifths of the population composed of economic dependents. (See Table 6–3 for a more complete comparison of developed and underdeveloped countries.)

Table 6–3
Salient characteristics of D.C.s and U.D.C.s

D.C.s	U.D.C.s
Low birth rates	High birth rates
Low death rates	Low death rates
Death rate lowered over long period of time	Death rates rapidly lowered since the 1940s
Nearly 1/4 of population under 15 years old	Nearly 1/2 of population under 15 years old
High literacy level	Low literacy level
Mainly urban	Mainly rural
High average income: affluence	Low average income: poverty
Great mobility	Little mobility
Primary occupations are government, skilled labor, white collar jobs	Primary occupations are farming, service, and labor
Extensive use of fuels and electricity	Small consumption of fuels
Many doctors and hospitals	Few doctors and hospitals

Table 6–4 shows the discrepancy in 1973 between population growth rates in underdeveloped countries or regions and developed areas. Developed countries characteristically have a very low growth rate. (For example, Canada has a rate of 1.2 percent per year; New Zealand, 1.7 percent.) In most European countries the growth rate is under 1 percent per year; in Hungary, Austria, and Ireland, the rate is under 0.5 percent. The United States and the USSR are growing at about 1 percent per year. Some areas even have a negative growth rate: East Germany has a −0.2 percent growth rate; Malta, −0.1 percent.

Contrast these data with those of the underdeveloped countries. In some Latin American countries and Mexico, the rate is about 3 percent per year. The population there will double in 24 years or less. Colombia, Ecuador, Venezuela, and Paraguay have a rate of 3.4 percent. Populations in these countries will double in 21 years! Some countries are growing even more rapidly. Gaza has a growth rate of 3.6 percent; Kuwait, 9.4 percent. About two-thirds of Kuwait's growth in population is due to migration into the country and one-third to birth and death rates.

In 1970 the percent ratio of population of D.C.s to U.D.C.s was 29:71. If the rate of increase is maintained, this ratio will be 25:75 in 1985. Those U.D.C. countries with the most rapid doubling rates will, of course, produce the greatest number of new people, mainly because the percentage of the population under 15 will have entered a reproductively active period between now and 1985. Today, there are over 1 billion children under the age of 15 in the U.D.C.s. Not that many people existed in the world until 1850! To appreciate how the ratio can be changed in so few years, contrast the doubling times of several major regions of U.D.C.s and D.C.s (Table 6–5).

Table 6–4
Population growth rates of some developed and underdeveloped world regions or countries in 1973

Some Underdeveloped World Regions or Countries	Growth Rate		Some Developed World Regions or Countries
Middle southern Asia	2.6	0.4	Northern Europe
Southwestern Asia	2.8	0.4	Western Europe
Southeastern Asia	2.8	0.7	Eastern Europe
Africa	2.5	0.9	Southern Europe
		1.0	USSR
Central America	3.2		
Caribbean	2.2	0.8	North America
Tropical South America	3.0	1.2	Japan
		1.7	New Zealand
		1.7	Temperate South America
		1.9	Australia

Source: *1973 World Population Data Sheet,* Population Reference Bureau, Inc., Washington, D.C., 1973.

Table 6–5
Population doubling times of D.C.s and U.D.C.s in 1973

D.C.s		U.D.C.s	
United States	87	Latin America	25
Europe	99	Africa	28
		Asia	30
		Oceania	35

Source: *1973 World Population Data Sheet,* Population Reference Bureau, Inc., Washington, D.C., 1973.

Probably, the D.C.s, although growing slowly, will eventually become much denser, but the threat is not imminent. Many U.D.C.s, however, are *now* in a crisis even at their present rates of increase. Soon their condition will inevitably worsen and reach disastrous levels. A look at the structure of the populations of U.D.C.s helps show why.

Causes for the plight of the U.D.C.s

You may wonder what specifically within each U.D.C. population accounts for the unhappy prediction of catastrophe. Many links contribute to the chain that holds some countries at the underdeveloped level. Considering population, however, the high birth and low death rates explain the problem and help point out why the populations of these countries increased so dramatically in only the last few decades. Figure 6–4

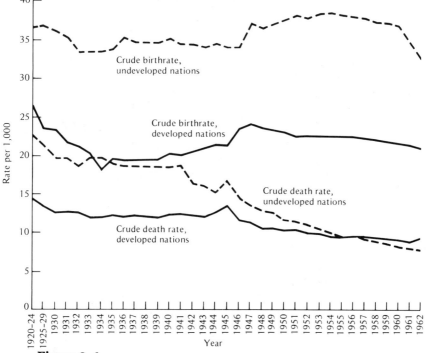

Figure 6–4
Trends in the crude birth rate and death rate in developed countries and underdeveloped countries show the abrupt decline in the death rate during the 1940s. (From D. J. Bogue, *Principles of Demography,* John Wiley & Sons, Inc., New York, 1969.)

shows clearly that, although the birth rate has stayed approximately the same in a few underdeveloped countries, the death rate has dramatically fallen, especially since World War II. Despite the death of millions and the far-reaching destruction, some benefits emerged from the war years. Science and technology were placed in crash programs to produce materials and devices that would hasten the war's end. Developments in medical science were especially important. Many new medical techniques and medicines saved the lives of thousands. These developments, used after the war on civilian populations, are now proving to be a mixed blessing.

Use of penicillin, for example, accounts for millions of the world's population being alive today. So does DDT, the insecticide, which is primarily responsible for Ceylon's population growth. Kingsley Davis reports on causes of the rapid decline of mortality in U.D.C.s, especially Sri Lanka (Ceylon). There DDT was used to control malaria, a disease that was the major cause of death. During 1933–1942, 1,736 deaths per million were reportedly due to this disease. Furthermore, many malarial deaths were not included in the total because they were reported as due to other causes. During the malarial epidemic of 1934–1935, doctors in Ceylon reported that 100,000 people died. According to Davis, about one-half of the deaths on the island were caused by malaria before the use of DDT.

In 1946, with help from the World Health Organization, Ceylon began to spray DDT to control mosquitoes. By 1949 the malarial mortality rate was reduced by 82.5 percent. Death from other diseases was also depressed, probably because of the destruction of other insect vectors and elimination of general debility owing to malaria (Fig. 6–5).

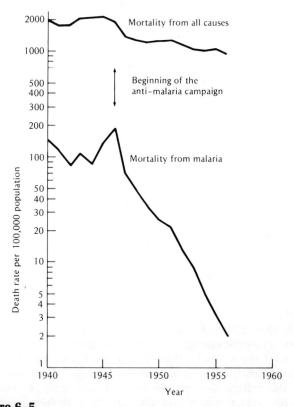

Figure 6–5
Data comparing deaths in Sri Lanka (Ceylon) in 1940–1957 due to all causes with those due to malaria point out the effectiveness of DDT campaigns in the 1940s. (From United Nations, *The Situation and Recent Trends of Mortality in the World,* New York, 1963.)

171 Population growth: The present and the future

In 1945 the death rate was 22 per thousand population. In 1954 it was 10.4 per thousand, a reduction of 53 percent in 9 years! Today, it is 8 per thousand. This amazing reduction of disease and mortality rate in so brief a period, well-illustrated with Ceylon, is also true of other U.D.C.s. For example, between 1940 and 1950 the death rate was depressed in Puerto Rico by 46 percent, in Jamaica by 23 percent, and in Formosa by 43 percent.

An analysis of the ages of the population within a country such as Ceylon or Puerto Rico tells clearly that it is mainly the youngest age group that benefits from such mass programs to eliminate disease. Consequently, the percentage of young people and children in the U.D.C.s is increasing.

Table 6–6
Population under 15 years in certain U.D.C.s and D.C.s

D.C.s	Percent of Population under 15		U.D.C.s
USSR	28		Asia
		40	Korea, Republic of
North America		43	Thailand
Canada	30	47	Phillipines
United States	27	41	Ceylon (Sri Lanka)
		46	Iran
Japan	25	45	Pakistan
		42	Turkey
Oceania			
New Zealand	32		Africa
Australia	29	47	Algeria
		46	Dahomey
Europe		44	Tanzania
Sweden	21		
East Germany	24		South America
Greece	25	42	Bolivia
Yugoslavia	28	47	Colombia

Source: *1973 World Population Data Sheet*, Population Reference Bureau, Inc., Washington, D.C., 1973.

Figure 6–6 contrasts the population age distribution in two U.D.C.s with two D.C.s. Table 6–6 compares the 1973 percentages of people under 15 in other D.C.s and U.DC.s. The increase in the percentage of the population under 15 has at least two significant implications when predicting the fate of the U.D.C.s. First, as illustrated with Ceylon, the rapid decline in deaths began in the late 1940s. The percentage of young people in the population of the U.D.C.s has steadily increased since that time. The first of this group are now in their early reproductive years. This means that, very likely, the percentage of babies born will continue to increase, be-

172 Chapter 6: The growth of the human population

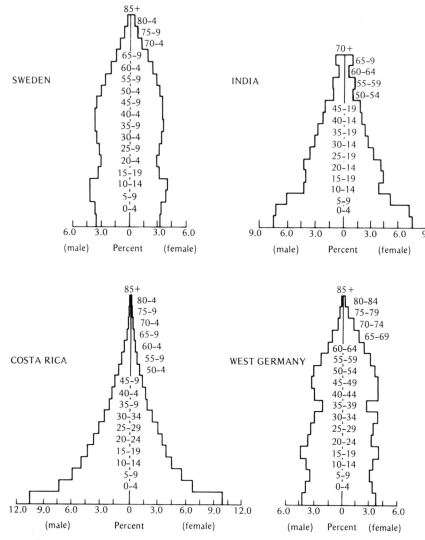

Figure 6–6
Differences in population composition can be seen in a comparison of the population pyramids from two developed nations (Sweden and West Germany) with those of two underdeveloped countries (India and Costa Rica). (From D. J. Bogue, *Principles of Demography,* John Wiley & Sons, Inc., New York, 1969.)

cause more people of reproductive age will produce more children. As time passes, an increasing percentage of the population will be young, and they will accelerate the trend by producing more babies.

Second, a greater percentage of young people in a population means a lesser percentage in the economically productive ages between 20 and 64. Consequently, a greater economic burden is placed on the U.D.C.s with a large percentage of children. Already the countries are straining economically. More young people add still more strain.

Table 6–7

Economically dependent population in D.C.s and U.D.C.s

	Age Group	North America	Latin America	Asia	Europe	Africa	World
	0–19	36.6	50.3	50.6	33.7	51.3	46.3
	65–70	9.6	3.8	2.7	9.5	2.6	4.5
Economically dependent	Total	46.2	54.1	53.3	43.2	53.9	50.8
Economically productive	20–64	53.9	45.7	46.6	56.9	46.2	49.1

Source: Adapted from Table 19, *The Future Growth of World Populations*, Population Studies 28, United Nations, New York, 1958, p. 37.

Table 6–7 compares several continents and the world population according to the economically dependent and productive age brackets. Note especially the contrast between the U.D.C.s and D.C.s.

Population density

One result of rapid population growth is an increasing density of people in all countries of the world but most notably in Europe and Asia. Table 6–8 shows clearly the densest areas in 1960 and the densest areas predicted in 1980. Figure 6–7 illustrates the density of population globally in 1967. Note that the less developed regions of moderate and high density will produce the greatest density increase in the next decade. The more developed regions will increase by 12.6 percent, but unhappily, the less developed regions will increase by 55.6 percent. As a result, the world's population will increasingly be located in countries that are already underdeveloped or, if not, face space limitations, such as Japan.

In Fig. 6–7 you can tell at a glance the percentage of the population in underdeveloped and developed countries in 1967 and projected for 2000 A.D. Figure 6–8 illustrates the projected growth of population in millions for specific areas for the year 2000 A.D. in contrast with the 1965 population.

In the past many significant events, notably wars and migrations, have occurred as the result of populations' becoming too dense to satisfy their needs. Migrations to meet demands for space and food are common

Table 6-8
Most densely populated areas of the world in 1960 and predictions for 1980

Region and Group	Density (persons/km²) 1960	Density (persons/km²) 1980	Population Increase between 1960 and 1980 %	Population Increase between 1960 and 1980 Persons/km²	Ave. Persons/km²
More developed regions of low density	8	11	32	3	
Australia and New Zealand	1.6	2.2	38	0.6	
Temperate South America	8	11	39	3	
Northern America	9	12	32	3	
USSR	10	12	30	2	12.6
More developed regions of moderate density	74	83	13	9	
Northern Europe	46	50	7	4	
Southern Europe	89	101	14	12	
Eastern Europe	98	115	18	17	
More developed regions of high density	167	193	15	26	
Western Europe	136	153	13	17	
Japan	252	300	19	48	
Less developed regions of low density	10	17	73	7	
Melanesia	4	6	41	2	
Middle Africa	4	6	46	2	
Southern Africa	7	11	67	4	
Northern Africa	8	14	76	6	
Tropical South America	8	15	88	7	
Eastern Africa	12	18	51	6	
Western Africa	14	24	74	10	
Southwest Asia	15	26	73	11	
Middle American mainland	19	36	91	17	55.6
Polynesia and Micronesia	20	38	89	19	
Less developed regions of moderate density	65	97	49	32	
Southeast Asia	49	81	66	32	
Mainland east Asia	59	77	30	18	
Middle south Asia	87	141	63	54	
Caribbean	85	136	60	50	
Less developed regions of high density					
Other east Asia	181	309	70	128	

Source: *World Population Prospects as Assessed in 1963*, United Nations, New York, 1966.

175 Population growth: The present and the future

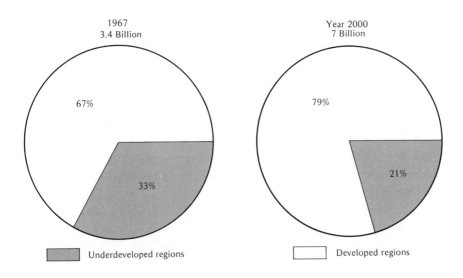

Figure 6–7
Comparison of population of underdeveloped and developed regions in 1967 with the population predicted in 2000 A.D. shows how the population distribution will shift. (From Agency for International Development, *Population Program Assistance,* Washington, D.C., 1968).

in Oceania. The Maori of New Zealand, for example, are descendants of inhabitants of central Polynesia who migrated by plan in the 1300s. Predecessors had discovered the island in the 1100s.

The South Indians, the Pallava, present a similar story. They established colonies in Oceania in the first century B.C. to avoid attacks of antagonistic Indian states and gain room for expansion. By the fifth century Java, Borneo, Sumatra, Cambodia, and the south Malay Peninsula contained Indian cities.

Possibly, too, the infamous Hun invasions into Europe in the fifth century A.D. may have been caused by a too dense population reaching out for new territories. In the Far East the Hsiung-nu tribes, who repeatedly invaded China from central Asia and provoked the Chinese to construct the famous Great Wall, may have been motivated by lack of space and food (Fig. 6–9).

More recently, the Crusade migrations in the eleventh century were most likely motivated to some extent by the possibility of gaining territory for the sons of families disadvantaged by the trend toward primogeniture, the inheritance by the oldest son of the family fortune.

The discovery of new lands in the fifteenth and sixteenth centuries when sea transportation became more dependable offered the opportunity for those deprived of space, food, or decent socioeconomic position to

176   Chapter 6:   The growth of the human population

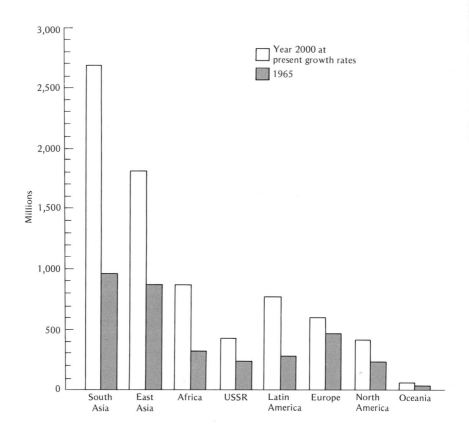

Figure 6–8
Comparison of the population of specific regions in 1965 and that projected for the year 2000 points out the unevenness of population increase in the world. (From Agency for International Development, *Population Program Assistance*, Washington, D.C., 1968.)

migrate to new opportunities. The rapid settlement of the New World resulted.

The aggressive attacks of Hitler's Germany in the 1930s and Hirohito's Japan in the 1940s are thought to be very much linked to the need for space (Fig. 6–10). Japan doubled its population between 1874 and 1937. Its pre-World War II population of 70 million was extraordinarily crowded on the small islands. Japan's population today is 103.5 million. Liberal birth control programs compensate for the population density, but how long Japan can endure within its present boundaries is a question.

China makes her bordering countries anxious with her frequent forays into their lands. China has in recent years invaded India and Tibet. As her population grows, history suggests that only more strife can be expected.

177 Population growth: The present and the future

Figure 6-9
View of the Great Wall of China, a superstructure possibly built in response to the frequent invasions into China of peoples from the Far East who were seeking space and food. (Wide World Photos.)

The prime concern is to what point the increasing density of some world areas can be pushed. Violence, internal uprisings, border disputes, and famines must be expected in the future. A large percentage of some populations, notably in U.D.C.s, are existing on inadequate caloric or protein diets. How long can this trend continue, and worsen, before some force reduces the population? (See the insert, "Robert Malthus.")

Robert Malthus
Prophet of doom?

Thomas Robert Malthus, son of a country gentleman, was born in 1766 (Fig. 6-11). He won prizes in English and Latin during his private education at Jesus College. At 22 he became a curate, but he seldom worked actively at ministering. Political economy was a deeper interest, and until his death in 1834 he held a post at the East India College at Hertford as England's first appointed professor of this subject and history.

Malthus's significant contribution was his book *Essay on the Principle of Population*, published in seven editions, the first when he was 32.

178   Chapter 6:   The growth of the human population

Figure 6–10
The attacks by Germany's armies on the rest of Europe in the 1930s were excused by Hitler as a national need for space. (Wide World Photos.)

In it he speaks out against two major schools of thought of the era: mercantilism and revolutionary utopianism. Officials of the state were intricately involved in economic enterprises. Flourishing business, even at the expense of agriculture, meant a strong nation–state. Wars or preparation for war meant strong business. Business hoarded for the rainy day when the nation–state would be involved in one of the frequent wars. Such was mercantilism, a close involvement of economics and politics.

To hoard people was also patriotic and economically sound. Large working industrial populations especially strengthened the government and economy, even though food might have to be imported. The then-new French Constitution read: "No one can be a good citizen who is not a good son, a good father, a good brother, a good husband." Although the prospect of overcrowding seemed imminent, the trend toward more people was still encouraged. A few saw that the concern for a quality of life was as important as an increase in population. Man, in fact, is capable of a utopia, they said, without inequalities of any kind, without war, without

Figure 6-11
In *Essay on the Principle of Population,* Thomas Robert Malthus in the nineteenth century predicted disaster for an England with an uncontrolled population. (Radio Times Hulton Picture Library.)

disease, without starvation. The state with the proper attitude could produce such conditions.

In response to these utopian remarks and mercantilist milieu, Malthus wrote his *Essay*. He claimed the following: (1) The quantity of food limits the size of the population. (2) As the means of substinence increases arithmetically, the population increases geometrically unless held in control by "checks." That is, food supply may increase in the order 1, 2, 3, 4, 5, etc., but the population will increase 1, 2, 4, 8, 16, etc., until sooner or later it is curbed. (3) The "checks," or controls, are major and obvious. They may be "preventive" or "positive."

The main preventive check is "moral restraint," which means postponement of marriage until the partners are financially sound, and no sexual intercourse before or during marriage. Some positive checks, or "vices," included homosexuality and birth control. Other positive checks, or "misery," included wars, starvation, disease, "unwholesome occupations, severe labor and exposure to the seasons, extreme poverty, bad nursing of children, great towns, excesses of all kinds." When examining the alternatives to population control, certainly abstinence is the most moral, he claimed. If this check is not chosen, the positive checks of vice and misery will take over.

Malthus opposed unimpeded population growth. He also claimed that man's social condition is tied to nature by his biological demand for sex and food. Furthermore, he said it is fulfillment of these needs that accounts for civilization. "To the laws of property and marriage, and to the apparently narrow principle of self-interest which prompts each individual to exert himself in bettering his condition, we are indebted for all the noblest exertions of human genius, for everything that distinguishes the civilized from the savage state." Maltus disagreed with the utopian writers of the era: "The structure of society in its great features will probably always remain unchanged."

Fear of overcrowding and concern over inequality in the social levels in early-nineteenth-century England made Malthus's *Essay* a best seller. Later, when emigration and better subsistence gave promise of a better existence, its popularity declined. Malthus's predictions did not come true, because he did not consider the possibility of greater food production through new agricultural developments, such as crop rotation, chemical fertilizer, better animal feed, and improved varieties of plants and animals. Neither did he realize the prospect of improved standards of living caused by industrialization and technology or recognize the impact on society of faster and more reliable means of transport. Nor did he entertain the consequences of birth control, because he did not consider it moral.

In the nineteenth century, unforeseen conditions prevented the demise of England as he predicted. However, today his prophecy of doom may be understood more clearly than ever before.

The trend toward urban living

As the crowds accumulate around the earth, more and more people will find themselves living in an urban environment. Between 1950 and 1960 the population of cities in the D.C.s increased 25 percent; in the U.D.C.s, 53 percent. Today, in Kenya's capital, Nairobi, growth is at a rate of 7 percent per year. That means that Nairobi adds more people per year than did Los Angeles between 1950 and 1960! Ghana's capital, Accra, is increasing by 8 percent per year; the Ivory Coast's capital, Abidjian, by 10 percent per year; the capitals of Zambia and Nigeria, by 14 percent per year! The migration to the cities is a search for a better and easier existence. Unfortunately, today in the U.D.C.s peasants arrive with no marketable skills. Consequently, they continue to live in as rural a fashion as possible but within the environment of the city. The consequent sociological problems are immense (Fig. 6–12).

Population growth in the United States

The problems faced by the underdeveloped countries are critical. Indeed, survival of many of their inhabitants, and perhaps of the coun-

Figure 6–12
Filthy shanties, the typical homes for many people in overpopulated regions such as Hong Kong. (WHO photo by Takahara.)

tries themselves, is the issue. At the moment they are more often concerned with maintaining existence than with the quality of life. The addition of approximately 1.2 million inhabitants to India each month is a calamity; the addition of nearly 170,000 people to the U.S. population each month is a serious problem of a different sort. In most developed countries, like the United States, the problem is maintaining a quality of life; they can probably support many more people than they now contain.

Extrapolations show the immensity of the developing problem. Increasing at 1 percent per year, the U.S. population by 1985 will be 241.7 million. At this rate, the population of 209.2 million in 1972 will double in approximately 70 years. To appreciate the demands made on the country, contemplate the changes necessary in the transportation systems to handle twice as many people; or the demands put on our natural resources, such as water or oil. Consider, too, the already serious problems of pollution, housing, privacy, recreational areas, and so on.

Many demographers are convinced that despite such an incredible increase in the population — which in a hundred years may pass 700 million — the economy of the country can absorb this huge population. Although the population will not go hungry, future generations will be born unaware of the quality of life common to their ancestors and forever lost to them. Remember, 1 billion people will produce a density similar to France and Poland now, and 3 billion people will yield a density similar to England, Belgium, or the Netherlands today.

Perhaps the predictions will not materialize. The U.S. growth rate, a result of the death rate and the birth rate, has unpredictably vacillated in the past. Perhaps the growth of the population will slacken in the future. New figures on our population growth seem to indicate that this is happening already.

The death rate

Exact data concerning the death rate over the past decades are not available for many countries and not for the United States before 1933. By that year all U.S. states established systems to record annual deaths. From 1900 until 1933, death statistics were gathered by the government from registration sources in selected parts of the country. Between 1850 and 1900, death data were gathered every 10 years at the census, but these data are not reliable. A few states compiled death statistics before 1900; for example, Massachusetts has done so since 1639. Extrapolation from information mainly from Massachusetts indicates a decline of the death rate in the whole country in the nineteenth century.

Until 1950 the death rate in this country declined notably, nearly 45 percent. Since 1950, however, the death rate has remained approximately the same, about 9.5 deaths per 1,000. Periodic outbreaks of disease, such as influenza in 1918, account for some variation in the general declining trend before 1950, but no explanation so far exists for the maintenance of a constant rate since 1950.

More specifically, the U.S. mortality trends in the twentieth century can be delimited into four periods: a period of rapid decline between

1900 and 1921 and another between 1937 and 1951, and a period of stability between 1921 and 1937 and another between 1950 and the present.

An analysis of the crude death rate in the United States according to age, sex, marital status, color, place of residence, and socioeconomic status produces a more detailed picture of the death rate through the decades. Compare the number of deaths per 1,000 inhabitants by age bracket in 1900 and 1964. Again the advances in medical care most likely account for the percent decline in the 64 years.

Data concerning the number of deaths by sex indicate that between 1900 and 1960 the death rate of American women dropped from 17.0 to 5.9 per 1,000 people, or 65 percent. For men in the same period the rate dropped from 18.6 to 9.5, only 49 percent. Furthermore, the death rate in 1960 was 61 percent more for men than for women. The reasons for the advantage of women over men are difficult to define. Biological resiliency and social acceptance of emotional release in women may partly explain their longer life.

Differences in death rates between whites and nonwhites are also well-recorded. Inequities in social and economic status are probably primary reasons for the poor living conditions and poor health care of nonwhites in the United States. In 1960, for example, $5,088 was the median income for the white population, whereas the median nonwhite income was $2,520. Educational opportunities are profoundly different, too. Since 1950 the percentage decline in mortality rate, approximately the same between 1900 and 1950, declined 15 percent for nonwhites and 9 percent for whites. Still, the nonwhite population dies at a faster rate than the white population, probably because they have less medical care and generally poorer diets.

Married people live longer than unmarried. In 1959 the death rate for men 54 to 55 years old was 8.3 per 1,000. For unmarried men in the same category the rate was 13.9 per 1,000. A similar trend is true of women. Reasons for this vary, and numerous jokes to the contrary, marriage generally does produce a healthy environment of regular hours, regular and healthful meals, companionship, and a longer life for men.

Place of birth, whether foreign or U.S., can also be correlated with death rates. In 1900, for example, death dates for foreign-born males and females in the United States were 18.9 and 18.4, respectively, whereas death rates for native-born males and females were 17.4 and 15.7, respectively. There were a number of reasons for such differentials in death rates. Certainly, the cultural training of the foreign-born, which frequently did not allow or encourage the recognition of symptoms of illness or the securement of treatment, was a factor, but so was the urban, sometimes ghetto, existence in which many groups were forced to live.

Since 1900 the death rate of foreign-born people has risen, whereas the death rate of native-born individuals has fallen. Such a trend can be interpreted on the basis of age. Immigration has fallen off. Many foreign-born are now elderly, and the death rate in advanced ages rises.

Place of residence, urban or rural, is an additional differential in analyzing the death rate. In 1960 the death rate in rural areas was 20 percent lower than that in urban areas. Interestingly, the larger the urban community, the higher the death rate. Iowa, Kansas, and Nebraska had the lowest death rates of the predominantly rural states in 1960. Occupation also affects death rates. Salary, socioeconomic position, lack of education, and level of occupation are all factors that contribute to a short life.

Many interrelating factors contributed to the crude-death-rate trends in U.S. history. All can be studied to help predict and explain the future death rate. The U.S. death rate will probably continue to decrease slowly as the social and economic conditions of deprived citizens improve as a result of better living conditions, heath, jobs, and education.

The birth rate

The highest birth rates in the world are in underdeveloped countries. In 1973 Dahomey, West Africa, for example, had a rate of 51. A few others were also high: Sudan, 49; Zambia, 50; Pakistan, 51; the Philippines, 45. The birth rates for developed countries are much less. Sweden's birth rate was only 13.8; Finland's, 12.7. The U.S. rate is 15.6. This birth rate, high for a developed country, is contributing to the overall increase in the U.S. population.

Data concerning the country's birth rate before 1900 are not reliable. No systematic attempts were made to gather information. Some evidence suggests that colonial America had a rate of about 55 births per 1,000 people. By 1820 the births began to drop, a trend usually witnessed when a country undergoes industrialization:

Year	Birth Rate
1840	52 per 1,000
1860	44 per 1,000
1880	40 per 1,000
1900	32 per 1,000

Birth rates in the twentieth century continued this decline but not always consistently. Generally, the downward trend persisted through the decades until World War II. Unquestionably, this general decline accompanied a greater national trend toward an increasingly technological society with all its benefits. The birth rate climbed slightly just prior to the war and reached 23 per 1,000 in 1943.

In 1944 and 1945 the rate declined. Then came the "baby boom," the period when the birth rate rose dramatically. Very likely, the return of men to their families and the promise of a plush economic era accounts for the increase in births. From 20.5 in 1945 the rate rose to 27 per 1,000 in 1947!

Such an upward trend differs from the trends observed in the other D.C.s at the same time. After a slight increase after the war, the birth rates then dropped off, a trend that continues today.

The baby-boom period, 1947–1957, maintained a birth rate of approximately 25 per 1,000. Demographers have studied the reasons the high rate was maintained in this country and not in other countries. The return to more normal conditions following the war caused many Europeans to make up for the marriage and child bearing that had been postponed. In the United States, however, many people not only "made up" for time lost to the war but married and began raising families at a much earlier age than before. In addition, the *size* of the family increased. By 1957, women in the peak reproductive years averaged 3.7 children each by the end of their reproductive years. Women in the 1930s bore an average of only 2.3 children. No doubt the favorable economic climate of the postwar years, never enjoyed before in the United States and unique to this country in the world, was a prime cause of the elevated birth rate.

By the 1960s, however, the baby boom ended. The birth rate began to fall from the 1957 rate of 25 per 1,000 people. In 1967 it was 17.9 per 1,000 people; by 1970, 17.6 per 1,000. Consequently, the actual number of babies born declined from 4,317,000 in 1961 to 3,555,000 in 1967. If the birth rate in 1967 had equaled that in 1957, approximately 1.5 million more babies would have been born that year.

Explanations for the termination of the baby boom and rapid decline of the birth rate since 1957 are difficult to discern. The wide acceptance of the birth control pill has been cited. However, the pill was not available to the general populace until 1960, and the effects of its use would not be felt until 1961. By then the birth rate had already been dropping for 4 years. A more reliable explanation may be that those who gave birth to their children during the baby boom had now completed their families and were out of the baby business.

Furthermore, because of the declined birth rate in the depression years, fewer women of prime reproductive age existed. Finally, women of the prime reproductive age in the late 1950s and 1960s have had lower birth rates than women at an earlier time. Two reasons may exist for this: the women are having fewer children per family than comparable women before; and couples are spacing their children much farther apart and so they have not yet completed their families.

The future

Most all of the factors mentioned seem to suggest that the U.S. birth rate may continue to drop. However, some reasons exist which indicate that the birth rate may even increase. For example, the plentiful postwar babies are in or about to enter their reproductive periods. Women 20 to 24 years old totaled about 5.5 million in 1960. In 1970 that number was *at least doubled*. Even with family-size limitation, possibly because of the availability of birth control, the increase in the total number of women bearing children is greater. Hence, the birth rate may rise.

As with the crude death rate, an analysis of the crude birth rate uncovers the differentials affecting population growth in this country. This information should allow for more knowledgeable predictions of its future growth, specifically from what group the population will be derived. For example, a fertility pattern can be detected upon analysis of the crude birth rate of the white and nonwhite segments of the population. The birth rate of the nonwhites in this country is higher than that of whites. In 1960 it was 32.1 per 1,000 for nonwhites and 22.7 per 1,000 for whites, a 41 percent difference. By 1963 a 43 percent difference existed, and the gap seems to be widening.

Greater fertility is most probably due to lack of knowledge and use of birth control methods. In addition, planning for the future has little appeal for those who have difficulty gaining the benefits of a technological society.

Religion is another differential that affects fertility rates, possibly the most significant. Data gathered by the Bureau of the Census in 1957 indicate that Catholics are most fertile and Jews least fertile. The Catholic Church's official stance that sexual intercourse should be for no reason other than procreation precludes the use of most modern birth control methods. Among Protestants, the lowest fertility rate is among the most liberal groups; the highest, among fundamentalist groups.

Finally, the birth rate can be analyzed by the criterion of socioeconomic status. As the level of education of the woman and the family income goes up, the number of children decreases. As with religion, birth control information and ability to purchase devices or pills probably influences the decline in the fertility rate. Only 70 percent of grade-school-educated women expect to use birth control some time in their lives; 95 percent of college women do.

The implications of these data concerning color, religion, and socioeconomic status of women are important. Can a democratic society, supposedly to run on the decisions of an able, educated populace, continue to run adequately, or in fact survive, when an increasing proportion of its inhabitants are in the lower socioeconomic, poorly educated, and impoverished groups?

The United States probably will not have difficulty accommodating its population changes in the near future. But the world as a whole is not so fortunate, and parts of the world already cannot manage to feed or house their numbers. The future can bring only more misfortune.

Is the population problem unique to mankind? Do other organisms have similar problems? How do natural populations normally retain a balance? These are questions that we shall ask in the next chapter when we examine populations of other organisms in the hope of gaining insight into our own population problems.

Summary

Let us review the main ideas in this chapter. Man is a success as an organism because he has devised ways to relate with his environment in a life-giving manner. As a consequence, through the ages man has become very abundant and widely distributed. However, today he has reached a point where his very success as a species, the product of thousands of years of evolution, may be working against him. For example, in the underdeveloped countries high birth rates and low death rates have produced a large annual increase in the population in the last decade. Consequently, these countries have been overrun with people very quickly and have been unable to develop an economy to keep pace with the growth. In contrast, the developed countries are more fortunate, because their populations are growing more slowly, mainly because their death rates and birth rates have been low for decades. This, coupled with economies that are more able to provide for the population, gives the developed countries a higher standard of living than the underdeveloped countries. What will happen as the global population continues to grow is an ominous threat. Most likely, even the developed countries will eventually be seriously affected.

Supplementary readings

Bates, M. 1962. *The Prevalence of People.*
Charles Scribner's Sons, New York. 283 pp.

Bogue, D. J. 1969. *Principles of Demography.*
John Wiley & Sons, Inc., New York. 917 pp.

Borgstrom, G. 1969. *Too Many: The Biological Limitations of Our Earth.*
Macmillan Publishing Co., Inc., New York. 368 pp.

Callahan, D., ed. 1971. *The American Population Debate.*
Doubleday & Company, Inc., New York. 380 pp.

Davis, W. H., ed. 1971. *Readings in Human Population Ecology.*
Prentice-Hall, Inc., Englewood Cliffs, N.J. 251 pp.

Detwyler, T. R. 1971. *Man's Impact on Environment.*
McGraw-Hill Book Company, New York. 731 pp.

Ehrlich, P. R. 1971. *The Population Bomb,* rev. ed.
Ballantine Books, Inc., New York. 201 pp.

Fisher, T. 1969. *Our Overcrowded World.*
Parents' Magazine Press, New York. 256 pp.

Harrison, G. A., and A. J. Boyce, eds. 1972. *The Structure of Human Populations.*
Oxford University Press, New York. 447 pp.

Hinrichs, N., ed. 1971. *Population, Environment, and People.*
McGraw-Hill Book Company, New York. 227 pp.

Leisner, R. S., ed. 1971. *Population and Food.*
William C. Brown Company, Publishers, Dubuque, Iowa. 83 pp.

Pyke, M. 1970. *Man and Food.*
McGraw-Hill Book Company, New York. 256 pp.

Wrigley, E. A. 1969. *Population and History.*
McGraw-Hill Book Company, New York. 256 pp.

Chapter 7

Man is a fraction of the animal world. Our history is an afterthought, no more, tacked to an infinite calendar. We are not so unique as we should like to believe. And if man in a time of need seeks deeper knowledge of himself, then he must explore those animal horizons from which we have made our quick little march.

African Genesis
Robert Ardrey

Characteristics of animal populations

In the early part of this century one of the greatest natural tragedies occurred on the 727,000-acre north rim of the Kaibab Plateau near the Grand Canyon in Arizona. Before 1906 this area supported nearly 4,000 mule deer and their competitors for the forage, thousands of sheep (200,000 in 1895), cattle, and some horses. In 1906 a Presidential decree set aside the area to protect and perpetuate the famous deer herd. Wolves, coyotes, and cougars, the natural predators of the deer and other grazing animals, were removed. Between 1907 and 1917, 600 mountain lions were killed, 74 more were destroyed between 1918 and 1923, and 142 more between 1924 and 1939. Three thousand coyotes were destroyed between 1907 and 1923, and between 1923 and 1939 nearly 4,500 more were eliminated. By 1939 the last wolf in the area was shot.

At the same time the cattle and sheep were removed. The increase in food, reduction of predatory pressure, and protection from hunting resulted in a huge increase in the deer herd. By 1918 deer had increased 10 times, and by 1924 the number of deer had risen to about 100,000, approximately 25 times what it was in 1907.

Then came catastrophe. In the winter of 1925–1926, the size of the deer population crashed until only about 40 percent of the population survived. By 1939 additional losses lowered the number to approximately 10,000 deer. Since then the deer population has fluctuated greatly, apparently influenced by the severity of winter, range conditions, and population density.

The decimation of the deer population on the Kaibab Plateau is a major example of interference with natural population without knowledge and concern for the consequences (Fig. 7–1). The reduction of predators, the increase of available food resulting from the elimination of sheep and cattle, and protection from hunting all contributed to the increase of the deer herd and its eventual catastrophic decline. Such a calamity can occur when man does not understand or appreciate the careful balance that organisms maintain with their environment in the natural world. The unregulated chaos due to the rapidly expanded deer population could have been avoided if information and sense prevailed.

Hopefully, from relating the story of disaster on the Kaibab Plateau and by examining in detail the characteristics of animal populations, especially the ways they are naturally controlled, some insight may be gained into how we, as another animal species, may propagate a life-sustaining relationship with our environment. This may be especially timely in view of the cancerous expansion of the human population.

With this in mind, we shall examine next the characteristics of populations, especially the ways they change, the controls on population size exerted both as a result of the density of the population and from outside the population, and examples of natural and predictable population changes in natural communities, known as *succession*.

Population change

A population of organisms can be defined as a group of animals or plants of any size that exists in a particular place at a specific time. Although such a population sounds static and unchanging, its most salient

Population change

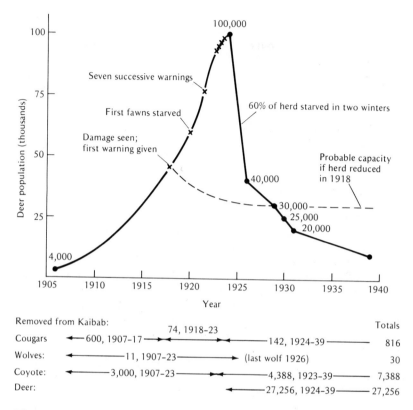

Figure 7-1
Reduction of predators and removal of cattle, sheep, and horses from the 727,000-acre Kaibab Plateau in Arizona helped "spike" the mule deer population, which led to an abrupt decimation of the herd, starting in 1925. Application of basic ecological principles could have avoided the catastrophe. (From A. S. Leopold, *Wisconsin's Deer Problem*, Wisconsin Conservation Bulletin 321, 1943.)

characteristic is change in size and internal composition. Change results from interaction between the nature of the individuals in the population and the environmental conditions surrounding the population. This interaction can lead to a new ratio of genes in the population and, hence, to transition of the physical and behavioral characteristics of a group of organisms with new characteristics. Charles Darwin wrote of this kind of population in his famous *The Origin of Species*, suggesting that it was the source of new kinds of plants and animals.

Such a change requires a great amount of time and is one of the most profound population changes to occur. Population changes requireing less time also occur constantly as a result of the interaction of individuals with the environment. In natural populations these changes are often life-sustaining for the group, although individuals may be sacrificed.

Occasionally, however, sudden changes are induced in a population or its environment which threaten all individuals with the most profound change of all — extinction.

Knowledge of the factors that affect animal populations is vitally important to man today because his population may be reaching a size and composition which threatens its existence.

Ingredients of population change

Analysis of the sources of change within a population reveals several characteristics of groups that cause it to change in internal structure and size. Among the most important are rate of birth, rate of death, and rates of immigration and emigration.

The birth rate, the rate of occurrence of new individuals, of most organisms is extraordinarily high. Some bacteria, for example, are able to divide every 15 to 20 minutes. Within 36 hours bacteria have the potential to cover the entire earth 1 foot deep. Then, $1/2$ hour later, bacteria would cover our heads! The bluegill is another example. This common freshwater fish often produces from 4,000 to 60,000 young fish in a single nest. Obviously, not all these offspring survive, but many do. Most organisms, as Darwin observed, produce numerous offspring. Such abundance of individuals usually helps maintain the population, but it can also threaten its existence. In any case, the coming of new individuals causes change in the group.

Changes within a population also are a result of the death rate, the number of individuals that die over a period of time. Obviously, the rate of death limits all populations, which potentially could increase to astronomical limits. The varied causes of death are usually so abundant in any natural environment that overproduction of individuals is usually easily counteracted. When overproduction is not managed, it is often to the detriment of the population. Because man has adroitly managed his environment through history, his population continues to expand — to threatening proportions.

Immigration, the movement of individuals into a population, and emigration, the movement of individuals out of a population, are two other sources of change of a group of organisms. The control of the size of the human population in various global areas historically has been controlled with these methods. For example, when Ireland's populace in 1843 suffered the scourge of potato blight, which reduced the available food, the Irish migrated to a new country. As a result, the population of the United States grew quickly as the population of Ireland dwindled to manageable numbers. Such emigration and immigration are natural causes of change, even in plant populations.

Population change through time

Birth rates, death rates, and the rates of emigration and immigration change in time the composition of individuals in a population as well as the size of the population. Such changes are often affected by environmental forces. The effect of environmental changes, such as exhaustion of food, can easily be observed in laboratory cultures of fruit flies in which the walls of the container allow no flies or food to enter or to leave the population. The tendency to expand is checked by the boundaries of the environment. Indeed, a population can be said to possess a peculiar kind of population tension: that is, a tendency to expand and grow opposed by environmental factors that discourage expansion and growth.

Many calculations have been made of the potential of populations to expand rapidly and plentifully if environmental impositions did not exist. For example, one entomologist calculated the quantity of houseflies produced if *one* individual female fly produces 120 eggs at each laying and half the eggs were females. The quantity of flies is staggering if we assume that all females born will reproduce. In a growing season of seven generations the number of offspring in the second generation derived from the one housefly would be 7,200 houseflies, and by the end of the seventh generation, over 5 trillion!

The English sparrow is another example. This bird, introduced into the United States in 1899, was calculated to produce 275,716,983,698 descendants in 10 years from each single pair of sparrows. At this rate, 575 birds per 100 acres could be expected by the early 1920s.

The tendency for populations to increase to incredible potential numbers is usually checked, however, because sooner or later environmental forces tend to counteract a population's "biotic potential," its natural tendency to expand. Most organisms in the wild are subjected to this pressure constantly. Only a few natural populations and organisms grown in the laboratories may tend to approach their biotic potential because of the temporary elimination of environmental inhibitions.

The English sparrow population in the United States illustrates the containment of a natural population. Although by 1920 it was calculated that several 100s of these birds could occur per 100 acres, only 18 to 25 birds per 100 acres could be found, less than 5 percent of the calculated quantity, thanks to factors in the new environment. Such population-limiting factors are known as "environmental resistances."

The sigmoid growth curve

The biotic potential of a population is a theoretical condition that organisms do not attain except temporarily at very early stages in their development or under artificial conditions. Nevertheless, the biotic potential of a population, as well as its actual expansion into a particular environment, are mathematical expressions that can be represented graphically.

As the population grows, the inhibitions of the environment become great. Consequently, on a graph, the actual increasing density of population plotted against time produces a curve representative of population growth of any organism. The curve is known as a *sigmoid,* or *logarithmic, growth curve* (Fig. 7–2).

As a population approaches its maximum size in its particular environment, the logarithmic curve will eventually begin to stretch into a flat and more stabilized line, indicating that the number of new individuals born

194   Chapter 7:   Characteristics of animal populations

in the population equals the number of individuals that die. In other words, the rate of birth, or natality, and the rate of death, or mortality, balance each other. Earlier, of course, natality exceeded mortality and accounted for the rise of the plotted curve.

However, even within a stable population in which significant expansion of the population size is prevented because the environment resistances are impinging on the biotic potential, the curve will fluctuate slightly.

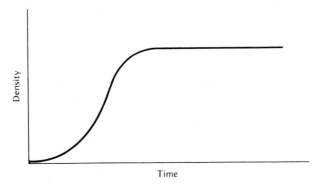

Figure 7–2
The sigmoid growth curve is typical of developing populations.

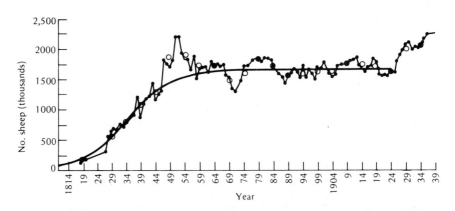

Figure 7–3
Growth curve calculated for the sheep population in Tasmania. Open circles represent the average quantity of sheep in a 5-year period; black dots represent the count from official records for each year. (From J. Davidson, *Transactions, Royal Society of South Australia*, vol. 62, pp. 342–346, 1938.)

This is because even stable populations have their "ups and downs," depending upon minor changes in the environment in which they find themselves. A minimal degree of fluctuation around a mean can be seen in the growth curve of sheep introduced into Tasmania (Fig. 7–3). More pronounced fluctuations can be seen in the growth curve of prey and predator, as would be expected. The prey–predator relationship will be discussed later in detail.

Thus, a natural population normally reaches a balanced relationship with its environment, so that, within a range, its size is maintained. However, environmental resistances can be removed as environmental conditions change. Consequently, the growth curve will again rise; or if environmental conditions become increasingly harsh, for example, if food or water becomes scarce, the population size may be reduced. When the enviromental resistances overwhelm the population's ability to maintain itself, the population dies out; on a graph, the line will fall.

The J curve

Populations that undergo a sudden stress, such as starvation, produce a growth curve that falls abruptly (Fig. 7–4). This curve is known as a J curve and results from the deaths of many individuals in the population within a very short time. The J curve, as well as the S curve, can be seen in the plotting of the Kaibab deer population through several decades (Fig. 7–1).

The growth of plankton — because it is intimately tied to the seasons of the year — also produces J curves. The temperature of the water changes seasonally. The turbulance that results distributes nutrients throughout the water and encourages the reproduction of plankton. Consequently, in May and June, and again near the fall, "blooms" of plankton occur. Then as the food material abruptly decreases, the blooms recede, producing J-shaped curves.

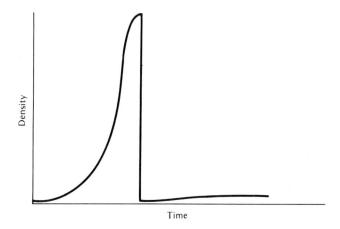

Figure 7–4
Typical J curve representing a population that is suddenly reduced, possibly due to starvation.

This information concerning animal populations is significant, especially when applied to the growth of the human population. The increase of human beings plotted against time yields the early stages of a sigmoid growth curve. However, the path of the curve in the future is uncertain. It may well level out if human beings reach some kind of balance with their environment, but it could drop abruptly if an imbalance with the environment persists. Insufficient food to nourish the billions could affect the human population as the lack of food affects the plankton. The kind of growth curve the human population will trace in the future is today one of the most serious problems facing mankind.

Resistance to population growth
Environmental resistance

The types of environmental resistance affecting a population can conveniently be categorized as factors independent of the density of a population or dependent on the density. *Density-independent* factors normally are those aspects of the environment that are nonliving, such as light, heat, temperature, moisture, and availability of nesting sites. This means that whether the population is dense or not makes no difference as far as these factors are concerned. *Density-dependent* resistances include other organisms, such as predators. Some factors that influence population size are difficult to classify, so the categories are not mutually exclusive.

Density-independent resistance

An algal "bloom" is an example of the size of a population controlled by density-independent factors. High temperatures and plentiful oxygen and nutrients resulting from spring and late summer turnovers of the water seem to cause the seasonal occurrence of abundant plankton, or "blooms," in lakes in temperate areas (Fig. 7–5).

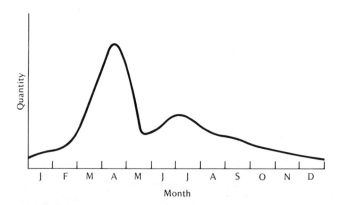

Figure 7–5
The fluctuating growth curve of algae throughout the year is due to the distribution of oxygen and nutrients in late spring and late summer. Changing air temperature causes the water to mix.

Temperature also appears to explain the fluctuating population of a species of heron *(Ardea cinerea)* in Great Britain between 1928 and 1950. Years with cold winters appear to support fewer herons during the warm months (Fig. 7–6).

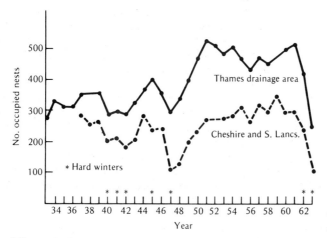

Figure 7–6
Fluctuations in two heron populations which appear to be tied to temperature. (From D. Lack, *Population Studies of Birds,* The Clarendon Press, Oxford, 1966.)

Another environment influence is moisture. Classical studies by biologist Paul L. Errington illustrate the influence of the quantity of pond water on the size of muskrat populations. In dry seasons, entrances to muskrat houses are left unsubmerged, and predators, such as foxes, gain easy access. If the water is low in the colder months, food is frequently unavailable. Consequently, the animals must leave the area, exposing themselves to predators as they migrate extensive distances on land.

Abundant moisture is also a problem. It inundates the muskrat burrows and houses, destroying great numbers of animals. Any of these extremes in moisture affects the size of the muskrat population.

More recent examples of an environmental control of insect populations is shown with the use of pesticides. The decimation of populations with these chemicals is well-documented (see Chapter 12).

Density-dependent resistance
Internal physiological control

Researchers only recently gained insight into the causes of control which appear to originate within the individuals of a very dense population. In these individuals alteration of internal physiology due to malfunctioning glands suppresses reproductive behavior and may even cause death.

The adrenal gland appears greatly involved in limiting populations subjected to the stress of high density. For example, house mice populations were allowed to develop from a few pairs in large cages with ample food, water, cover, and nesting material. As the populations grew in size, infant survival and birth rate declined. Compared with control groups, productivity of the experimental female mice was reduced at least 31 percent. Adrenal glands in the males and females in the dense populations were larger and heavier than normal. The development of the sex glands was delayed. Increasing levels of stress due to increased density apparently were reflected in the adrenocortical hypertrophy and suppression of reproduction. Experiments now suggest that similar changes in the glands occur in mice at the bottom of the pecking order in populations not especially dense. This appears to be due to the stress induced by too much harassment by mice higher in the hierarchy. Individuals higher in the peck order do not show the symptoms.

Similar physiological reasons may influence the several-year cycles seen in some animals. In the north, the lynx, the great horned owl, the goshawk, and their prey, the snowshoe hare, all evidence a 9- to 10-year oscillation in population density. Lynx skins collected by the Hudson Bay Company since 1845 indicate the cycle (Fig. 7–7).

The regular fluctuations seen in lemming populations may be caused indirectly by the glands. These animals reach a population peak every 3 or 4 years and, if the population is exceptionally dense, will migrate many miles, often ending their trek by dropping into the sea. In some Norwegian towns the migrating hordes are so great that dogs and cats do not bother to pursue them. Predators of the lemming, such as snowy owls and arctic foxes, also evidence a 3- to 4-year cycle.

Although the prey-predator relationship must influence the oscillation, some evidence suggests that endocrine changes at high densities, at least in the snowshoe hare, help maintain the cycle. These usually lethal changes compose the syndrome called "shock disease."

Shock disease in the snowshoe hare occurs in the late winter and early spring following a summer in which the population reaches an extraordinarily high density. Behavior changes fall into one of three categories: (1) convulsions, initiated by sudden running motions of the legs, extension of the hind legs, and a retraction of the head and neck (some animals initiate convulsions on landing after a sudden leap into the air); (2) extreme lethargy; and (3) a sequence of both behaviors. Internal symptoms include transformations in the liver and spleen. The liver becomes soft, emits a sweet unpleasant odor, and becomes darker than normal. The spleen becomes brownish, shows little blood, and is covered with a fibrous capsule. Hemorrhaging of the brain, kidney, thyroid gland, and adrenal gland also may occur. Hypoglycemia and a reduction in the liver glycogen also occur. These symptoms are thought to be influenced by the exhaustion of the adrenal and pituitary glands. The secretion of hormones from these glands presumably is excessive under the stress conditions of a dense population, and the glands degenerate.

Some factors in the natural environment may help induce shock disease in dense populations, according to the mammologist John Christian. These factors may include scarce and inadequate food, the low tempera-

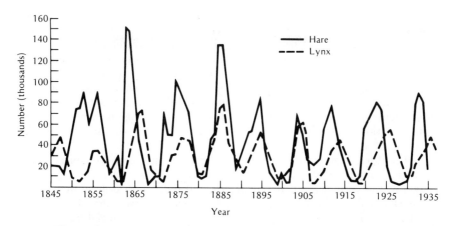

Figure 7-7
The changes seen in this lynx and rabbit population is a classic example of oscillation in population density involving prey and predator animals. (From E. P. Odum, *Fundamentals of Ecology,* 3rd ed., W. B. Saunders Company, Philadelphia, 1971.)

tures of winter and early spring, increased muscular exertion during more extensive food forays, fights, inadequate shelter, and the onset of reproductive activity.

Other examples of internal control over population size can be cited. For example, aphid populations, during seasons when living conditions are optimal, normally produce only wingless females, which produce offspring parthenogenetically (from unfertilized eggs). However, as the population becomes dense, the competition increases. Sexually reproducing winged females develop, which migrate from the area.

Physiological changes have also been seen in dense populations of several species of locust, especially *Locusta migratoria*. As populations become dense, the locusts darken, develop longer wings, and store fat. Eventually, when the population is most dense, the locusts will emigrate from the dense area in great swarms.

Another example of physiological changes induced by great density is the fungus fly, *Oligarchs paradoxus*. These flies reproduce parthenogenetically, remaining immature as long as food is plentiful. However, as the density of the population increases and the food supply runs out, competition among members of the group becomes intense. Then winged adults develop, which emigrate to other areas.

Territoriality

As we have pointed out, some animal populations appear to be limited by the onset of behavior that curbs population growth. The causes of such behavior remained uncertain until very recently when evidence of physiological control was uncovered.

However, some species of animals appear to limit their population density through mechanisms not yet determined to be physiological. No definite aberration in size of glands or amounts of hormones in the blood have yet been detected, although their behavior effectively limits population size. Until further research discloses such physiological controls, these social behavior patterns must be discussed separately.

One of the commonest behavioral adaptations to maintain a manageable density is the establishment of a geographical area for carrying out life activities, such as feeding or breeding. This area is known as the "home range." If all or part of this range is carefully defended against any other members of the species, or perhaps just against members of the same sex, the area is known as a "territory." Defense of such an area is known as "territoriality."

Territories serve many functions. For example, some songbirds stake out a territory for mating, nesting, and feeding. Before the arrival of the female in the spring, the male songbirds will defend an area with singing and occasionally fighting. Later, when the females arrive, they join in the defense. Still later, when nesting activity is over, the territorial boundaries are neglected.

Some birds, such as hawks, use the territory only for mating and nesting, and some, such as the woodcock, use it only for mating. Other birds set up territories only for feeding or roosting. Fish, another animal group that is customarily territorial, often determine areas for feeding or breeding. Some mammals, such as the muskrat, antelope, seal, vicuña, and deer, also develop territories (Fig. 7–8).

The factors that determine the size, shape, or location of the territory are not completely known for most animals, although several characteristics of territories have been identified. Among them are the following: as the mating season wears on, territories often become smaller as more animals develop areas; most territories are roundish except along waterways; some territories are known to move because the animal moves, as illustrated by the Canada goose; and the location of territories of birds, such as the American robin, which returns to the same general area year after year, often varies.

A result of territoriality is the prevention of the too-dense population. Another result can be the prevention of underpopulation. Excess animals for which no territory exists are known to take up residence outside the territorial areas and move in when room develops, as work with the Australian black-backed magpie, for example, shows. Males of this species, while in the fringe group, do not even develop sexually.

Animals actively defend their territorial boundaries. Experiments with stickleback fish show that if a male stickleback in a test tube is placed within the territory of another male, he becomes visibly inactive and apparently subdued, especially when the "owner" of the territory attacks, as he usually does. The fish in the test tube when in his own territory, however, is vigorously aggressive.

Despite such aggressive behavior, few animals of the same species actually kill each other. Normally, the great show of aggressiveness will chase rivals away. If a defender has met his match, usually he will relinquish his right to the territory to the victor without bloodshed.

Predation

One environmental factor that limits the size of some populations, such as the deer on the Kaibab Plateau, are predators. A *predator* is any

Figure 7–8
Two buck deer with locked antlers end their battle for territory. (National Audubon Society. Photo by W. L. Miller.)

animal or plant that feeds on other animals or plants, the prey. Such a relationship between prey and predator is a delicate one, because obviously, if the predator too effectively destroys its prey, he himself is faced with extinction.

However, stability usually characterizes a prey–predator relationship. Predators seldom become so plentiful as to exhaust the prey species, and the prey species never become so numerous as to overwhelm the predator species unless drastic changes occur in the environment. Rather, there is a dynamic balance between the two populations. The relationship between the deer and its predators on the Kaibab Plateau before the entrance of man, cited early in this chapter, illustrates well a naturally balanced prey-predator relationship.

The maintenance of the balance probably depends to some degree on the repeated immigration of individuals of the prey species into the area. In 1934 biologist G. F. Gause showed this in his laboratory with two protozoans, the ciliate predator, *Didinium,* and the prey, *Paramecium caudatum.* In the controlled experiment the prey appeared always to be eliminated by the predator. However, when Gause repeatedly introduced additional prey organisms, the prey did not die out.

Although predation is not always defined to include the relationship of a first-order consumer on a producer, R. L. Smith, author of *Ecology and Field Biology,* uses this interaction as seen with the prickly-pear cactus (*Opuntia*) and its predator, a cactus-feeding moth, to illustrate the significance of immigration in preventing prey and predator populations from dying out. Early in the 1800s the prickly-pear cactus was introduced from America into Australia as an ornamental plant. It escaped cultivation and spread rapidly, eventually covering 60 million acres in Queensland and New South Wales. Then a South American cactus-feeding moth was introduced. The moth devoured the cacti in such quantities that only a few greatly separated colonies survived. Consequently, the moth declined, surviving in only a few areas, and the prickly-pear cactus reestablished itself. Soon, however, the moth would immigrate to these areas and reduce the cactus population.

Such a relationship of prey and predator can be shown graphically. The life curve of either population can be represented with a typical sigmoid curve. However, instead of leveling off, each curve will normally show undulations. The relationship of the prey–predator curves can be seriously interrupted by changes in the environment or by the immigration of new organisms into the community where the two populations exist. One of the most disrupting forces has been man. Among the best illustrations is man's removal of the predators of the Kaibab deer.

Another disruption by man of the prey–predator relationship is illustrated by the use of insecticides. For example, chemicals were used to kill cyclamen mites that damage strawberries, but they produced an unpredicted effect. The insecticide decimated a second kind of mite, the predator of the cyclamen mite. As a result, the cyclamen mites, freed of their predators, quickly reinfested the strawberry fields and caused even more damage than before the application of the insecticide.

Much is still to be learned about many aspects of the prey–predator relationship. For example, predation not only drops off when the prey become very scarce but also when the prey become abnormally dense. Some biologists think the cause to be that the population is satiated, but other explanations may exist. More research is needed.

Parasitism

An extreme form of predation is *parasitism*. The size of a parasitic population is dependent on the availability of the host. Numerous animal groups exist which are made up almost entirely of parasitic members; almost all animal groups contain some parasitic types. Ubiquitous parasites are the fungi, bacteria, viruses, and roundworms.

Parasites live on the surface of their host or within the host. Parasites that live on the surface of their host, *ectoparasites,* often have special clinging structures, such as suckers, sticky surfaces, and sucking or biting mouthparts, to withdraw fluids from the host; lice, mites, leeches, and many fungi are ectoparasites. *Endoparasites,* which live within the host, have specialized structures to exist within that environment. Many have devices, such as hooks, which help to maintain their position on the internal surfaces of the host.

The structures, functions, and life cycle of a parasite are usually greatly specialized. Most systems, such as the digestive system, are very abbreviated, because many functions are assumed by the host. In addition, the stages of the life cycle are related to the parasite's dependence on the host. For example, tapeworms are eaten in meat, where they are lodged in the muscle (Fig. 7–9). After growing to adulthood in the digestive cavity of the predator, they release eggs, which pass into the environment in the feces. Grazing cows ingest the eggs. Larvae from eggs hatched in the digestive cavity of the cow move into the blood system and eventually lodge in the muscle of the animals. Here they encyst and begin the cycle again if the meat is eaten.

Competition

Competition is a major density-dependent limiting factor. The closer organisms are genetically, the more they compete for food, water, space, light, and other factors that are necessary for existence. Consequently, competition controls the size of a population and the type of organisms it contains; a role of competition is at the heart of Charles Darwin's theory of natural selection, which explains how new kinds of organisms, or species, originate.

Competition as a factor that controls population size is easily observed and has been the subject of much research. For example, in 1934 Gause placed two species of paramecia, *Paramecium aurelia* and *P. caudatum*, in a tube containing bacteria, their usual food. *P. caudatum* died out

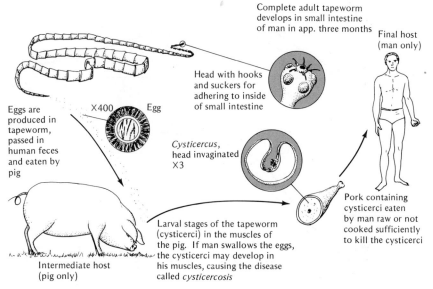

Figure 7–9
The life history of the pork tapeworm, *Taenia solium*, illustrates well the special adaptation of many parasites.

(Fig. 7–10). In another experiment Gause placed *P. caudatum* with a third species, *P. bursaria*. Both species reached a stable population. Gause discovered the reasons for the survival of *P. caudatum* when with *P. bursaria* and its demise when with *P. aurelia*. He found that *P. caudatum* and *P. aurelia* fed at the same level in the culture and, hence, made similar demands for food, whereas *P. caudatum* and *P. bursaria* fed at different levels and did not compete.

The behavior of two species of flour beetles was explained similarly. When the two populations were placed together, one died. However, when a population of the beetle that died was placed with a third species, both species survived. Examination revealed that one of the beetles took up residence inside the wheat grains and the other outside. Consequently, the two species did not compete for food and other essential environmental materials. Such evidence suggested that the conclusions Gause drew from the experiment with paramecia were correct: if two species attempt to meet identical needs in the same environment, only one will survive. The environment that supplies the unique ingredients for a species survival Gause called an *ecological niche*. Today, the concept of only one organism occupying a niche has come to be known as *Gause's principle,* or the *competitive exclusion principle*.

More support for Gause's principle is illustrated with the weed called bed straw. When two closely related species, *Galium saxatile* and *Galium sylvestre*, are grown in mixed stands on calcareous soil, *G. saxatile* loses out to *G. sylvestre*. Both species are apparently competing for the same

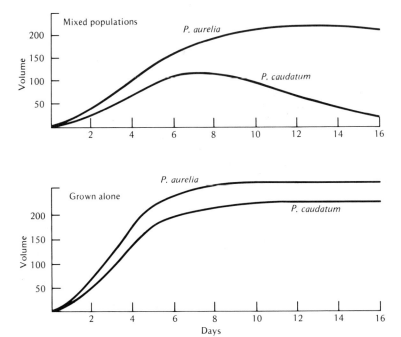

Figure 7–10
Two species of the protozoan *Paramecium* thrive in isolated cultures but are unable to survive together because they make similar demands on their environments.

environmental factors, and *G. saxatile* has, for some reason, an advantage over *G. sylvestre*. Curiously, though, when both are grown in acid peat, the reverse is true.

In addition to laboratory examples, organisms living in the wild illustrate the principle. The myrtle warbler, the black throated green warbler, and the blackburnian warbler all have similar feeding habits, yet all can survive together in the same forest. Studies show that Gause's principle is not contradicted, however, because each species feeds in a particular layer in the trees. Hence, each has a separate niche (Fig. 7–11).

The finches that Charles Darwin observed on the Galápagos Islands also illustrate niche specialization. As noted earlier, distinct species of finches occur there, presumably all derived from a common ancestor. Yet six species are ground feeders and have heavy beaks for crushing seeds.

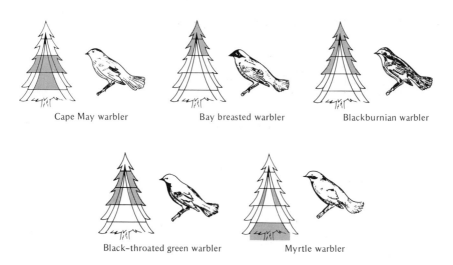

Figure 7–11
Studies revealed that several warbler species could live in the same habitat because each occupied only a small portion of it. (From R. H. MacArthur, *Ecology,* vol. 39, pp. 599–619, 1958.)

Eight are perching birds, which have lighter and narrower beaks for eating insects. (A very specialized insect eater is the cactus finch, which actually removes a spine from the cactus and, holding it in his beak, probes into cracks of the cactus stem for food.) All these finches live side by side and apparently avoid competition because of their specialized feeding habits (see Chapter 4).

Succession
Community succession

Until now we have considered the general characteristics of single populations. In nature, of course, many populations are found in the same geographical area. These interacting populations are known as *communities.* Some communities undergo changes that can be identified in the changes found in the composing populations. The causes of changes are sometimes difficult to ascertain, but certainly they do not appear to be due to climate, because over many years the range of climate conditions changes little. Apparently, the populations effectively modify the physical environment within the community. Thus, conditions support populations which in turn alter the environment so that different populations can exist. The environmental and populational changes extinguish some species and encourage the emigration of others. These population interactions in a community are not easy to study; yet despite difficulties, significant observations concerning naturally occurring changes have been made. As a result, for some areas the sequence of population changes is even predictable. Such a predictable series of community changes is known as *natural succession.*

Natural succession in any area usually shows several trends. These include:

 1. A rapid succession of the early stages, but a slower succession of stages as the community ages.

2. An increase in the diversity of species of plants and animals up to the older stages of succession. At maturity the number and types of species tend to stabilize.
3. As the number of species increases, the relationships among them become increasingly complex; that is, the food webs become more complicated.
4. The total amount of organic matter increases as the community ages, eventually increasing very little when the community approaches maturity.

At which point the community ceases changing, that is, is "mature," or at the "climax stage," is debatable. Some ecologists hold that all communities in an area inexorably progress toward maturity, a stage recognized by a unique collection of species and environmental conditions. Any community not represented by this unique collection of plants and animals characteristic of the climax stage is still changing and will reach that point eventually.

Other ecologists do not agree. Rather, they maintain that any particular collection of species characterizing a community is a result of local environmental conditions which vary through time, such as the nature of the soil, the velocity and direction of the wind, and the quantity of water. Consequently, the plants and animals that exist there must also vary but not necessarily in a predictable succession. They regard a community as a continuum of development, characteristically not having a beginning or an end. There is no certainty that any of the communities would attain a "mature" flora and fauna; rather, they would stay in an "immature" stage indefinitely. Consequently, a mature, or final, stage of development must be defined in terms of the very particular local environmental conditions. In short, the maturity in a natural succession is defined in absolute terms by one group and in terms quite relative to the environmental conditions by the other.

Important, too, is the effect on the sequence of stages of a severe environmental alteration, such as prolonged lack of rain, a fire, erosion, or human activities such as scraping away topsoil for building, or bulldozing trees for a road. Any influences such as these can revert the area to an earlier stage of succession and, hence, retard its development. A community in its mature stage, however, is least affected by these sudden environmental changes. This is because any disruption destroys only a few of the many ecological niches established there. A disruption of a younger stage of succession is often extremely devastating to the community.

Whatever the proper interpretation of natural successions of populations, whether absolute or relative, in studies of abandoned fields, the shore of Lake Michigan, bare rocks in mountainous areas, and in lakes and ponds in the temperate areas, a predictable sequence of populations can often be identified.

In a field no longer cultivated, pronounced population changes can be identified over the decades. Immediately after abandonment, small annual plants, usually recognized as weeds, appear. These give way in a few years to perennials, mostly grasses, which send out well-established root systems. In 10 years or so bushes and even small trees appear. Eventually, in areas that originally supported forests, deciduous and evergreen trees reappear. The animal population also changes as the flora changes (Fig. 7–12).

Another well-known example of natural succession is the dune area that borders Lake Michigan. Shrinkage of the lake over the years leaves younger and younger sand dunes. As the lake continues to shrink, the aging dunes change. Today, several stages of development can be found there. These include:

1. The beach near the water's edge, which supports little life mainly because of the sandy nature of the soil and the erosive action of the water.
2. The middle beach, which is ordinarily dry, although occasionally washed by waves from severe winter storms.
3. The upper beach, which supports some plants, although vegetation is still sparse. Driftwood decay adds organic matter to the soil.
4. The dune, in which beach grasses hold the soil and provide an environment for sand and tiger beetles, grasshoppers, and burrowing spiders.
5. The cottonwood community, built on sand but which supports trees, mainly quick-growing cottonwoods. The trees hold the soil and resist the shifting action of the wind.
6. The jack pine–juniper community, which supports those trees and bearberry bushes on soil that contains much humus.
7. The pine community, which includes jack pines.
8. The oak community, which has oaks, sugar maples, and beeches, and a very rich humus soil with many forest floor animals, such as snails, crickets, sawbugs, earthworms, and ants. This community may take 1,000 years to form from the original sandy shoreline.

Another example of natural succession can be seen in rocky areas. The effects of water and lichens break the rocks into smaller pieces, which, in time, produce just enough soil to support some easily accommodated flora. The death of those plants soon adds humus to the soil and provides a substratum for the support of even more demanding plants.

A final example of natural succession is seen in small ponds and lakes. A natural tendency is for the pond to fill with eroded soil and decaying organic material. Eventually, the pond environment changes drastically and alters the kinds of organisms that can live in the water. In a young pond, for example, bass and other game fish thrive, but as the pond ages, low-oxygen-demanding fish, such as carp and catfish, take over. Man, of course, can hasten the natural succession in a lake with water pollutants, a topic to be discussed later.

Primary succession

Natural succession can also be identified in an area that has never been occupied by a community, such as the lava rock newly exposed after volcanic eruption. The series of population changes seen in such virgin areas is known as *primary succession.*

Primary succession has been studied on the island of Krakatoa near Java in the East Indies. In 1883 a violent volcanic explosion equal to a

10,000-megaton H bomb blew 6 cubic miles of Krakatoa rock into the air. Only a beach completely coated with ashes and rock remained of the island. A few fungal spores or soil plants or animals may have survived in protected crevices, but all apparent life was destroyed.

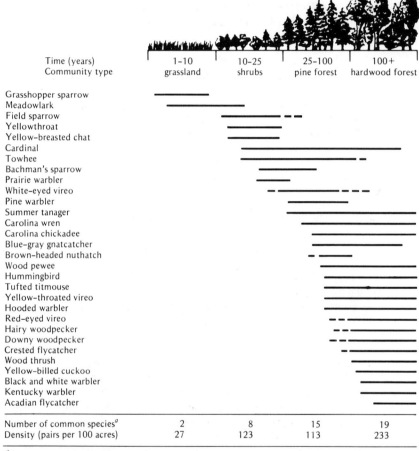

Figure 7–12
Succession in abandoned farmland in the southeastern United States illustrates the rapid changes that are undergone by some ecosystems. (From R. H. MacArthur and J. H. Cannell, *The Biology of Populations,* John Wiley & Sons, Inc., New York, after E. P. Odum, *Fundamentals of Ecology,* 3rd ed., W. B. Saunders Company, Philadelphia, 1971.)

No life was seen for 9 months after the explosion. Then a spider was sighted. After 3 years, 11 species of ferns and 15 kinds of flowering plants were found. After 10 years, cocoanut trees grew near the shore, and wild sugar cane and orchids grew in the vegetation covering the island. By 1903, 25 years after the explosion, 263 species of animals had relocated on the island, including 16 kinds of birds, 2 kinds of reptiles, and 4 species of land snails. Most of the other animals were insect species. After 50 years, dense forests sheltered 47 species of vertebrates, including 2 kinds of rats and many birds and bats.

The source of new species probably was the neighboring island. The wind carried the seeds of many plants, including grasses and orchids. Birds probably brought seeds, too.

Plant and animal life on Krakatoa Island today is abundant and well-established. However, it still is undergoing change, because all the original flora and fauna has not returned, and the relationship between populations on the island is not yet in balance.

In the examination of Krakatoa, as well as the other topics in this chapter, one point should be very clear: that populations of organisms are intimately related to their environment and that any factor that affects one will affect the other. Out of the relationship comes a natural balance, as seen, for example, in the control of the snowshoe rabbits when the environment could no longer provide their needs.

Human populations are also tied to their environments, as seen often in previous chapters. Their size, and in fact their existence, depend on the maintenance of a healthy relationship with their surroundings. A human population too large for its area sooner or later will be reduced in size. In Chapter 8 we shall examine the ways the human populations can be sanely controlled so as to avoid the kinds of disasters seen in this chapter occurring in populations of other organisms.

Summary

In this chapter we have examined (1) the characteristics of animal populations, most notably how they change; (2) the factors affecting the growth of populations; (3) the population density-dependent and density-independent controls of population size; and (4) succession, the natural and predictable changes in populations of a community.

Supplementary readings

Boughey, A. S. 1967. *Population and Environmental Biology.* Dickenson Publishing Co., Inc., Belmont, Calif. 108 pp.

Boughey, A. S. 1973. *Ecology of Populations,* 2nd ed. Macmillan Publishing Co., Inc., New York. 182 pp.

Hazen, W. E. 1970. *Readings in Population and Community Ecology.* W. B. Saunders Company, Philadelphia. 421 pp.

Keith, L. B. 1963. *Wildlife's Ten Year Cycle.* University of Wisconsin Press, Madison, Wis. 201 pp.

MacArthur, R. H., and J. H. Connell. 1966. *The Biology of Populations.* John Wiley & Sons, Inc., New York. 200 pp.

Rosenzwig, M. L. 1974. *And Replenish the Earth.* Harper & Row, Publishers, Inc., New York. 304 pp.

Slobodkin, L. B. 1961. *Growth and Regulation of Animal Populations.* Holt, Rinehart and Winston, Inc., New York. 184 pp.

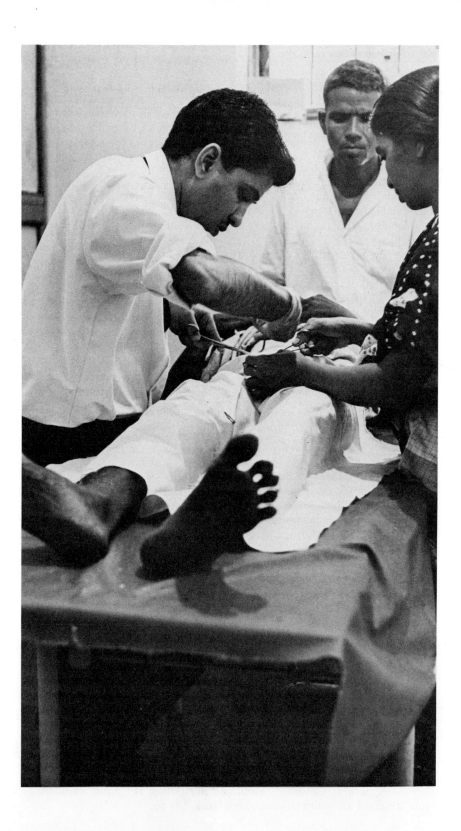

Chapter 8

Man cannot look to any natural process to restrain his rapid growth. If the growth is to be slowed down, it must be by his own deliberate and socially applied efforts.

Wild Heritage
Sally Carrighar

The control of the human population

In the previous chapter several controls were discussed which occur naturally in animal populations to limit their size and help maintain a life-sustaining relationship with the environment. Unfortunately, none of these controls appears dependable for the containment of man's population. Other controls must therefore be sought. Deliberate birth control appears to be an answer. In this chapter we shall examine the history of birth control, religious and political attitudes affecting its use, and methods of control.

Birth control in the past

Birth control and family planning have been practiced for centuries. Aristotle in the fourth century B.C. and the gynecologist Soranus of Ephesus in the second century described contraceptive methods. Very widespread knowledge and practice of birth control, however, may have occurred only as recently as the nineteenth century, although in 1797 an Englishman, Jeremy Bentham, pleaded ineffectively for acceptance of contraception. *Illustrations and Proof on the Principle of Population* by Francis Place in 1882 drew attention to birth control as a more appropriate way to contain population growth than abstinence, proposed earlier by Thomas Malthus.

In the 1800s works on contraception circulated in Europe, but broad dissemination of contraceptives and information in the United States was impeded, mainly by the Comstock Law, adopted by Congress in 1873. This law, named after Anthony Comstock of New York, a zealous antivice crusader, classed birth control information as obscene and forbade its distribution through the mail. "Little Comstock Laws" were subsequently passed by many states. In 1890 the Tariff Act prevented any birth control books or pamphlets from being imported into the country.

Early in this century the situation began to improve. In 1912 Margaret Sanger (Fig. 8-1) published a series of articles pleading for the use of birth control methods to help free women from the burdens of unwanted child rearing. In 1916 she opened the country's first birth control clinic in New York, for which she was arrested and served 30 days in jail. In 1917 the new National Birth Control League, later to be renamed the Planned Parenthood Federation of America, organized the birth control clinics that were sprouting up around the country. In 1921 Dr. Marie Stopes opened England's first birth control clinic.

In the last few decades greater progress has been made. The Planned Parenthood Federation has become very influential in helping couples plan and space their children and in instructing people in birth control methods. In the 1950s Margaret Sanger joined with other early pioneers in birth control movements in other countries and formed the International Planned Parenthood Federation (IPPF). The IPPF is supported by private contributions and has membership in more than 40 nations on 5 continents. Today, many other organizations contribute to birth control research and education. These include the Population Reference Bureau, the Population Crisis Committee, Zero Population Growth, and the Population Council.

Figure 8-1
Margaret Sanger, founder of the birth control movement in the United States. (Courtesy of Planned Parenthood Federation of America, New York.)

Religious positions on family planning and birth control

The Bible's admonition "to be fruitful and multiply" has, through history, influenced opinions concerning family planning and birth control in the Western culture. The ancient book of the Jews, the Talmud, although it also allowed for sexual abstinence during periods of scarce food, encouraged large families. Today, the orthodox Rabbinical Alliance of America condemns birth control techniques used by the husband but allows the wife to prevent conception if her health demands it. Conservative Jews and Reformed Jews are more liberal, and in the 1960s the Central Conference of American Rabbis proclaimed that parents can decide the appropriate number and spacing of the children.

Before the twentieth century many Protestant groups allied with Roman Catholicism to oppose birth control in general. Now Protestantism, in general, approves of birth control. The Episcopal Church, for example, gave full approval to parents in 1958 to decide the size and spacing of their family, although in 1908 and 1920 it was on record as opposing contraception. Roman Catholicism, on the other hand, has maintained to the present time that all birth control methods except rhythm are unnatural and, hence, immoral because they prevent the natural purpose of intercourse, the procreation of children.

For example, the development of oral contraceptives has recently caused a great deal of controversy over birth control in the Catholic Church. In 1958 Pope Pius XII approved of the use of a hormone pill to inhibit conception if such treatment was necessary for some medical reason other than simply to prevent conception. Since that time, however, many church moralists have interpreted the Pope's words very liberally, because they claim the Pope's statement includes psychological problems accompanying prospective pregnancy. The Second Vatican Council was greatly concerned with this controversial issue. The council made clear that the attitude between man and wife and the likelihood of the proper rearing of their offspring is as important as the children's birth.

Incidentally, the effect of the Vatican's position on the size of Catholic families as well as the size of populations in Catholic countries has been documented. A study done in 1941 concerning Catholic couples in Indianapolis indicated that 57 percent of all Protestant couples had 2 or fewer children, but only 52 percent of Catholic couples had 2 or less offspring. Also, the average number of children born to Protestant couples was 2.19, to Catholic couples, 2.74. More recent studies show that young Catholic couples are less influenced by the Vatican's position and that their family size compares favorably with similar Protestant groups.

In contrast to religions in Western countries, the Eastern religions have no strong objection to birth control or family planning. Hindus stress the religious importance of giving birth to a son, but the Buddhists emphasize the detachment of one's self from earthly passions. Infanticide and abortion are not encouraged. Other kinds of contraception are accepted, however.

Family planning in the developed countries

The earliest and most active birth control and family planning proponents were citizens of developed countries. Some of the earliest publications on contraception are products of activists in England and the United States. In the 1920s and in the previous decade many birth control clinics were opened throughout the United States, thanks primarily to Margaret Sanger, and in England to Dr. Marie Stopes.

One of the first developed nations to open government-supported birth control clinics was Sweden in 1930. Great Britain developed clinics in 1932. The United States began several state public health programs in the 1930s which by 1942 were assisted with federal money.

Eventually, dissemination in the United States of birth control devices and information was no longer prohibited. In 1970 the U.S. Supreme Court declared the Comstock Law an unconstitutional invasion of privacy.

In general, in the developed countries there has been both social and governmental liberalization of birth control and family planning. Several developed countries, the United States, for example, have taken official stands on the necessity of population limitation and have aided underdeveloped countries with family planning and birth control programs. Within the United States a more liberal attitude is evidenced by an increasing liberalization of abortion laws and the American Medical Association's declaration in 1964 that birth control is an essential part of responsible medical practice. In recent years an obstacle to the use of birth control in the United States has been the concern of the minority groups, such as the blacks, that birth control and family planning programs may be a scheme of the majority groups to limit the size of the minority populations in the country. Only constant effort to sincerely erase the inequities in the treatment of the different ethnic and racial groups in all ways in this country will eventually eliminate such a concern.

Family planning in the underdeveloped countries

Many underdeveloped countries, alreay burdened with too many people, support the greatest rate of population increases. In addition to attempting to feed the present population, many of the underdeveloped countries are also hoping to curtail population growth. Consequently, they have initiated family planning and birth control programs.

Developed countries have helped many of the underdeveloped countries initiate family planning programs. The U.S. government, for example, in 1964 offered several nations technical assistance in population control programs as part of foreign aid plans. In addition, many private organizations have committed their time and money, such as the Ford and Rockefeller Foundations and the Population Council, an organization founded in 1953 especially to study rural population problems.

One of the first countries to inaugurate a national population control was India in the mid-1950s. Lack of adequate administration and doctors accounted for a slow and inauspicious beginning. The Indian program, administered mainly through clinics, primarily advocates IUDs (intrauterine devices) and male and female sterilization. Chemical and other mechanical techniques are less favored, and oral contraceptives are least popular.

Pakistan, a country with a high annual growth rate of 3.3 percent and a population increase of 20.6 million between the years 1965 and 1970, also has inaugurated extensive birth control programs. Pakistan's goal is to supply family planning aid through the present channels in hospitals, rural clinics, and dispensaries. The national budget has increased greatly to accommodate the more extensive birth control plans. In 1965 the family planning directorate was given a 5-year budget of 300 million rupees, which equals an average annual allotment of approximately 12 cents per person. A major thrust to the birth control program is the recruitment of midwives, who will be paid to insert IUDs under medical supervision.

China inaugurated a national birth control plan soon after India. Although federal interest seemed to wane in the late 1950s, by 1962 China expressed new concern in a national birth control program. Apparently, by the 1960s all methods of contraception were available, including the Pill, and a policy of late marriage at a minimum age of 25 for women and 30 for men was in force. The government has strongly encouraged the birth of two children per family with no less than 3 years between the births. Probably, the employment of women, free child care, and national health plans support the program. Although data on the population growth of China are difficult to obtain, the Population Reference Bureau in 1972 estimated that that nation's population was close to 786.1 million people, an annual increase rate of 1.7 and an increase of 64.6 million people between the years 1960 and 1970. The Chinese population represents close to one-fourth of the human beings on the earth. The effectiveness of their birth control plan in view of their high annual increase and already large population is impossible to assess because of the difficulty in obtaining information.

Family planning programs in Korea and Taiwan seem to be fairly successful. South Korea adopted a national family planning policy in 1961 and by 1964 had inserted 112,000 IUDs and by 1965, 233,000 IUDs, more than hoped for. Many birth control devices are manufactured in Korea. Hopefully, by 1980 social changes within the country as well as a more extensive birth control program will reduce the rate even more.

In Taiwan, although that country does not have an official population policy, annual goals for the insertion of IUDs have been repeatedly attained, a success that is attributed to the effectiveness and the indirect help provided by the U.S. Agency for International Development (AID). Oral contraceptives are now being introduced.

Many Middle East countries, including the United Arab Republic, Tunisia, Turkey, and Morocco, have adopted birth controls in the 1950s and early 1960s.

Many African countries now also have inaugurated birth control programs. Official plans for attacking the problem exist in such countries as Kenya, which now operates about 50 clinics; Egypt, which launched its program in 1965 and proceeded to develop it quickly until the war with Israel began in 1967; Morocco, Uganda, Liberia, and other countries. Although national policies have not been inaugurated in a few of the countries, planned parenthood movements have met with great success.

Unfortunately, many countries in Africa still have very high birth and death rates. Many authorities feel that strong interest in birth control programs will not come about until the death rate is lowered. This is unfortunate, because what is needed is a concomitant reduction in the death and birth rates.

After World War II, Japan experienced a rapid and profound drop in its birth rate mainly because of the legalization of abortion. Now with the development of new birth control methods, Japan still enjoys a low birth rate. Japan has also reduced the quantity of abortions. Consequently, the country is generally thought of as the nation to be emulated by developing nations, because it nearly neutralized the growth of its population in a brief time. In 1972 Japan had an annual growth rate of only 1.2 and an increase of only 5.5 million people between the years 1965 and 1970. Success of the birth control program in Japan is attributed to an intensive and effective education program and a national establishment of clinics.

Latin America is perhaps the slowest region to develop birth control policies, although in 1973 it had one of the highest annual population growth rates. Middle Latin America in 1973 had an annual rate of 3.2 percent, tropical South America, 3.0 percent. Many of the countries have no official family planning policy. Part of the reluctance to accept birth control methods most likely is due to the strong effect of Roman Catholicism in these countries. Nevertheless, some help is being administered by private family planning organizations in almost every nation of Latin America, often under the sponsorship of the International Planned Parenthood Federation and local contributions. As a result, family planning services will probably rapidly increase in a short time in these countries. The Pill is becoming increasingly popular, although many of the traditional birth control techniques are still publicized. The Population Reference Bureau has opened an office in Bogotá, which helps to disseminate information about family planning and population growth to all of Latin America. Furthermore, the Panamerican Health Organization has put into effect a major population program and encourages the study of Latin America's population problems. In addition, AID is contributing a great deal of the technical help. The realization in many Latin American countries that high growth rates are serious obstacles to achieving overall planning goals encourages rapid development of population control programs.

In general, the development of family planning programs around the world is astonishingly rapid despite the frequent and overwhelming problems of administration, finance, distribution of contraceptives, and lack of skilled help. The programs, in general, have not really exploited all possibilities. For example, in many countries abortion has not been sufficiently embraced as an effective birth control method, although in some parts of a country or region, notably Japan, the Soviet Union, and eastern Europe, evidence undeniably exists that abortion is an effective and rapid birth control technique. Also, the lack of concern about the male's role in birth control methods causes neglect of the potential control of birth by male contraceptive devices. Difficulties due to lack of strong and extensive organization result in a lack of effective communication and education as well as a lack of adequate data on the accomplishments of birth control programs.

It seems clear in a review of the programs of many countries that substantial reductions in birth rates can be accomplished with birth control methods. However, the question is: In those countries with a large population and a high population growth rate, can birth control be administered extensively and quickly so that disaster in the near future can be avoided? Birth control techniques, along with improved agricultural methods and other attempts to feed the world's human beings, could well head off a global disaster. One of the significant factors that may help avoid this calamity is the development of even newer birth control devices. Hope rests in such predictions as the following: The U.S. National Academy of Sciences calculates that if child bearing in developed nations were reduced by 50 percent over a 30-year period, 40 percent could be added to per capita income in these countries. The United Nations predicts that if Latin America could reduce its population growth from 2.5 to 2.0 percent per year, in 2 decades the chronically deprived one-third of its people would be absorbed into the productive work force.

Methods of birth control

Margaret Sanger, the organizer of the birth control movement in the United States, once helped attend a young mother who was recovering from a self-induced abortion. When warned that another attempt to abort might be her last, the mother asked how to prevent pregnancy. The doctor sternly admonished her with: "You can't have your cake and eat it too, young woman. Tell Jake to sleep on the roof." The woman later died from another abortion attempt. That was in 1912. The obvious need for birth control information inspired Mrs. Sanger to expand her family-planning activities.

The need for control of population around the world is much more serious today than in Margaret Sanger's time. Yet, despite much effort and money, the initiation and maintenance of programs to curtail births have met with difficulty in some parts of the world. The problem lies in the limitations of available birth control techniques and devices (Table 8–1). Described below are most of the contraceptive methods and devices now used, their limitations, and the contraceptives hoped for in the near future.

Foams, suppositories, creams, and jellies

Among the least effective methods of birth control are the vaginal foams, suppositories, creams, and jellies (Fig. 8–2). These are simple methods often used by couples wishing merely to reduce the probability of pregnancy rather than to prevent it totally. Each is a chemical placed in the vagina before intercourse; the chemicals kill or immobilize sperm on contact and block the cervical opening with a thin film so that living sperm cannot swim up the uterus into the Fallopian tubes to fertilize the ovum. The foam spermicide can be purchased in an aerosol can or in tablets. The jellies, foams, and creams are applied with a plastic applicator. The suppositories and foam tablets are inserted manually into the vagina. Despite the high rate of failure to prevent pregnancies, these methods are used widely because they are inexpensive and simple.

Douche

Douching, or flushing the vagina with water, is a very ancient and not especially effective birth control method. Only approximately 7 percent of American married couples depend on its effectiveness. In other countries, such as France, it is common, however, and the bidet, a basin with an upright jet of water, is a common lavatory fixture there. Bulb syringes are also used.

219 Methods of birth control

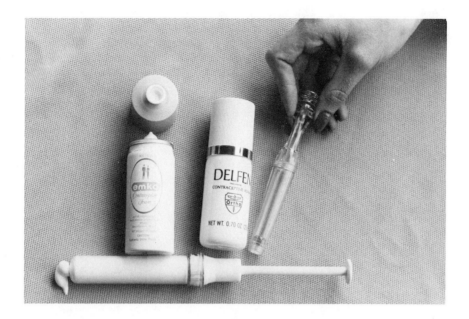

Figure 8–2
Vaginal foams, creams, and jellies are sometimes effective, but they are not among the foolproof methods of birth control. (Courtesy of Planned Parenthood–World Population, Des Moines, Iowa.)

Douching, as a preventive to conception, must be done immediately after intercourse. This requirement places the burden for birth prevention solely on the wife and, at least for this reason, makes it unpopular. Furthermore, the technique does not necessarily flush out the sperm or kill them, even when a douching solution of vinegar, alum, or soap is used. That is because 3 minutes after sperm is deposited in the vagina the cells are frequently already migrating into the uterus. Douching is probably the least-effective birth control method, but if nothing else is available, it should be used.

Rhythm

Rhythm is probably more effective than the birth control methods so far discussed, but not much. This technique is based on the biology of ovulation (Fig. 8–3). (See the insert, "Hormonal Control of Ovulation.") An ovum is available for fertilization only approximately 2.5 days each month; a sperm is viable only approximately 48 hours after ejaculation. The whole point of rhythm is to avoid getting the sperm and ovum together during these times, a feat that accounts for this method's lack of success in preventing pregnancy.

Table 8-1
Characteristics of common birth control methods

Methods	"The Pill"	Intrauterine Device (IUD)	Diaphragm with Jelly or Cream	Foam, Jelly, or Cream
How does it work?	Prevents egg's release from woman's ovaries.	It is not known exactly how the IUD prevents conception.	Blocks and prevents sperm from reaching egg.	Prevents sperm from reaching egg.
How reliable is it?	More effective than any method other than sterilization.	Highly effective. Protects about 97 of 100 users.	Highly effective. Protects 95 of 100 users if used consistently and properly.	Foam: medium effectiveness. Creams and jellies: low effectiveness.
Are there problems with it?	Must be taken only when prescribed by a doctor. Some women should not take the Pill. All women should have a medical examination before taking the Pill.	Cannot be used by all women. Sometimes the body "pushes" it out, or it punctures the uterine wall.	Must be fitted by a doctor. Some women find it inconvenient.	Must be used just before intercourse. Some find it messy and inconvenient.
Side effects?	May cause clotting, weight gain, or internal bleeding. If this occurs, see a doctor.	May cause cramps or bleeding, spotting. If this occurs, see a doctor.	None.	Causes some irritation in rare cases.
Advantages?	Convenient, reliable, and not messy. Does not interfere with sex act.	Always there when needed, yet not felt by either partner.	Not felt by either partner.	Can be purchased at drugstore.
Prescription needed?	Yes.	Yes.	Yes.	No.

Methods of birth control

Table 8-1 (continued)

Methods	Condom (Rubber)	Condom and Foam (Used Together)	Rhythm, or "Safe Time"	Sterilization: Vasectomy (Male), Tubal Ligation (Female)
How does it work?	Prevents sperm from reaching egg.	Blocks and destroys sperm.	Intercourse only during woman's "safe time."	Closing of tubes in male prevents sperm reaching egg; closing of tubes in female prevents egg from reaching sperm.
How reliable is it?	Highly effective, particularly when used during foreplay and when removed with care to prevent spilling the fluid in the woman's vagina.	Highly effective.	Low effectiveness; 25 of 100 women become pregnant.	Highest effectiveness. No one should have the operation unless he or she is sure no more children are desired.
Are there problems with it?	Objectionable to some. Condom may break or tear.	Objectionable to some. Condom may break or tear.	Difficult to be sure of the safe time if menstrual cycle is irregular.	All surgery has some risk, but new procedures make these operations simple and safe.
Side effects?	None.	Causes some irritation in rare cases.	None.	None. There is no loss of sexual desire or ability.
Advantages?	Condoms offer excellent protection against venereal disease and can be purchased at drugstore.	Condoms offer excellent protection against venereal disease and can be purchased at drugstore.	Little if any religious objection to this method.	Many feel that removing fear of pregnancy improves sexual relations.
Prescription needed?	No.	No.	No.	Yes.

Source: From *Family Planning: Methods of Contraception*, U.S. Department of Health, Education and Welfare, Washington, D.C., 1973.

The main difficulty is knowing precisely when the ovum is mature, so as to avoid intercourse. Women who have extraordinarily regular menstrual periods can usually calculate the "safe" periods before and after the 2.5 days of "unsafe" time. Nevertheless, many cautious women reserve

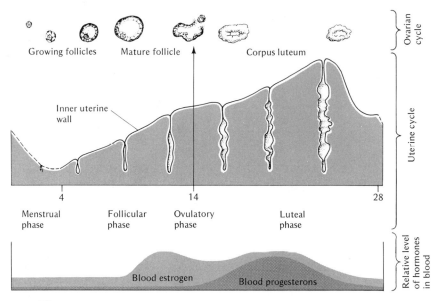

Figure 8–3
This diagram of the menstrual cycle illustrates the condition of the uterine wall, hormones secreted, and time of ovulation.

even more days of abstinence before and after the 2.5 day period. However, 15 percent of women do not menstruate regularly and, hence, cannot predict their safe periods. Furthermore, the method is not effective immediately after childbirth, because of the usual irregularity of menstruation then. Only after the third menstrual period is the technique again safe, even for those women on a regular cycle.

The method became more reliable when doctors discovered that a woman's temperature rises approximately one-half to seven-tenths of a degree between 1 and 2 days after ovulation and until menstruation begins (Fig. 8–4). Once the temperature rise is ascertained, intercourse can be resumed with little fear of conception. However, it is an essential nuisance to take the temperature each morning immediately after awakening with a special thermometer, the basal body thermometer. This device reads from 96 to 100, and the intervals are in tenths of a degree. Color tapes have recently appeared on the market which may simplify the detection of ovulation. The tapes change color when the acidity of the vigina varies at the time of ovulation. Another technique involves a written record of the beginning and end of the menstrual cycle for 12 months. Calculations by a doctor from these data will make the safe period more certain.

For Roman Catholicism to officially endorse rhythm as the only

223 Methods of birth control

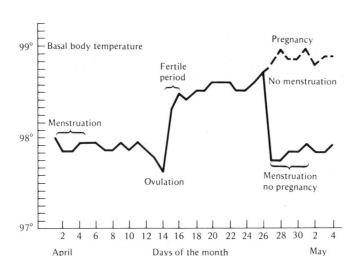

Figure 8–4
A sudden rise in body temperature during ovulation is warning of the fertile period. (From *Biology and Its Relation to Mankind,* by A. M. Winchester. © 1969 by Litton Educational Publishing, Inc. Reprinted by permission of Van Nostrand Reinhold Company.)

morally acceptable birth control method is especially unfortunate at a time when the need for population limitation is so pressing in the U.D.C.s, some of which are predominantly Catholic.

Hormonal control of ovulation

Hormones within the female cause the production and release of an ovum, known as *ovulation*. The source, kinds of hormone, and their effects are appreciated best with a close look at the hormonal activity of a woman who ovulates regularly every 28 days.

Four kinds of hormones are primarily involved: the *follicle-stimulating hormone (FSH)* and the *luternizing hormone (LH)* produced by the anterior pituitary gland at the base of the brain, and *estrogen* and *progesterone,* products of the ovary. At the beginning of the 28-day period, the pituitary begins secreting FSH. This hormone, as its name indicates, stimulates in the ovary the development of the ovum within its cellular enclosure, the follicle. As FSH increases in the body, the follicle containing the ovum continues to grow and soon begins to secrete estrogen.

The estrogens, accumulating in the body, produce three effects. First, they initiate vascularization of the uterine wall preparatory to implantation of the fertilized egg. Second, they inhibit production of FSH from the anterior pituitary. Third, they stimulate the anterior pituitary to produce LH. Consequently, as the uterus wall grows, the quantity of FSH falls in the body, and the amount of LH increases. By the fourteenth day, abundant LH in the body provokes ovulation by a mechanism not yet thoroughly understood.

As the released ovum migrates down the Fallopian tubes toward the uterus, hormonal activities continue. The increasingly abundant LH changes the old follicle, turning it into a yellow mass of cells well supplied with blood vessels. This yellow body, the *corpus luteum*, continues to secrete small amounts of estrogen and begins to secrete the hormone progesterone.

Progesterone has three functions. First, it continues thickening the walls of the uterus, begun earlier by the estrogens. Second, it continues to inhibit the production of FSH begun by estrogen and, consequently, to prevent another ovum from developing. Third, it inhibits the secretion of LH from the pituitary. Consequently, the *corpus luteum* is not maintained nor is its product progesterone. When the amount of progesterone falls, the amounts of FSH and LH increase. Each of the four hormones drops to its lowest level at the end of 28 days. Then, a lack of progesterone causes a disintegration of the thick and vascular uterine wall. Blood flows from the broken blood vessels. Menstruation begins, which marks the first day of a new menstrual cycle (Fig. 8-3).

Until now, the description of hormonal events assumes that fertilization has not occurred. If the ovum is fertilized and implanted in the uterine wall, the level of progesterone is kept high, resulting in inhibition of LH and FSH and new follicular development. The high progesterone level is maintained during pregnancy by a secretion from the developing placenta similar to LH, which preserves the *corpus luteum*, the source of progesterone.

Diaphragm and cervical cap

Both the diaphragm and the cervical cap function as barriers to the uterus. To be effective, both must be fitted by a physician. The diaphragm is a thin cup of rubber, 2 to 4 inches in diameter, bounded with a flexible metal ring (Fig. 8-5). It fits over the opening between the vagina and uterus. An ill-fitting apparatus can be dislodged during intercourse, and sperm can swim into the uterus. Therefore, frequent checkups to ensure proper fit are essential to be certain of its effectiveness. Spermicidal creams or foams decrease the likelihood of contraception.

The cervical cap is a small metal or plastic cup that fits over the opening to the uterus, the cervix. The cap is more difficult for some women to insert than the diaphragm. Once in place, however, it is a very effective device. Spermicidal foams or creams help its effectiveness.

Coitus interruptus

Coitus interruptus is mentioned in the Bible and is very likely the oldest form of birth control. No apparatus is needed; the penis is withdrawn from the vagina moments before ejaculation. The method provides an effective contraceptive, but it has drawbacks. First, the success of the method depends on exceptional timing, a factor not always uppermost in the mind during intercourse and, hence, difficult to depend on. Even one drop of semen may be dangerous, especially since the first 10 drops of semen have been shown to contain the greatest quality of sperm. Second,

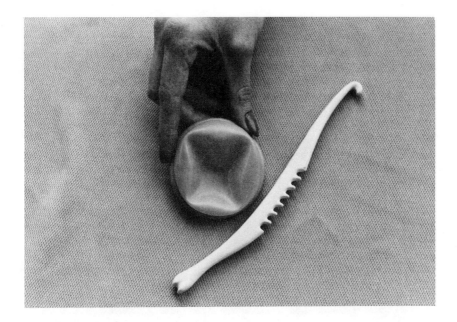

Figure 8–5
The diaphragm (shown with applicator) is one of the most effective birth control devices. (Courtesy of Planned Parenthood–World Population, Des Moines, Iowa.)

it makes the sex act difficult for many couples to enjoy because of the anxiety created by their preoccupation with withdrawal moments before orgasms. Finally, the technique minimizes the woman's involvement and satisfaction in intercourse. Withdrawal and male orgasm can occur before the woman achieves an orgasm, thus preventing her final satisfaction.

Condom (prophylactic or rubber)

The condom is usually a small latex rubber sac about $1^{3}/_{8}$ inches in diameter and approximately $7^{1}/_{2}$ inches long (Fig. 8–6). It weighs less than $^{1}/_{20}$ ounce and folds into a small wad; hence, it is convenient to carry in wallet or purse. Rubbers are fitted onto the erect penis and catch the semen on ejaculation. Some are constructed with a small bulge or reservoir at the end to accommodate the ejaculate, and some may be lubricated. Some rubbers are not latex at all but sheep intestine. These are called "skin" condoms and, because they are much thinner than rubber condoms, do not interfere greatly with the pleasure of intercourse, a serious drawback of the latex condom for some men.

Condoms occasionally are defective and will leak sperm. This was more of a problem in the past than now, however, since the U.S. Food and Drug Administration set standards in 1938. As a result, this agency makes periodic spotchecks of the devices. Substandard products are confiscated, and a penalty of 3 years in prison and $10,000 fine can be imposed for repeated violations. Consequently, condoms are generally reliable today.

One prime advantage of the condom is its effective prevention of

226 Chapter 8: The control of the human population

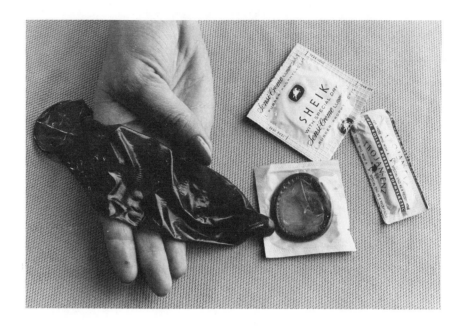

Figure 8–6
The condom, with sperm reservoir at the tip, is a very effective birth control device. (Courtesy of Planned Parenthood–World Population, Des Moines, Iowa.)

venereal disease. The condom when used correctly completely covers the penis and prevents it from directly coming in contact with the vagina. Consequently, any venereal disease carried on the muscous membranes of the reproductive organs cannot be transmitted. Authorities fear that with the increased use of the contraceptive pill, venereal disease has spread rapidly.

The Pill

The Pill has been hailed as the answer to every maiden's — and matron's — prayer. It is an extraordinarily effective contraceptive. The Pill operates through the hormonal system of the female and affects the reproductive organs by preventing ovulation (Fig. 8–7). It is composed of two hormones, estrogen and progestin, a chemical similar to the progesterone naturally found in the woman's body. A pill is taken on the fifth day following the onset of menstruation and every day thereafter up to the twenty-fourth or twenty-fifth day. By the twenty-eighth day menstruation begins, often with less discharge and discomfort than experienced when not taking the pill. Five days after the menstrual period begins, the pill cycle is begun again. It is vital that the woman does not neglect to take a pill each day. If she does, ovluation could occur, and fertilization result.

Specifically how does the Pill work? Recall that the normal menstrual cycle is regulated by four hormones, two from the pituitary and two from the ovary. (See the insert, "Hormonal Control of Ovulation.") Estrogen and progestin repress the secretion of FSH and LH from the pituitary, just as estrogen and progesterone naturally do in the body. Low amounts of FSH and LH prevent the development of the follicle containing the egg, and low amounts of LH prevent discharge of the egg from the follicle. In fact, even though a small amount of FSH allows the ovum to develop, it never matures.

Besides repressing the development and release of the ovum, the Pill alters the lining of the uterus, the *endometrium,* so that it is not recep-

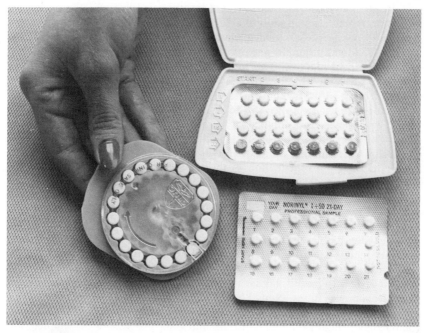

Figure 8–7
The Pill, to be effective, must be taken according to schedule. (Courtesy of Planned Parenthood–World Population, Des Moines, Iowa.)

tive to the fertilized ovum. Also, it affects the nature of the mucous which fills the cervix, so that the sperm cannot penetrate. Finally, the endometrium is prevented from developing. Consequently, when the pills are stopped and menstruation occurs, the bleeding is less pronounced than before. The Pill has several effects, then, all of which help prevent a pregnancy from developing.

A form of the Pill called the "sequential" is used also. This pill is really two kinds. One, mainly of estrogen, is taken the first 14 to 16 days of the period. The other, estrogen and progestin, is taken the last 5 to 6 days. The sequence of pills attempts to imitate the natural occurrence of the hormones in the woman's body.

The unified Pill has been shown to be an extraordinarily effective contraceptive. In tests wih Enovid, the oldest brand and most studied of the pills, only 3 pregnancies have resulted of 14,083 women who used it over 116,000 cycles, a rate of pregnancy of 0.028 percent. Other brands, introduced to the market more recently, have a similar rate of effectiveness.

Brands of the sequentials are less effective, with a pregnancy rate 5 percent. The difficulty with "sequentials" is that omission of even one pill may induce ovulation, because the quantity of estrogen that represses ovulation is not very high. Furthermore, these pills do not alter the lining of the uterus to make it difficult for implantation, nor do they make the cervical mucous impenetrable to sperm as do the "unified" pills.

Recently, the Pill has been greatly criticized because of the occurrence of side effects, some of which are said to cause death. Since the early days of testing the Pill, some side effects were known. Included is weight gain in 14 to 50 percent of the users, nausea, intermittent menstrual bleeding or "spotting," blotchy facial pigmentation of the skin (chloasma or melasma), and migraine headaches. The more serious effects may include leg vein inflammation with blood clotting. Most difficulties can be alleviated by changing to a pill brand that has a different dose of estrogen or progestin.

Blood clotting, the most serious possible side effect, needs a closer look. Two reports from England first announced the possible incrimination of the Pill in causing pulmonary embolism (blood clots in the lung tissue) or cerebral thrombosis (clots in the brain veins). Data presented to the Food and Drug Administration in the United States by representatives of eight drug companies indicated that clotting deaths among Pill users was not above the normal range. A more detailed study in 1968 confirmed these data for users of the Pill in the United States. More studies are under way.

Some evidence suggests that the quantity of estrogen in the Pill is the cause of blood clotting, and that reducing the quantity of estrogen would reduce the death rate caused by clotting. The evidence is not conclusive that the Pill increases the danger of emboli in the body. As R. W. Kistner points out in *The Pill: Facts and Fallacies about Today's Oral Contraceptives* (Delacourte Press, New York, 1969), we simply need further studies that incorporate standardized laboratory tests, common definition of terms, standardized record-collection forms and data-processing techniques, and reliable rates of blood clotting in specific and general populations stable for some time.

The effects of using the Pill over a long period of time are yet to be catalogued because it was introduced so recently. Scientists, aware that the human body is a delicately balanced chemical system, are constantly alert for any apparent long-range damage including major problems such as cancer. So far, no serious symptoms have been detected, but only time will tell.

Intrauterine devices

The intrauterine device (IUD) effectively prevents, by a mechanism still unknown, the fertilized egg from implanting in the uterine wall. The IUD is a plastic or metal object inserted by a physician into the uterus, where it usually remains until removed (Fig. 8–8). Over 6 million women have used these devices with considerable success over the past 5 years.

Data support the effectiveness of the IUD. Only 1 percent of women using IUDs expel them spontaneously (approximately 5 percent of the others must have the device removed because of bleeding and discomfort). Furthermore, the longer a woman wears the device, the less her chances become of getting pregnant. The first year the pregnancy rate is approximately 2.4 percent; the second year, 2.0 percent; and the third year, about 1 percent.

Although the exact method of the preventing pregnancy is not known, some doctors believe that the presence of the IUD hastens the descent of the egg in the tube. Others think that the uterus may be stimulated to secrete a chemical that inactivates the zygote. Research will, no doubt, soon reveal the truth. Already known is the remarkable effectiveness of the double plastic coil over other shapes. Recent research on a T-shaped IUD wrapped in copper wire indicates that the copper increases the effectiveness of the device and reduces expulsion. The IUD is an ideal contraceptive for those women who are leery of the effects of the pill, or who find other techniques too distracting. One disadvantage is its tendency to dislodge in approximately 1 percent of women. Periodic checks are essential to make certain that it is still in place. The new copper-wrapped IUD is said to be as effective as the pill, although more data are needed to confirm this.

In 1974 serious effects were discovered by users of some forms of IUDs. The Dalkon shield, for example, is no longer used because it was found to puncture the uterine wall and cause excessive bleeding. As time goes on, serious problems with other IUDs may be found.

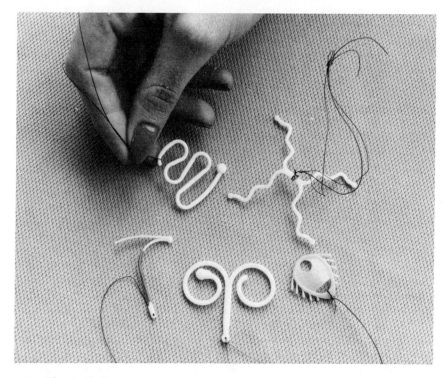

Figure 8-8
Intrauterine devices (IUDs) are effective birth control devices, although why they work so well is still a mystery. (Courtesy of Planned Parenthood-World Population, Des Moines, Iowa.)

Sterilization

Sterilization of either the man or woman is a 100-percent-effective birth control device. Both operations are becoming increasing common. The operation on the female, the *tubal ligation,* or *salpingectomy,* is the more serious. It is often performed on a consenting woman while she is undergoing abdominal surgery for other reasons. Anesthesia and hospitalization are required. The Fallopian tubes (Fig. 8–9) are snipped and tied, so the ovum released from the ovary is reabsorbed and does not reach the uterus.

The operation on the male is much simpler and is increasing in popularity. It is brief, does not require general anesthesia, and can be performed in an outpatient clinic. The doctor merely cuts and ties off the *vasa deferentia,* the tubes through which sperms move from the testes out of the body during ejaculation (Fig. 8–10). Consequently, the sperm cannot be ejaculated and are reabsorbed by the body cells. The testes are untouched. All glandular fluids that normally accompany the sperm and make up a good portion of the semen are still produced and ejaculated. Consequently, all sexual functions, including orgasm and ejaculation, are maintained.

In sterilization of both male and female, the hormonal balance is maintained. Sperms or ova are still produced; they just never get together. The body continues its regulated activities. Reversals of the operations can be done but not in 100 percent of cases. About a 50 to 80 percent reversal is possible for men; for women, approximately 52 to 66 percent.

Abortion

Abortion has a long history. It is a very old birth control method and probably the one most used today. A Chinese record written 4,600 years ago prescribes the use of mercury to abort. The Greeks advocated it. England labeled it a misdemeanor in 1803; the United States, in 1830. Today, nearly two-thirds of the world (based on population) prohibits abortion completely or allows it only for very limited medical reasons.

Because of the legal stringency, the method of aborting varies according to how it is performed, clandestinely or legally. Abortions under a doctor's care usually occur no later than after 12 weeks of pregnancy. Two tools are used, the dilator and the curette. The operation is appropriately known as "dilation and curettage," or D and C. Several dilators, of increasing size, are used to stretch the cervical and canal of the uterus. When the canal is adequately stretched, the inside wall of the uterus is scraped with the curette, and the uterine contents removed, often with suction from a vacuum pump. The whole process takes only a few minutes. Abortion performed late in pregnancy is accomplished by slitting the abdomen and uterine wall and removing the fetus. This is called a *Caesarean section.* An injection of a very concentrated salt or sugar mixture into the uterus also effects expulsion of the embryo in 2 to 3 days.

Abortion is performed illegally in various ways. In early pregnancy a tube is inserted into the uterus. Air seeping into the womb provokes bleeding, contractions, and eventually expulsion of the fetus, although the placenta may not dislodge. Consequently, many abortions not performed in a hospital eventually bring the patient there anyway.

Before 1973 rigid abortion laws prohibited widespread abortion as a

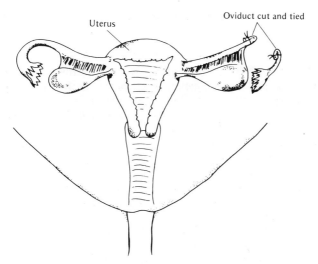

Figure 8–9
Sterilization of the female involves the cutting of the oviducts.

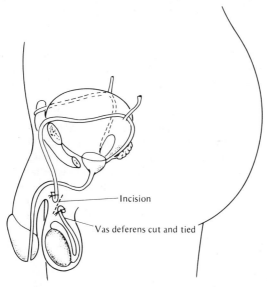

Figure 8–10
A vasectomy requires the severing of the *vasa deferentia*.

method of population control in the United States. Most states allowed abortion only if a continued pregnancy threatened the mother's life, a criterion proposed by the American Law Institute. This was interpreted to mean that adequate reasons for abortion included substantial risk to the mother's physical or mental health, the chance that the child might be born with a physical or mental defect, or rape or incest as the cause of the pregnancy. Then in January 1973 the Supreme Court ruled that no state can influence decisions made between a woman and her doctor during the first 3 months of pregnancy. Consequently, the state laws were superseded, and abortion could be obtained more easily.

Previous to 1973 the number of abortions each year in the United States was estimated to be about 1 million, and in the world, about 30 million. At least one-half of these abortions were probably illegal. The number of deaths in the United States resulting from abortions performed in unsanitary conditions with unskilled help has only been estimated. One study of the causes of deaths of women due to complications during pregnancy or childbirth in New York City indicated that 25 percent of the white women, 49 percent of the nonwhite women, and 56 percent of the Puerto Rican women died from abortion. The data reflect the socioeconomic position of the women and the consequent degree of medical attention.

The legality of abortion in foreign countries varies. Rigid prohibition exists in Africa; much of Asia, except Japan and China; southern and western Europe; and the rest of the Western Hemisphere. Liberal laws exist in China, the Scandinavian countries since the 1930s, Japan since 1948, and Britain since 1958. The number of legal abortions has naturally risen in these countries. In Hungary, the number of abortions per year now exceeds the number of births. This information, along with a decreasing birth rate, indicates that abortion is accepted as a means of birth control. The evidence that abortion can be effectively used as a birth control device influenced Romania and Bulgaria to revise their abortion laws to prevent too abrupt a decline in their birth rates.

Countries with rigid rules seem to encourage illegal abortion. Apparently, they are most common in Latin America, especially among the urban poor. A study indicates that in Chile 20,000 illegal abortions occur per year, as compared to 77,440 births.

You may wonder about the reason for the rigid laws that surround this method of birth control. The issue is an ethical one: When is the human embryo human? It is this issue that implicates the churches and accounts for the resistance to liberalizing abortion laws in Catholic countries, for example. At one time, the Catholic Church believed that the soul was infused in an embryo several weeks after conception. Today, the church believes that a human being is initiated at conception. This belief allows for baptism of the fetus any time it dies. The life of the mother and child are, therefore, equally valuable, so a pregnancy cannot be terminated, even if the mother is threatened. Consequently, abortion is not permissible for any reason at any time.

Other religions look upon abortion differently. Judaism never worried about the inception of soul in an embryo or the identity of the embryo separate from its mother. Protestants, in general, feel similarly. If not approving of it as a birth control method, they do not absolutely forbid it.

Nevertheless, legalization of abortion is slow in coming. L. Lader, in his book *Abortion* (Bobbs-Merrill, New York, 1966) believes that it is because "all Protestant denominations are trapped in the same Catholic web of animation and the impenetrable obscurities of ensoulment, though cloaked in more modern technology. Though rejecting the theory of soul entering the fetus at a given moment, Protestantism has nevertheless developed its own mystique built on the phrase 'protection of the sanctity of life.'" Perhaps society will eventually consider abortion as merely another birth control method and institutionalize it.

Evidence suggests that the public is becoming more accepting of abortion. A poll taken in 1965 indicated that a considerable majority of the public approved of abortions if the mother's health was in danger, and a lesser majority approved if the woman was raped or if the child might be born defective. Only 13 percent sanctioned abortion for unmarried women, and only 8 percent for any woman who desired it. In 1967 the results of a Gallup Poll showed greater acceptance of abortion: 28 percent for any woman who desired it. In 1967, also, the House of Delegates of the American Medical Association formally endorsed more liberal abortion laws, and in 1968 the American Public Health Association took a similar stand. With increasing liberation of the woman in our society, increasing pressure will be exerted in the press, in schools, and in the legislative bodies for laws that give the pregnant woman and her doctor the authority to determine the propriety of an abortion.

Contraceptives in the future

Today, the safest, surest, and cheapest contraceptive is the Pill. Nevertheless, it has drawbacks, primarily the side effects and the need to take it on a strict schedule. Needed still is an improved birth control method or, more realistically, several methods. Today, foundations such as the Ford Foundation, recognizing the need for research in human fertility and contraception, are supplying millions of dollars throughout the country and abroad on research in these areas. Researchers are approaching the problem from many directions.

A better pill

Doctors found that many of the side effects experienced by women taking the Pill could be eliminated by reducing the amounts of estrogen and/or progestin. Breakthrough bleeding is still too frequently a problem to herald the new "minipills" as a success at present.

Another interesting variation is a pill with only small amounts of progestin and no estrogen. This pill suppresses pregnancy as effectively as an IUD, even though ovulation occurs. Precisely why pregnancy does not occur is unknown, but perhaps it is due to a speedy descent of the ovum to the uterus where it disintegrates because of an unprepared uterine wall. The cervical mucous may also be unreceptive to the sperm.

Long-lasting hormones

Research is especially active on this type of contraceptive. The advantages are numerous, including the lack of necessity to administer the contraceptive frequently, such as daily or at the time of intercourse. The product studied most so far is a combination of estrogen and progestin taken by mouth 24 hours before menstruation or on the eighth day following menstruation. The side effects are similar to those of the pill. Research on even longer-lasting injections is also under way.

Another method that shows promise is the injection under the skin of a porous capsule filled with hormones. The hormones diffuse out slowly from the capsule and give long-lasting contraception. The capsule, made of Silastic, a material used in the surgical reconstruction of Fallopian tubes, is completely harmless. Removal of the capsule will release all inhibition to ovulation. So far it has been tested only on animals. A similar method involves "micropellets," small doses of hormone mixed with an adhesive that controls release of the drug. The pellets are injected under the skin by needle.

Postcoital pill

Although originally hailed as the most promising birth control pill, research has encountered many difficulties in perfecting it. In 1966 two doctors from Yale Medical School announced a pill that could be taken after intercourse to prevent pregnancy. Today, only one compound, ORF-3858, is safe and effective on laboratory monkeys, animals whose reproductive cycles closely resemble man's. The exact function of ORF-3858 is unclear. Perhaps it causes the uterus to contract and, hence, prevents the egg from implanting. Perhaps it speeds up the descent of the ovum down the Fallopian tube. To prevent pregnancy, however, it must be taken not once, but 6 days, one pill per day during the third week of the menstrual month. Perhaps dosages can be reduced to one pill to be taken near the end of the cycle.

IUDs

IUDs, despite the numerous advantages for many women, have a few drawbacks. For example, IUDs tend to come out of the uterus. New shapes are being tried to avoid this problem. One experiment utilizes new plastics that are warmed to pliability and then inserted into the uterus, where they take the uterine shape. Another technique is to increase the retention of the IUD by inserting it immediately after an abortion or birth. Still another impregnates hormones in the device, producing a two-fold protection.

Contraceptive methods for men

Research progresses on an injection or pill that would control male fertility. The difficulty lies in the close physiological relationship between sperm production and hormonal balance. All spermicidal drugs so far tested appear to have undesirable side effects. The effects of one product lowered the sperm count, but the drinking of small amounts of alcohol then caused the eyes to turn bright red, flushed the face, and accelerated the heart beat.

Vaccination

A technique that may have the greatest promise as an effective contraceptive is an antifertility vaccine. This procedure, for example, could involve sensitizing a woman to her husband's sperm so that the sperm are inactivated in the woman's reproductive tract. Evidence suggests that some women naturally develop antibodies to their husband's sperm and, hence, are infertile.

Research in the future will undoubtedly produce new and effective birth control methods. The sad and alarming fact is that in some countries effective birth control is long overdue and, in fact, is forever too late. The U.D.C.s could have used contraceptive help decades ago. Today, new contraceptive methods will be developed in the D.C.s. Every effort must be made to hasten their development, mainly because the impoverished and uneducated masses in the U.D.C.s are desperately in need of help.

While contraceptives are being developed, efforts must go on to feed the people of the world who are already born, and those that will be born.

In the next chapter we shall examine ways in which the population of the world may be fed.

Summary

Birth control, especially via abortion, has a long history, and the religions of the world have differing views toward it. Widespread use of diverse methods is a recent development and is more common and, therefore, more successful in D.C.s than in U.D.C.s Methods of birth control presently used include foams, suppositories, creams, jellies, douching, rhythm, the diaphragm, the cervical cap, coitus interruptus, the condom, IUDs, the Pill, sterilization, and abortion. Contraceptions in the future may include a more effective pill, longer-lasting hormones, a postcoital pill, more effective IUDs, and improved pills for the male.

Supplementary readings

Callahan, D. 1972. "Ethics and Population Limitation." *Science*, vol. 175, pp. 487–494.

Demarest, R. J., and J. J Sciarra. 1969. *Conception, Birth, and Contraception: A Visual Presentation.* McGraw-Hill Book Company, New York. 129 pp.

Granfield, D. 1969. *The Abortion Decision.* Doubleday & Company, Inc., Garden City, N.Y. 240 pp.

Guttmacher, A. 1963. *The Complete Book of Birth Control.* Ballantine Books, Inc., New York.

Hardin, G. 1969. *Population, Evolution, and Birth Control: A Collage of Controversial Readings,* 2nd ed. W. H. Freeman and Company, San Francisco. 386 pp.

Hardin, G. 1973. *Stalking the Wild Taboo.* W. Kaufmann, Los Altos, Calif. 216 pp.

Harkavy, O., and J. Maier. 1970. "Research in Reproductive Biology and Contraceptive Technology: Present Status and Needs for the Future." *Family Planning Perspectives,* vol. 2, no. 3, pp. 5–13.

Lader, L. 1966. *Abortion.* Bobbs-Merrill, Indianapolis, Ind. 212 pp.

Lader, L. 1973. *Abortion II: Making the Revolution.* Beacon Press, Boston. 242 pp.

Mishan, E. J. 1970. *Technology and Growth: The Price We Pay.* Praeger Publishers, Inc., New York.

National Research Council, Committee on Plant and Animal Pests. 1968. *Principles of Plant and Animal Pest Control,* vols. 1 and 2. National Academy of Sciences, Washington, D.C.

Seiler, P. 1963. *The New Handbook of Modern Birth Control.* Dell Publishing Co., Inc., New York. 157 pp.

Westoff, L. A., and C. F. Westoff. 1971. *From Now to Zero: Fertility, Contraception, and Abortion in America.* Little, Brown and Company, Boston. 358 pp.

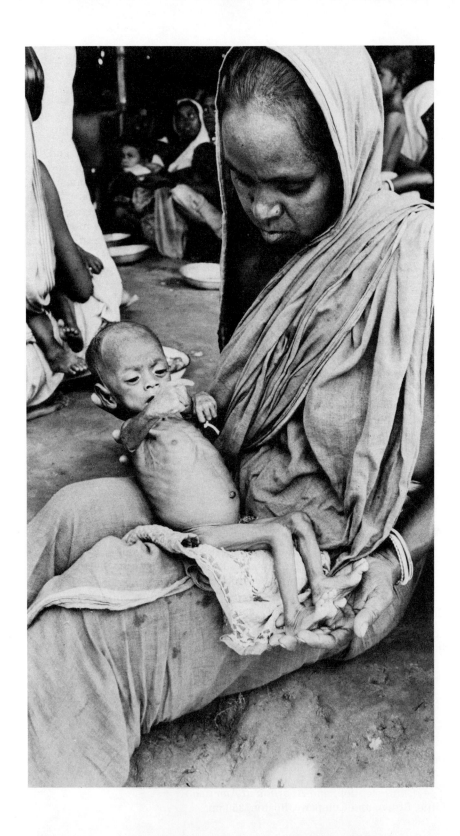

Chapter 9

A starved or once starved man has all his human capacities impaired: starvation breeds men who are permanently diminished.

The Hungry Future
Rene Dumont and Bernard Rosier

Feeding the world

By the year 2000 A.D., close to 7 billion people will inhabit the earth. Reckoned from 1970 this represents a doubling of the population in a mere 30 or so years. Clearly, the needs of the population will also multiply. Food supplies will have to increase to two to three times what they were in 1970 to adequately feed the world in 2000 A.D. Supplying adequate food may well represent the greatest challenge that has ever confronted the human race. We might not meet it.

In 1970 food supplies were already seriously inadequate in some parts of the world, mainly in the underdeveloped countries. In fact, an individual died, directly or indirectly because of a lack of food, every 8.6 seconds. Every 60 seconds, 7 people died. Every 50-minute class hour, 350 people died, and in a 60-minute lunch hour, 420 people died. In one day 10,000 people died. In one week, 70,000, and in one year, 3.6 million deaths were due to insufficient diets.

To better understand the reasons for this alarming number of deaths, let us look at the dietary requirements of human beings as compared to the typical nutrition of people in many U.D.C.s, describe some nutritional diseases that result from inadequate diets, and examine a few of the many proposals to feed the increasing number of people in the world.

The nutritional requirements
The food groups

The U.S. Department of Agriculture has devised criteria for maintaining an adequate diet: each day eat foods from four carefully defined groups in quantities dependent on age and sex. The defined food groups are milk products, meat products, vegetable and fruit foods, and cereal grain foods (Fig. 9–1).

Milk products (group I) contain a good proportion of some of the most important nutrients. The basic molecules, fats, and carbohydrates are plentiful, as are proteins, especially those incorporating the amino acids tryptophan and lysine. The minerals phosphorus and calcium, and vitamins A, B_6, B_{12}, and thiamine are abundant, too.

Meat products (group II) are high in protein, of course. This group includes all animal flesh, as well as eggs, and high-protein legumes and nuts. The protein in these foods contains a highly balanced and constant proportion of amino acids and is consequently a dependable source of materials for body growth and maintenance. The amino acids methionine and lysine are less plentiful in nuts and legumes. Nevertheless, they remain valuable sources of protein. Meat foods supply a great many minerals, such as phosphorus, sulfur, potassium, sodium, and especially iron. Vitamins of the B complex are especially abundant, and vitamin A is plentiful in liver.

The vegetables and fruits (group III) are the major sources of vitamin A. This group includes most parts of plants, except mature seeds, which are considered part of group IV, the cereal grain group. Besides vitamin A, the foods in this group supply a great amount of iron, as well as other minerals and the B-complex vitamins. Many of the green vegetables, such as kale and broccoli, supply a considerable amount of calcium. The dark-

239   The nutritional requirements

green leafy vegetables and many fruits are notable for their supply of iron. Others are rich sources of potassium. The citrus fruits supply plentiful ascorbic acid, vitamin C.

Cereal grain products, some of the oldest cultivated plants, compose group IV. Because of low production cost, plant adaptibility, abundant

Figure 9–1
In the United States, a developed country, foods from four groups — milk products, meat products, vegetables and fruits, and cereal grains — are commonly consumed by each person each day. (Courtesy of the National Dairy Council.)

yield, and good storage properties, cereals are a staple today in the diets of the majority of peoples around the world. Rice is most common in the East, wheat in America and Europe, and corn in Central and South America.

Cereal grains have abundant protein, iron and phosphorus, and B-complex vitamins. Unfortunately, the amino acid makeup of the proteins is not as complete as in animal protein. Wheat, rice, and corn are deficient in lysine, corn is deficient in tryptophane, and rice is low in threonine. Nevertheless, the cereal grains contribute to the diet valuable amounts of many other nutrients needed by the human body, except vitamin A, vitamin C, and calcium.

Calories

To maintain a healthy condition any human being must regularly take in certain amounts of food from the four basic groups. Carelessness concerning the kinds of food eaten will result in an inability of the body to function properly, manifested in abnormal growth or general debility over a period of time, an insidious "hidden hunger."

Beside the body's need for the proper assortment of basic kinds of molecules for proper functioning, it also demands a constant amount of energy for its metabolism. The energy needs vary according to individual activity, sex, and age of the person, but whatever the needs, if they are not met, the individual starves. Energy required for metabolism is taken into the body as molecules of the food from the four food groups. Molecular activity releases the energy from these molecules for any synthesis or molecular breakdown intrinsic to proper metabolic activity.

Energy is measured in calories. Food samples, burned in oxygen, yield heat. The amount of released heat which raises the temperature of 2.2 pounds (1 kilogram) of water from 15 to 16° centigrade is called the large calorie, kilocalorie, or "Calorie." (This unit is not to be confused with the small calorie, or "calorie," a unit 1/1,000 as large, used by physicists.) The quantity of Calories usually needed to maintain an individual is generally no less than 2,250 and usually no more than 3,000 per day. The President's Science Advisory Committee placed the average world caloric requirement at 2,354 per person per day. Table 9–1 gives an idea of the common caloric requirements for various activities.

The global food problem

The problem that faces a great number of people, especially in underdeveloped countries, is sufficient provision of foodstuffs to supply the Calories needed and an adequate assortment of foods to provide the proper molecules for growth and maintenance of the body. The United Nations Food and Agriculture Organization estimates that an average of 2,420 Calories per person per day is now produced. However, inequitable distribution and loss due to spoilage reduces the Calories available to many people in some countries. The President's Science Advisory Committee Panel on the World Food Supply estimated in 1967 that 20 percent of the

Table 9-1
Caloric requirements for various activities

Activity	Number of Hours	Calories per Hour	Total Calories
Sleeping	8	60	480
Sitting, reading, eating, driving, card playing	6	90	540
Standing	6	150	900
Hiking and brisk walking	2	180	360
Occasional sports (swimming, golfing, tennis)	2	270	540
			2,820

Source: M. A. Bernarde, *Race against Famine*, McCrae Smith Co., Philadelphia, 1968, p. 23.

inhabitants of underdeveloped countries were not furnished with adequate Calories per day to exist, and that 60 percent had diets that significantly lacked nutrients, usually protein (Table 9-2). Consequently, the quantity of undernourished or malnourished people, mostly children, in the world is at least 1.5 billion. Some authorities claim many more. Many of these unfortunates succumb directly or indirectly to this deprivation of food. Although 3.6 million deaths per year were cited earlier to be the result, some authorities boost this estimate to 10 or 20 million.

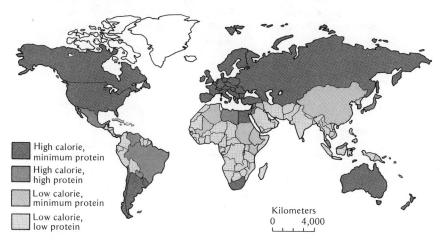

Figure 9-2
The areas of the world where hunger exists. (From *Population Resources and Environment: Issues in Human Ecology*, 2nd ed., by P. R. Ehrlich and A. H. Ehrlich. W. H. Freeman and Company, San Francisco. Copyright © 1972.)

Table 9-2
Recommended daily intake of basic nutrients

I. Milk Cheese and ice cream can replace part of milk	Two or more servings (1 cup or 8 oz each)	Protein Calcium Riboflavin Vitamin A
II. Meat Beef, veal, pork, lamb, poultry, fish, eggs Dry beans and peas and nuts can be used as alternatives	Two or more servings (at least 2 oz each)	Protein Thiamine Riboflavin Niacin Iron
III. Vegetables and fruits Should include (a) a dark green or deep yellow vegetable at least every other day for vitamin A (b) a citrus fruit or other fruit or vegetable rich in vitamin C daily (c) other fruits and vegetables, including potatoes	Four or more servings ($1/2$ cup each)	Vitamin A Ascorbic acid Iron Thiamine Riboflavin Niacin Trace minerals
IV. Bread and cereals whole grain, enriched, or restored	Four or more servings (1 slice bread or $1/2$ cup cereal)	Thiamine Niacin Iron Protein Riboflavin
Other foods — sugars, fats, oils, including butter or fortified margarine for vitamin A		Calories Vitamin A

Source: G. A. Goldsmith, *Nutritional Diagnosis,* Charles C Thomas, Publisher, Springfield, Ill., 1959, p. 12.

The situation worsens each year (Fig. 9-2). The populations of the countries that already have difficulty feeding their people usually have the greatest rate of population increase. The President's Science Advisory Committee Panel in 1967 estimated that, to feed the world population in 1985 at the 1965 consumption rate, the quantity of food must increase by between 43 and 52 percent, depending on the constancy of the fertility rates. Some countries, with especially rapid increases of population, will have even higher percentage increases of caloric intake. Projections for 1985 at the 1965 consumption rate will be a 88 to 108 percent increase needed for India, a 92 to 104 percent increase for Brazil, and a 118 to 146 percent increase for Pakistan.

Demands for specific foodstuffs, as well as Calories, will also increase. The President's panel estimates a 50 percent increase in world need for vitamin A and a substantial increase in demands for calcium and other vitamins.

Nutritional diseases and their results

The dietary demands in the future, certainly by 1985 when the world's population may have increased to 5 billion, will be extremely difficult to meet in view of the few years that man has to decide how to supply the food and develop the procedures to do it. If man cannot channel his efforts effectively, the world will soon be faced with rampant global starvation.

In some countries today we can already see the effects of insufficient availability of calories and certain kinds of foods, notably protein. The physiological and morphological symptoms of malnourishment and starvation are severe. Outlined below are some of the common nutritional-deficiency diseases.

Goiter

Goiter is one of the principal diseases of the world that are caused by inadequate nutrition. West Africa, Ethiopia, Central America, South America, southeastern Asia, the Himalaya area, the Near East, and other iodine-poor areas are frequent sites for its occurrence. The World Health Organization estimates that 200 million people in the world suffer from the disease. In eight states of Mexico one-fifth or more of the inhabitants suffer from goiter; in Basutoland, 40 percent; and in East Cameroon, 25 percent.

Goiter is a disease of the thyroid gland, a gland located in the throat region. Here 2,500 times more iodine is found than in any other part of the body. If the intake of iodine is insufficient, the gland enlarges or hypertrophies in its attempt to supply adequate combined iodine to the body. The endemic goiter is usually the only public health consequence of inadequate iodine. In severe cases, however, cretinism and thyroid cancer occur. Cretinism, a disease occurring in infancy when the mother is severely deprived of iodine during pregnancy, is a disease characterized by low basal metabolism, lack of muscular tone, dryness of skin, affectation of bone development, and severe mental retardation. Except for the effects on the nervous system, the symptoms of inadequate iodine intake can be improved by eating animal thyroid glands. The addition of iodine to salt sold over the counter prevents the widespread occurrence of goiter.

Hypochromic anemia

This disease is the result of inadequate intake of iron. The red pigment of the blood, hemoglobin, is reduced, and the red blood cells microscopically appear pale, or hypochromic. Weariness and shortness of breath are more readily seen symptoms. Infants, adolescent girls, and pregnant women are most frequently susceptible to anemia.

Iron must come into the body from foods from all four food groups eaten daily, because only small amounts exist in the food in any one group (see Table 9–2 for the daily requirement of iron and its sources).

Nyctalopia and xeropthalmia

Diseases of the eye, such as nyctalopia, night-blindness, and xerophthalmia, loss of sight due to prolonged dryness of the cornea, are diseases that result from a deficiency of vitamin A. Night-blindness is an early sign of vitamin deficiency. Sight is lost in very dim light because the eye pigment needed to see, visual purple, cannot regenerate itself after exposure to light, a normal function in individuals with adequate vitamin A. In Indonesia a single hospital has recorded approximately 1,000 cases of vitamin A–deficiency blindness for each of the last 10 years.

Prolonged deficiency hardens or keratinizes the epithelial tissues of the body and is especially deleterious when the eye epithelium is attacked. First symptoms include itchiness and redness of the eyes and sensitivity to bright light. The cornea then becomes dry and inflamed because the tear glands no longer function to wash away the bacteria. Eventually, severe cloudiness and infection soften the cornea and produce permanent blindness.

Rickets

Rickets, a child's disease, results from the lack of vitamin D. It is rare in tropical countries, although it occurs in southern Africa and northern India. It is more frequently found in children in northern climates and urban areas, especially in the winter months when the sun cannot irradiate the skin frequently enough to promote vitamin D synthesis in the body. Heavily pigmented children are more susceptible than light-skinned children, an observation that allows speculation on the origin of races (Chapter 4). Fifteen percent of children in Damascus and Cairo hospitals have rickets.

Generally, the symptoms include a lack of muscle tone; restlessness; "pot-belly"; weakness; enlargement of ankles, wrists, and knees; bowing of the legs; soft fragile bones; and bulging of the forehead. Early skeletal deformations persist through life.

Deficiency of vitamin D in adults produces osteomalacia, "adult rickets." It is especially prevalent in the Orient in pregnant or lactating women. Calcification of the bones seriously lags behind other metabolic processes. Consequently, the bones soften and become deformed. Walking is difficult. Frequently, the bones fracture.

Beri-beri

Beri-beri has long been a common disease in those areas of the world where polished rice is a dietary staple. It is especially common today in southeastern Asia and in the Philippines. A recent survey in Burma indicated that 40 percent of 2,000 pregnant women and nursing mothers had the disease. In 1956, 24 percent of the adults in northern Thailand suffered from polyneuritis, a symptom of beri-beri. It is endemic in Assamard and Bengal, India. Severe deficiency of one of the B-complex vitamins, thiamine, produces beri-beri. A diet of polished rice is the most common cause of the disease, because the thiamine exists in the outer brown layer of the grain, which is removed. The disease is characterized by gastrointestinal problems, muscular weakness, and paralysis of feet and legs, enlarged heart, and emaciation. It is often fatal.

Ariboflavinosis

Inadequate quantities of the yellow B-complex vitamin, riboflavin, causes one of the most common deficiency diseases, ariboflavinosis. Symptoms of this disease include morphological changes in the tongue and lips and impairment of vision.

Pellagra

The deficiency of another B vitamin, niacin, is believed to have been at one time the leading cause of mental illness and death in the United States. It produces the disease pellagra. In Spain, northern Egypt, Yugoslavia, and Africa it is still a serious disease, although in the United States it is becoming less common, presumably because of better diets and education. Early symptoms are minor, but late symptoms can be fatal. Progressively, they include a characteristic dermatitis in which the skin becomes cracked, red, and scaly; diarrhea; and mental aberrations, including delusions of persecution and dementia.

Kwashiorkor

Kwashiorkor is seldom observed in the United States but is common in some underdeveloped countries. In Africa, 25 percent of the children between 1 and 4 years of age have this disease. One of the largest recent outbreaks occurred in 1969–1970 in Biafra (Fig. 9–3). The disease is caused by a severe lack of protein or by ingestion of protein, such as some plant proteins, that lacks certain amino acids. Kwashiorkor victims usually eat little or no animal protein.

Most commonly, the disease occurs in babies soon after weaning when the child is taken off the breast milk and put on a high carbohydrate diet. Symptoms include lack of growth, retarded development, apathy, lack of appetite, diarrhea, edema, and changes in hair texture and color. Many children, in addition to those who have Kwashiorkor, have "prekwashiorkor," a mild but debilitating form of protein malnutrition. One

Figure 9-3
Kwashiorkor was a common ailment of young children in Biafra in 1968. The disease is seldom seen in the United States.

investigator suggests that 100 cases of pre-kwashiorkor exist for every case of kwashiorkor.

Kwashiorkor is treated by feeding reconstituted nonfat dry milk or vegetable protein concoctions. Improvement can be seen in days, and recovery in a very few weeks.

Marasmus

This term is applied generally to the symptoms produced as a result of inadequate proteins and calories in the diet. Carl Taylor at Johns Hopkins University describes marasmus in Indian children in his article "Population Trends in an Indian Village" (*Scientific American*, July 1970):

> They had overt marasmus: a wasting condition characterized by wizened, shrunken features, like those of tiny old men, and sluggish matchstick extremities. The condition (which the villagers call sukha, or "drying up") starts with the common practice of "breast starvation." Village mothers think that as long as the breasts produce any milk they need offer no other food until children can start the coarse adult diet, and they also prefer prolonged breast-feeding because they know it is associated with decreased fertility. The synergism of the resulting undernutrition and common and normally nonfatal infections during the weaning period is our most frequent cause of death. These babies had had recurrent diarrhea and respiratory infections.

Death

The most severe result of inadequate diet is, of course, death. From 3.5 to 20 million deaths in the world per year are attributed to the lack of food. Most of those who succumb are children. The President's Scientific Advisory Committee believes that moderate protein–calorie malnutrition affects at least 50 percent of the 1-to-4 age group in underdeveloped countries. Interestingly, the mortality rate in the United States for this age group is approximately 1 per 1,000, whereas in some U.D.C.s where food is scarce, it is between 10 and 30 per 1,000. Increased susceptibility to diseases has been reliably correlated with poor diets, especially in children. Deaths due to measles in 1960 were 100 times greater in Chile than in the United States.

Death as a result of poor diet may indeed be the merciful end to a life begun in an environment without adequate food. Yet many children, although subject to the ravages of inadequate diet, including increased susceptibility to diseases, do survive. What are the long-range effects of prolonged malnutrition to a child who survives? Only recently have researchers become concerned with this problem. The results of their inquiries are not encouraging, because malnutrition in the early years of a child's life appears to affect development, especially of the central nervous system.

Mental underdevelopment

In the past, field workers concerned with nutrition have spent most of their time preventing starvation. Today, some are concentrating their attention on the effects of malnutrition. Discoveries concerning the effects of malnutrition on the mental development of young children are especially disturbing.

The nutrition of babies in U.D.C.s allows for ready study of the effects of malnutrition. Babies normally are fed breast milk until approximately 6 months of age. Then, by supplementing the diet with small amounts of animal milk and solid food, such as rice, the infant is weaned. Meat and vegetables, staples in the adult diet if available, are notably absent from the child's menu, mainly because it is believed that such foods are inappropriate for babies. In many countries diet taboos also limit the diet. Furthermore, a limited quantity of food often means caloric deprivation.

Frequently, then, between 6 and 12 months of age the child's resistance to disease is seriously lowered. The combination of malnutrition and disease can be seen in characteristic growth patterns of the children in certain areas. During the first 6 months the breast-milk diet is adequate, and the babies grow normally. After weaning, however, the babies' growth rate is depressed, and a level of development common to better-fed children is forever lost.

Beside the physical debility associated with inadequate diet, poor mental development is also known. Unfortunately, the acute effects of malnutrition occur precisely at the time when the child's brain is developing fastest. By 3 years of age, the human brain is 80 percent developed, although the child has attained only about 20 percent of his height. Researchers have attempted to measure the amount of brain damage. Most laboratory studies concerned with tissue development have been done with experimental animals. Lipids, one of the main molecules in brain tissue, in poorly fed pigs fall greatly below the usual concentrations and do not increase even when the animals' diets are improved. Similar tests with rats produced similar results.

Experiments on dogs also show physical and behavioral effects of malnutrition. Two litters of puppies from poorly fed mothers were treated differently at birth: one was fed a high-protein diet, the other a low-protein diet. Two other litters from well-fed mothers were treated similarly. The puppies from the underfed mothers were small at birth, walked atypically, and suffered head tremors. The puppies fed the better diet soon lost the nervous disorders and began to resemble the puppies fed well throughout the experiment. The puppies from well-fed mothers who were fed a low-protein diet soon developed stiff legs and head tremors. Some puppies suffered convulsions.

The worst symptoms were seen in the puppies born of underfed mothers and kept on a low-protein diet. These animals became extremely irritable and lost interest in their environment. Their legs became stiff. Some became convulsive and died. Those that survived improved at about 12 weeks but never regained normality although the ratio of brain weight to body weight appeared similar to normal animals (Fig. 9–4).

Figure 9–4
Two 6-week old puppies, the smaller one born of and nursed by a malnourished mother, the larger born of and nursed by a well-fed dog. (From B. S. Platt and R. J. C. Stewart, *Developmental Medicine and Child Neurology*, vol. 10, no. 1, 1968. Courtesy of R. J. C. Stewart.)

Measuring the effect of hunger on the human brain is more difficult. Intelligence as measured with IQ tests and head circumferences have been used as criteria. One test that involved both criteria was performed in Cape Town, South Africa. Twenty severely malnourished children and 20 well-nourished children comparable in sex and age have been studied since 1955. The average circumference of the heads of the undernourished has remained approximately 1 inch smaller than the well-fed group. Furthermore, electroencephalograms of the undernourished group show more aberrations known to be related to mental disorders. These children also perform less well in school than the control children.

Solutions to the world's food problem

Short of preventing the birth of children, a topic discussed previously, what can be done to alleviate the hunger problem that now exists in some parts of the world and to head off the threat of global starvation in the future? Many preventive measures have been suggested. However, probably no single solution to the problem exists, because areas of the world with economic, political, social, and geographic uniqueness demand different solutions. Outlined below are a few of the more serious proposals.

Food fortification

The causes of nutritional diseases are nutritional deficiencies. With proper fortification of foods many of the diseases would never occur, and the crippling effects of an impoverished diet could be prevented. The prevention of dietary diseases is not easy, however. Economically and politically, it is difficult to distribute the supplements only to those people who need them. Education is no more successful. Most successful has been the addition of nutrients to foods already acceptable and regularly eaten by most of the population.

Data gathered from several places in the world testify to the success of food enrichment. The fortification of white bread with thiamine, riboflavin, and iron in the United States resulted in a startling decline of pellagra cases, for example. In 1941, 1,836 deaths from this disease were recorded. In 1966, 16 years after enrichment of bread was initiated in the country, only 21 people died from the disease. Similar data testified to the reduction of beri-beri in Newfoundland after the B-complex vitamins were added to the diet. Thiamine added to rice in the Philippines reduced by two-thirds the deaths from this disease in one year. Currently, nutritionists are interested in adding other nutrients to foodstuffs. It is well known, for example, that an inadequate amount of even one amino acid will prevent the manufacture of some proteins within the body. Lysine deficiency is common in many countries in which the dietary staple is wheat. Consequently, protein deprivation is observed, even though other amino acids are present, because only one-half of the protein in wheat is utilized by the body when lysine, one of the protein-building amino acids, is scarce. Lysine added to food commonly eaten would prevent malnutrition in many children. A similar problem exists in those countries which use a great deal of corn, such as Central and South America. Corn is deficient in lysine and tryptophane, and hence, much of the potential nourishment in corn is lost to the human body.

U.D.C.s, because of the instability of their governments, find it more difficult to put broad food-enrichment programs into effect, even though some dietary supplements are extraordinarily cheap. The government-owned Modern Bakeries in India have fortified bread with 0.2 percent lysine for some time. However, only now is fortification of other wheat foods being considered, even though bread with inexpensive lysine has been shown to have 33 percent more usable protein than unfortified bread.

Trytophane, however, is a much more expensive amino acid than lysine, costing approximately $20 per pound. Threonine costs approximately $7.50 per pound. Consequently, fish meal or oilseed meal, inexpensive foods containing these amino acids, may be the best sources of these essential molecules.

Some civilizations have mixed foods routinely. The American Indians, and people of the Andes today, mix corn flour with bean flour; southeast Asians and Japanese add fish to rice; and the Chinese for centuries have added Chinese cabbage and soybeans to rice. In this way the proteins in the various foods supplement each other and prevent deficiency diets.

New foods

The meal that remains after the oil has been extracted from soybeans, peanuts, and cotton seeds contains between 40 and 50 percent protein. This meal is perhaps the world's cheapest source of abundant protein. It costs only $0.08 to $0.12 per pound on the U.S. scale compared to $0.61 to $0.71 for dry milk solids and $1.00 to $2.00 for animal protein. In the poor countries more than 20 million tons are produced each year and fed mainly to animals or used as an inefficient protein fertilizer. However, certain artificial food mixtures have been synthesized using these materials, such as Incaparina, CSM, Pro-nutro, Fortifex, and Multipurpose Food (MPF).

Incaparina is a mixture of corn and cottonseed protein. It is promoted by INCAP, the Institute of Nutrition for Central America and Panama. Unfortunately, it has been only moderately well received, probably because it is tasteless. Mothers with young children are encouraged to buy the product. The relatively high protein content is sometimes improved by adding lysine, as in Guatemala. Another food mixture is CSM, composed of 70 percent gelatinized corn, 25 percent soybean, and 5 percent nonfat dry skim milk. An Indian variant, Bal Ahar, is made up of 70 percent wheat and 30 percent peanut protein concentrate fortified with lysine.

To have widespread acceptance, these foods must be attractive and palatable as well as nutritious. Poor texture and flavor prevent or discourage their use. Now, research has produced techniques to introduce texture into foods. Soy proteins spun and blended with other proteins and the defattening of soy products are common methods in use today. Research is still needed to improve the taste of many products, although with proper application of modern nutritional chemistry, this should be an easy problem to solve.

The biggest problem faced by industries sponsoring these foods is the unwillingness of people to adopt the new products. The new protein beverages are most quickly acceptable. They appeal to most segments of a population and can be used in place of cold soft drinks and, in some cases, even hot drinks. Vitasoy, a soybean protein beverage, has taken over one-fourth of the soft drink market in Hong Kong. Saci, a caramel-flavored beverage sold in Brazil as part of a pilot project of Coca-Cola, has the nutritional content equivalent to milk and is produced from materials indigenous to that country. The Monsanto Company has produced a banana-flavored drink which it makes and sells in Guyana. Swift & Company in Brazil and Pillsbury in El Salvador plan to market protein beverage powders soon.

The development of these products gives hope to the otherwise dreary prospect for countries burdened with large populations and shortages of foods, especially protein. Much credit for the development of new food products must go to AID, the Agency for International Development, although some pioneering experiments were done without this organization's help. AID has provided grants to private manufacturers to support initial attempts to put new food products on the market in countries where needed.

The availability of these products to a country's inhabitants who need them most is still a major problem. A nutritious and attractive product does not necessarily mean that children will eat better. Aaron Altschul, special assistant for nutrition improvement to the U.S. Secretary of Agriculture, comments in his article "Food: Proteins for Humans" (*Chemical and Engineering News,* Nov. 24, 1969):

> Most of the new foods will be available to and used first by people who probably do not need them, but the first thing to do with a new food is to establish it in the market place — to make the food attractive and desired, possibly giving it some sort of status appeal. Then the food can ultimately be cheaper than conventional foods or more readily accessible and can extend more rapidly to lower income markets. New foods will not eliminate the desire for animal foods, but they will temper the disappointment of not having enough of such foods and will make it possible to stretch the available supply a little further.

Fertilizer

Fertilizer increases crop yields, especially of the new varieties of crop plants to be described later. Most of the new breeds of crops efficiently convert the fertilizer into grain. One pound of nitrogen applied to old strains adds approximately 10 pounds to the grain yield, whereas nitrogen applied to the new plants adds up to 20 pounds to the yield.

Although fertilizer has proved to be essential in the production of high grain yields, especially of the new strains, the underdeveloped countries have great difficulty in obtaining an adequate supply. They find it difficult to develop fertilizer plants themselves, and their farmers, lacking credit, find it difficult to buy the product even when it is imported. Furthermore, many farmers still do not understand or accept the benefits of fertilizer. In many countries, this problem of education is the greatest difficulty and has not yet been attacked effectively.

At the moment, the production anywhere in the world of enough fertilizer to help food production in the U.D.C.s would be a major problem even if the economies and internal organization of the U.D.C.s would demand it. At the moment, India uses only approximately $\frac{1}{100}$ the fertilizer used in the Netherlands. It needs much more to grow crops optimally, but if India were to use fertilizer at the rate the Netherlands is using it today, the amount used by India alone would total nearly one-half of all the world's present production.

Japan is an example of a country whose agricultural productivity increased astonishingly after widespread acceptance of fertilizers. Japan uses more fertilizer than all of Latin America, Africa, China, or India, although as the Paddock brothers point out in *Famine, 1975!*, Japan's cultivated area is only 5 percent of China's and 4 percent of India's. It has been said that if India used fertilizer at the rate of Japan in 1963, India's requirements would have exceeded the amount produced by the entire world in that year.

Some U.D.C.s have been encouraged to develop their own fertilizer plants. During the mid-1960s, India and Pakistan and a few other countries received loans from AID, the same organization that supported the development and distribution of synthetic foods. Consequently, fertilizer plants within these countries are springing up, and the cost of the products is being reduced. Concurrently, as an educational service the Food and Agricultural Organization (FAO) of the United Nations ran thousands of field demonstrations showing the benefits of fertilizer application. Because of this, the use of fertilizer in U.D.C.s is rising at approximately 16 percent per year. For example, Brazil used fertilizer at approximately the same rate from 1960 to 1966, then doubled the quantity used in the next 3 years. Between 1966 and 1969 India increased her use of fertilizer more than threefold.

Today, new methods of production have been developed which will reduce the production cost of fertilizers and make them even more available. An efficient process now exists for the synthesis of ammonia from atmospheric nitrogen. Large centrifugal compressors reduce the initial cost outlay and the operating cost of the manufacturing plant by one-third. New processes for the production of phosphate and potash have also been developed.

Despite the increased availability of fertilizers, the major problem of unequal global distribution will probably be around for a long time. Use of fertilizer has nearly reached a saturation point in developed countries, where studies of its application and effect are longstanding, but U.D.C.s desperately need a great amount of fertilizer to help feed their starving populations. Presumably, they will need increased amounts in the near future. International arrangements are still not adequate to accommodate the transfer of excess fertilizer from parts of the world where it is abundant to those world areas that have a desperate need.

The use of fertilizer is not a panacea, however. The fact remains that more fertilizer spread on the soils of underdeveloped countries means increased yields only if high-grain-producing crops are grown and water and good weather are available. Excessive use of fertilizer frequently results in pollution problems because of the run-off into nearby waterways (Chapter 11). Fertilizer use alone cannot solve the problem of starvation in many countries, although it may help alleviate some of the problem.

Irrigation

Irrigation of land is another technique that may significantly increase food production around the world (Fig. 9–5). Its use to increase crop yields is long-standing, probably before recorded history. Canals, ponds of stored water, wells, dams, and terracing were all commonly used by ancient societies to bring water to their crops. Remains of extensive irrigation works have been found in Italy, Iraq, India, China, and Egypt, and parts of some are still in use today. Dams on the Nile, the Euphrates, and the Tigris were extremely important in the development of former cultures. One of the largest ancient dams, probably built some 2,000 years ago, is part of a huge irrigation project on the Ninkiang River in China.

Figure 9–5
Left: Expansive fields in California bordered by irrigation pipes and ditches. Right: Luxuriant cabbage, the product of irrigation of the fields. (Courtesy of Imperial Irrigation District, El Centro, Calif.)

In the last 100 years or so, irrigation has been developed on a massive global scale. By 1900 several large dams, notably in India and Egypt, were completed to serve thousands of newly cultivated acres, and they helped increase the amount of irrigated land throughout the world during the nineteenth century as much as sixfold. In this century dam building has not slackened. Georg Borgstrom, a noted nutritionist, predicts that by the year 2000 at least 600 million acres will be irrigated and that the twentieth century may properly be called the Irrigation Century.

The damming of flowing water, notably floodwater, to produce irrigation ponds or lakes has been the commonest irrigation procedure through history. Such ponds not only provide water during the dry seasons of the year but also prevent the previously unpredictable flooding of the floodplain early in the year and free more land for cultivation. In the 1950s China expanded her irrigation projects and produced another 100 million acres of fertile fields, an area equalling the total tilled land of Canada. A more recent dam project is the Aswan Dam of Egypt (Fig. 9–6). The area irri-

Figure 9–6
Machines and muscles together built the Aswan Dam. (WHO photo by P. Almasy.)

gated by this dam, equal to ½₀ of the United States, will increase Egypt's cultivated land by about 15 percent. Unfortunately, when the dam is completed, about 1975, Egypt's population will have grown nearly 35 percent, a rate of population growth that will prevent the dam from contributing greatly to the feeding of Egypt's people. The large dam project in China did not head off disaster either. Despite herculean efforts in dam building, China suffered a crop loss in the late 1950s.

The total amount of land in the world that could be irrigated and converted to productivity is difficult to estimate. The FAO estimates that, all political, economic, and social obstacles aside, possibly 650 million more acres could be irrigated in India, 117 million in Burma, 17 million in Thailand, and 5 million in Ceylon.

Another common irrigation method is the tube well, a closed perforated cylindrical pipe or shaft driven into the earth and operated with an electric pump. This device is inexpensive to install and, consequently, is increasingly popular in U.D.C.s (Fig. 9–7). The new wells cost from $1,000 to $2,500. Tube wells can be installed in a few weeks at most.

Figure 9–7
Another tube well being installed to inexpensively supply water to a farm in Pakistan. (Rolls-Royce [Composite Materials] Ltd.)

Farmers in Pakistan over a 5-year period in the 1960s installed 32,000 private wells. Private enterprise in India between July 1968 and June 1969 installed 76,000 tube wells, and the Indian government installed another 2,000. The immediate and plentiful supply of water provided by the tube wells contrasts with that supplied by storage ponds and dams. Because of their small initial expenditure, lack of governmental control, and the immediate use of water to grow the new seeds, tube wells may replace construction of dams in some areas of the world. Unfortunately, tube wells lower the water table in some areas, and this may tend to limit their use.

Although crop production usually increases dramatically with irrigation, supplying water artificially does have complications. For example, many large dams produce ponds that are filled with silt within a few years. Unused dams and ponds are common in South America, where farmers have been forced to abandon their lands when the ponds filled up after only 10 years. A major concern of the contemporary Aswan Dam builders in Egypt is that it, too, will suffer this fate.

Irrigating very arid regions has further complications. Because of the low rainfall in many irrigated areas, irrigation water evaporates and leaves salt deposits on the surface of the land. An accumulation of salt makes the areas untillable. Irrigation has produced millions of acres of salty desert in Pakistan, Egypt, Greece, and South America. Mesopotamia, India, and China very likely had this problem in ancient times, and in Pakistan today, the new area gained through irrigation has almost been canceled through losses caused by deposited salt (Fig. 9–8).

Figure 9–8
Struggling rye grass will never become robust, growing on very saline soil produced by extensive irrigation. (Courtesy of Imperial Irrigation District, El Centro, Calif.)

Another complication, the production of an impervious layer of soil just beneath the crop root line, results from irrigating with dammed water stored in ponds or lakes. This subterranean layer of hardpan prevents the downward movement of water. Hence, the water accumulates and eventually forms small shallow marshes. Hardpan and marshes have been found in the irrigation terraces of the Incas in the Andes and other places where irrigation has been practiced for some time. Recently, the phenomenon has been observed in the extensively irrigated Rio Grande Valley and the Imperial Valley of California.

The formation of this hard soil is initiated in the damming of the water eventually to be used in irrigation. Dammed water prevents floods and, thereby, frees the normally inundated land for cultivation throughout the growing season. Consequently, more than one crop can usually be produced. To ensure high crop yields, much fertilizer is frequently applied. In addition to supporting a longer growing season, the fertilizers also replace the organic material or silt which settles out of the water standing in the pond and which previously naturally fertilized the land when it was covered by flood waters. The numerous small particles of abundant fertilizer in the irrigation water soon are trapped in topsoil, usually just beneath the root layer, and form a hard impenetrable soil layer, or "hardpan."

The very tiny pores that remain in the soil after the deposit of minerals cause further difficulty. The pores effect a capillary action with the help of the evaporation of surface water which pulls water from beneath the ground toward the surface. Eventually, the formation of the hard layer, the capillary action, and the surface evaporation of water combine to complicate and frequently prevent the prolonged production of crops on irrigated land.

The problems described above are most common to those irrigated lands that have sparse rainfall during the growing season. But irrigated areas with rainfall have problems, too. In these lands salt gathering on the surface and in the top layer of soil is washed away to a great extent by rainwaters. But then the problem exists of the disposal of such salty water, or brine. The construction of ditches to drain the water away from the immediately irrigated area is only a partial solution. Its ultimate disposal can still present a problem. Near Baghdad, the water drained from irrigation fields is said to be saltier than ocean water. In fact, $\frac{1}{10}$ of the tilled land there has been devoted to huge ditches for drainage. And in Southern California in the Imperial Valley, close to 2,000 miles of drainage ditches have been constructed to direct the salty water away from the fields. In the Nile Valley, irrigation is a large enterprise, but one-third of the water used in irrigation projects is used to drain away the salt.

Still another complication accompanies the use of irrigation in some parts of the world. Evidence now shows that irrigation projects are influential in causing the spread of some diseases such as *schistosomiasis*. This disease is due to a flukeworm which in its early stages bores through the human skin into the bloodstream. (See the insert, "How Blood Flukes Cause Schistosomiasis.") General debilitation, susceptibility to other diseases, and frequently death result from infestation.

Vital to the completion of the life cycle is the completion of a developmental stage within the snail. The reason for the increase of schistosomiasis in some countries, for example, Egypt, is attributed to the presence throughout the year of the large snail populations in irrigated areas. Previous to the building of large dams the snail population existed only in flood times. When the waters dried up, the snails died. Now, however, the soil never dries out, so the snails persist through the year. Consequently, more flukes are present.

Another disease, called *river blindness,* is spread by a fly in whose blood is carried a parasite that is spread to man. The parasite causes blindness. Tropical western Africa is particularly afflicted. Constantly moist conditions encouraged by the irrigated areas seem to be essential for the development and spread of this pest.

The factor most seriously limiting irrigation around the world, however, may well be the water supply itself, particularly in underground reserves. Borgstrom of Michigan State University claims that Europeans use 3 times the amount of water that is returned to accessible reserve and that people in North America use twice the returned amount. For example, the recent increase in India of tube wells, which tap the underground water supply, may well hasten the exhaustion of easily available underground water in that country.

Desalinization, the desalting of sea water, has been proposed as a way to provide an ample quantity of fresh water for irrigation of crops. Because of the supposed inexpensiveness of nuclear power in the future, desalinization plants to be run with this energy are considered to be especially economical. Some desalinization plants have already been built around the world. Many of them can produce millions of gallons of water per day, but the cost of the water per 1,000 gallons ranges from $0.75 to $1.00 or more. Because of this cost, the desalted water is most sensibly used for drinking water, not irrigation.

The initial expense of building a desalinization plant and the annual cost of operation is great. One unit, for example, proposed to produce 1 billion gallons of water per day, is estimated to cost nearly $2 billion initially and nearly $1.5 million to operate annually. Such expense makes it even difficult for the developed countries to contemplate the construction of desalinization plants on a large scale and nearly prohibits construction in underdeveloped countries. Although cost is the big obstacle in considering desalting of ocean water as an abundant source of fresh water, other problems loom large, too, such as economically securing the ocean water and distributing the purified water.

Three scientists associated with Resources for the Future, Inc., in Washington, D.C., recently said after exploring the feasibility of desalinization plants in this country and the world that: "The conclusion is inescapable; the full and true costs of the desalting projects, now and for the next 20 years, are at least one whole order of magnitude greater than the water to agriculture.... It is impossible to bring planned costs and prospective values for agriculture together or even close."

Desalinization plants, especially those powered with nuclear energy, may contribute to the production of additional food in the future. At present, even for years to come, their contribution will be small. Hence, desalinization cannot be looked to now as a major factor in solving the world's food problems.

How blood flukes cause schistosomiasis

Three species of blood flukes, tropical human parasites, cause schistosomiasis. All live in the bloodstream and affect either the function of the urinary bladder or the intestine. Besides man, hosts include cattle, sheep, goats, dogs, and cats.

Figure 9–10
Boy with distended stomach, a victim of schistosomiasis. (Courtesy of G. A. Noble, California State Polytechnic College, San Luis Obispo, Calif.)

cost for this development would be near $80 million. Sample projects designed to increase the yield per acre in several underdeveloped countries vary in cost from $32.00 per acre in Guatemala to $973.00 in Kenya. Total cost ranges from $8.1 million to $29.2 million, respectively.

The complication of putting more land into cultivation is great because it demands a large input of money. P. R. and A. H. Ehrlich point out that just to open new land to feed people added to the population annually would probably cost approximately $28 billion. Consequently, although the increase of food production through the increase of cultivated land appears to be a sound plan, the money to initiate such projects in many countries makes use of the plan unrealistic. Furthermore, the soil found in tropical rain forests is not very productive. It contains only small quantities of materials needed for the growth of agricultural plants. This is primarily because the heavy amount of precipitation in these areas constantly washes away many nutrients from decaying vegetation. In addition, since the tropical forest does not have a season of dormancy such as found in temperate areas, the nutrients that remain are very quickly picked up by the constantly growing vegetation. Finally, when many tropical forest soils are exposed to sun and oxygen, as when the forests are cleared, chemical changes occur which convert the soils into *laterite,* a material so hard and dense that when cut and shaped it can be used as a building material. Thus, it is not only expensive to clear tropical forests for agriculture but also impractical in that the soils after a very short period of time will not support crops.

Reduction of food loss

According to the 1967 report of the President's panel, 55 tons of grain would be available to feed the hungry in the world if just one-half of the estimated loss of food grains was prevented. This is enough grain to provide a diet adequate in total Calories for 500 million additional people.

Grains of one kind or another provide approximately 70 percent of the food of Asian peoples. Yet surveys in India show that losses due to storage reduce the available amount of cereal and oil seeds by 10 percent, pulses (seeds of legumes, such as peas or beans) by 30 percent, and rice by 12 percent. The UN estimates the loss of stored cereal grains in six Latin American countries caused by insects to be as high as 35 percent. AID claims that the losses of stored grains in Brazil are as high as 15 to 20 percent, and pests claim approximately 33 percent of Africa's harvested cereal grains. The worldwide loss of all cereal grain, pulses, and oil seeds is 10 percent, according to the Food and Agriculture Organization.

In 1952–1954 a tremendous population of rats destroyed 90 percent of the rice, 20 to 80 percent of the maize, and about one-half of the sugar cane in two Philippine provinces. Some of the loss was caused by inadequate transportation to distribute the harvested materials. Lack of refrigeration accounts partially for the spoilage of 50 percent or more of the harvested fruits and vegetables and approximately 15 to 25 percent of all meat, poultry, and fish and dairy products in some U.D.C.s One Indian state loses 25 to 50 percent of vegetables and fruits, enough material to supply an additional 3 ounces of vegetables or fruits per day per person in that state.

Processing techniques also account for a great loss of food material before it reaches the consumer. One-third again as much rice could be produced in some Indian areas if currently used rice mills were replaced with more up-to-date machinery. In some places up to 50 percent of the sugar in sugar cane is lost because of the use of primitive cane crushers.

Despite these data, little money goes into research on loss of grain, vegetables, or fruits because of pests, storage, processing, or distribution. Money is put instead into programs to increase the yield per acre. In 1968, for example, India spent $265 million to import fertilizers, 800 times the amount spent on rat control. Controlling pests, of course, also has problems, particularly when chemicals are used. In a later chapter we shall consider the possible consequences of using chemical controls. Clearly, one of man's quickest routes to preventing starvation in the world may be to invest immediately in research talent and money to improve the storage and distribution of food.

Food from the sea

Since World War II, the fishing industry has grown greatly. Total average annual gain in aquatic yields since 1948 equals 9 percent. The production of no other food commodity has increased so suddenly.

263 Solutions to the world's food problem

Between 1956 and 1965 the annual average increase of products taken from the sea was 13 percent.

The apparent endless supply of sea products suggests that the oceans may be a source of food for the burgeoning billions around the world. Despite the increasing harvests from the sea, the total biological material reaped is very small. In fact, until now only $\frac{1}{10}$ of the total surface of the sea is fished, and nearly all of this is near the coasts of the Northern Hemisphere. Furthermore, only a few kinds of fish and mollusks are harvested regularly.

Recently, the fishing industry has expanded its exploitation of the ocean. The high seas are now more commonly fished, and coastal fishing industries have motorized their equipment and increased their efficiency. Much expansion is due to the construction of specialized boats, new synthetic materials, and new techniques such as refrigeration. Prior to World War II, fish were preserved by salting or drying. However, although salting is still commonly used, freezing and canning are increasingly popular. It is estimated that soon 50 percent of all fish caught will be either frozen or canned, rather than eaten fresh.

New lightweight, noncorrosive, knotless plastic nets and the echo sounder have greatly increased the efficiency of the fishing industry. The echo sounder provides an immediate impression of the bottom of the fished area and locates schools of fish. Still another device is the air-bubble curtain. Fish apparently consider the bubble walls impenetrable and, hence, can be herded and directed into nets. Ultrasonic waves can be used in this fashion, too, as has been shown in the anchovy fishing industry. Suction pumps are used to efficiently remove fish from nets into boats.

The countries most responsible for the introduction of new fishing techniques are the Soviet Union, Japan, China, and Peru. The Soviet Union was the first to use sonar equipment to locate fish and factory ships, boats that contain processing, freezing, storing, and fileting equipment. Consequently, fish caught by factory ships are ready for consumption when the product is delivered to shore. As a result, the USSR's fish production has been especially rapid in the last 10 years, increasing about 7 percent per year. Japan's fishing has expanded since 1965. Japan has formed joint fishing companies in several countries, for example, Indonesia, Ceylon, Ghana, Kenya, and most Mediterranean countries, and enjoys a worldwide industry. However, because of strong ocean currents that bring nutrients to the surface of its coastal waters, Peru is the major fishing country of the world. Its fishing industry has had an annual growth rate of more than 20 percent for more than 10 years. One of this country's biggest fish industries is the manufacture of highly proteinaceous fish meal. It is mainly exported to countries that can afford it. Approximately two-thirds of the entire production of fishmeal goes to western Europe, slightly less than one-third to the United States, and a small amount to Japan. In all cases it is mainly used as a food for animals.

No doubt the increased tapping of the ocean's resources would provide additional protein and possibly would be a source of help for the hungry people around the world. It has been pointed out, for example, that if the yearly production of fish meal alone were 3 times that of 1964 production, that is, brought up to 9 million tons instead of 3, and if this material could be used for food for human beings, it would satisfy the protein needs of 300 million people each year. Immediate and obvious difficulties, however, include the acceptance of the material in the countries that need it and the maintenance of an ample stock supply of fish in the oceans.

At least one scientist, J. H. Ryther of the Woods Hole Oceanographic Institute, believes that the maximum yield of utilizable fish from the ocean may well be near 100 million metric tons. Others believe that possibly 150 million metric tons could be harvested. Sixty million tons were removed from the sea in 1967, and by 1980 it is estimated that 70 million tons will be harvested. The number of years man can exploit the sea will soon come to an end at this rate, even if precautions are taken to ensure proper conditions for the existence of the fish now in the sea. Pollution of the ocean and overfishing may be serious problems, as seen only too well in the sad story of the destruction of the whale. (See the insert, "The Death of the Whaling Industry.") The depletion of the fishing industry in some countries even now would be disastrous. Japan, for example, secures 1.5 times more protein from her fishing industry than from her agriculture.

Fish farming may help increase the production of fish from the sea. One of man's most significant steps in his development thousands of years ago was the domestication and confinement of animals and the consequent transition from a hunting to an agricultural culture. Perhaps a similar development hinges on the cultivation of the sea and freshwater organisms. Carp, milkfish, tilapias, as well as certain oysters and clams, have been cultivated for some time, and other species may lend themselves to farming, or mariculture. However, this enterprise on a large scale demands a great deal more knowledge about the marine environment, because dependable culture demands far more control of pests, fertilizing, and temperature than is now possible.

The death of the whaling industry

The reckless and foolish hunting of whales by the whaling industry illustrates how easily and suddenly man can destroy species of living organisms. It also suggests what might be the fate of marine organisms if the sea were looked to as a main source of food for the world.

The whale has been hunted for many years because of its nutritious meat and blubber and excellent oil, valuable as a fuel for lamps and as a lubricant. In the eighteenth and nineteenth centuries, wind-driven whaling ships hunted five whale species, exploiting four almost to extinction. Today, with more efficient hunting and shipping techniques, eight species of whales are commercially hunted, again almost to extinction.

The catch of whales between 1933 and 1966 and the number of barrels of whale oil produced indicate a serious decline of the whale species in a brief time. In 1966, although almost twice as many whales were harvested as in 1933, the whale oil yield was only about 60 percent of that produced in 1933. The data testify to the decreasing catch of the large whale species, such as the blue whale, because of their decimation and the increased catch of whales of smaller species and young large whales.

In December 1946, 17 nations concerned with the decline of the whaling industry and conservation of whales formed the International Whaling Commission. The representatives from the nations on the commission agreed to be responsible for the compliance of their countries with the decisions of the group. Unfortunately, inspection or enforcement of decisions proved difficult.

A report by a committee appointed by the commission in 1963 predicted unhappily that the whaling industry must decrease its annual whale harvest and honor a quota or that many whale species would be in jeopardy. Apparently, in all cases the whaling industry ignored the advice of the commission and continued the unregulated harvest of the whales.

Since the commission's report in 1963 it has several times suggested harvest quotas, but the commission's advice has usually gone unheeded. Consequently, the depletion of the whale resources in the seas today has forced several countries out of the whaling business. Only two countries, Russia and Japan, are still heavily involved in the industry.

Today, advances in technology greatly increase the efficiency of the few remaining whalers and further threaten the whale's existence. Helicopters from ships detect the whales and then quickly direct boats to the area. Boats are sometimes equipped with sonar to better locate the prey. Strong nylon ropes are attached to the harpoons which sometimes bear explosives. After the kill, mechanical pumps inflate the whales with compressed air so as to easily float them ashore.

The exploitation of the whale over the last centuries, most notably over the last few decades, is a result of human greed and foolishness. Regrettably, the whale story is merely one of several that illustrate the exploitation of marine species. Some authorities claim that increasing world population will be fed adequately if the sea is more fully exploited. This solution may be "pie-in-the-sky" if the limits of the oceans resources are unrecognized and management procedures not instituted.

Food from algae and yeast

Algae and fungi, the group to which yeasts belong, are included in the plant group the Thallophytes, because they do not possess leaves or stems or roots. Although similar in many significant ways, algae differ from fungi in that algae are chlorophyllous and photosynthetic. The size and habitats of the Thallophytes are broad and varied. Some forms are unicellular; others are multicellular and large. Some of the brown algae reach hundreds of feet in length. Algae and fungi have been used for food for generations. The soldiers who accompanied Cortez to Mexico witnessed natives selling cheese-flavored bread made of algae. People along the coast of Scotland and on other North Atlantic islands and many Oriental peoples use algae today for food as they have for many years in the past. Today, the Japanese harvest 5,000 metric tons dry weight of red algae annually from the sea.

The high nutritional value of algae and fungi coupled with new technology suggests to many that they may be a prime source of food in the future. Two algae, *Chlorella* and *Scenedesmus,* have been researched (Fig. 9–11). The misconception that algae such as Chlorella were more efficient photosynthesizers than land plants probably originally accounts for their investigation. Additional impetus for the study of algae comes from space programs. Because algae consume carbon dioxide and produce oxygen and abundant protein, researchers supposed that Chlorella could help meet an astronaut's biological needs in a closed spaceship. Although much more research is needed before Chlorella can play a significant role in space exploration, considerable research has been done on the nutritive value of yeast and algae, especially Chlorella. Scientists have found that the composition of Chlorella can be modified according to the medium in which it grows. With an adequate source of nitrogen, Chlorella and other species of algae develop a 40 to 60 percent composition of protein. Although the two amino acids cystosine and methionine are not plentiful, all the essential amino acids do occur. Treonine and lysine are most abundant.

Nutrition studies with rats indicate that Chlorella can provide adequate nutrition for satisfactory weight gain. Chlorella and Scenedesmus have been fed to human beings in various experiments. Although food in which the algae is used liberally becomes greenish and frequently has a disagreeable taste, studies have shown that human beings can use algae as the main source of protein without ill effects.

The high percentage of protein and lipids (23 to 76 percent) in the algae when grown in appropriate media suggest that these plants may well be used extensively as food in the future. However, to grow Chlorella in quantities in the laboratory requires a concentrated source of carbon dioxide, minerals, nitrogen, light of the proper intensity, the proper temperature, and continuous movement of the medium for equal distribution of the nutrients. The cells are harvested with centrifugation, after which the algae are air-dried or freeze-dehydrated, and then ground into a powder. Experimental studies suggest that an acre of cultured algae could produce 40 tons annually of dried algae, which is about 20 tons of protein. An acre of soybeans in contrast produces only 0.75 ton of dried food and 0.25 ton of protein. Pilot projects have been attempted to determine the feasibility of culturing algae artificially. Unfortunately, studies show that the cost is high. One pound of decolorized dried algae costs about $1.12.

Fungi, especially mushrooms and yeast, have been used in the diet for centuries. Yeast is used in baking and in the production of alcohol around the world. However, until Germany supplemented soups with powdered yeast during World War I, yeast was probably not used directly as a food. In the future, because of inexpensive production and proved high nutritional value, yeast may be used more extensively as a dietary staple.

Production is not expensive because yeast can usually be grown on materials commonly considered waste, such as chips of wood or molasses. Yeast can also be grown on oil. In this sense it may prove to be a significant antipollution organism, because it removes the long-chain carbons, the *paraffins,* which make oil viscous, especially in cold climates. One ton of yeast protein can be produced from 60 tons of crude oil that contains paraffin. Based on the figures for oil production in 1966, if all the oil-refining companies cooperated, 20 million tons of pure protein could be produced which would provide about 25 grams of pure protein per person per day if distributed among underfed people of the world. Of course, the world reserves of oil are being depleted rapidly, so any long-range production of protein in this manner may be unrealistic (see Chapter 13).

Breeding new varieties

Of the 20 or so amino acids out of which proteins are made, 6 are supplied in protein from animal sources. Plant proteins, on the other

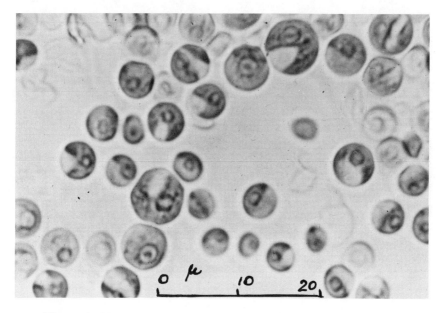

Figure 9–11
The one-celled *Chlorella* holds promise as a food source. (Courtesy of M. Calvin, University of California, Berkeley, Calif.)

hand, are seriously deficient in certain needed amino acids. Corn, for example, is deficient in lysine and tryptophane. Fortifying plant products with amino acids essential in the human diet is one solution to this deficiency problem. Another is to breed new species of grain plants which will genetically produce the necessary amino acids.

Today, as a result of prolonged research, a high-lysine corn has been produced. Edwin Mertz and his colleagues at Purdue announced in 1963 the discovery of the gene opaque-22, a high-lysine producer. Today, descendants of this variety are being grown in those countries where corn is a dietary staple, such as sub-Saharan Africa and Latin America. The effect of the corn on human beings is yet to be documented, but the effect on animals is already noted. Hogs in Colombia gained weight twice as fast as those fed unimproved corn.

Besides increasing the amino acid content of some plants, crop breeders have developed plant varieties with other new characteristics (Fig. 9–12). Out of these experiments may come the greatest hope for feeding the world's billions. In many developed countries, notably the United States, *and particularly* in the agricultural land-grant colleges, researchers have been working for decades to create new varieties of plants and animals. Consequently, because improved varieties have been around a long time, the people of this country have enjoyed better diets than most of the world's population.

Figure 9–12
A new strain of rice (right) grown next to an old strain. Note short stalks, fine leaves, and heavy yields of new strain. (Courtesy of International Rice Research Institute, Philippines.)

In most U.D.C.s agriculture research stations have a briefer history. Consequently, until recently few improved crop and farm animals were developed that were accommodated to their geography and climate.

For example, in the 1940s Mexico's average wheat yield was 11.5 bushels per acre (Fig. 9–13). Corn yields were one-fifth of the average produced in the U.S. corn belt at that time, a meager 8 bushels per acre. Only recently, since 1940, has Mexico attempted to develop new varieties of grain. But despite its brief history, Mexico's grain-breeding program is a success story. When Mexican officials realized the seriousness of the problem that faced their country in 1940, they enlisted the aid of the Rockefeller Institute to begin a research program aimed at improving Mexico's crop yields. Initial efforts were concerned with the collection of diverse kinds of corn and wheat varieties from around the world. These plants were tested in plots, and from them elite varieties were selected. Extensive studies identified the specific problems of disease and growth facing the farmer: potent races of the black stem rust; frequent droughts; impoverished soils; low grain yields; luxuriant foliage, but no increase in grain as a result of fertilization; "lodging," or falling over. Gathering information on corn was complicated because of an incredible abundance of varieties to examine.

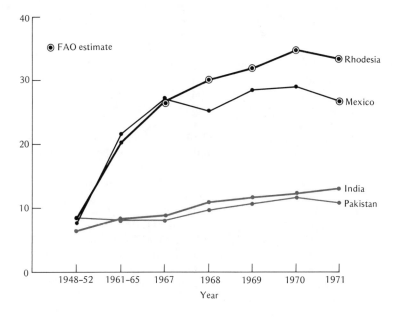

Figure 9–13
Wheat yields in Rhodesia, Mexico, India, and Pakistan. (From Food and Agricultural Organization of the United Nations, *Production Yearbook*, vol. 25, Rome, 1971.)

Careful selection, cross breeding, and testing eventually yielded improved disease- and pest-resistant varieties. As early as 1955, widespread cultivation of these crops met Mexico's grain needs. By 1968 national average wheat yields were more than 40 bushels per acre, 4 times the average recorded in 1943. Corn production had doubled in 25 years. Yields of other crops also involved in the improved-breeding program increased. Potato yields, for example, were 3 times that in 1950. Today, an all-Mexican National Institution for Agricultural Research continues the highly successful work begun by the Mexican Government and the Rockefeller Institute.

Encouraged by the success of Mexico's enterprise, Colombia in 1950, Chile in 1955, and India in 1956 initiated programs with the help of the Rockefeller Institute. These programs are still young, but the results of their research have already had an impact on agricultural productivity around the world.

Now, international centers of research continue research on food crops. These centers provide teams of experts to investigate problems and conditions that involve better food production in individual countries. Continuity of research through years, depth of research, rapid exchange of information and materials, education, and training are all benefits of the international organization not necessarily available in the home country. The development of international centers is a major step in cultural evolution and, hopefully, provides models for the solution of major world difficulties other than hunger.

One of the most productive international centers is the International Rice Research Institute (IRRI) at Los Baños, Philippines, jointly initiated by the Rockefeller and Ford Foundations in 1961. Here all aspects of the rice plant are studied. Varieties of rice from all over the world have been collected. Conditions that prevent abundant rice yields are studied. Recently, a new improved plant rice variety IR8 was developed as a result of a 4-year study. The new rice variety has a short stiff stem, few erect leaves, abundant and nutritious seed production, resistance to rice blast disease, and an insensitivity to the photoperiod, a set amount of light needed before flowering, which allows rice planting at any time of the year and in more areas of the world. Most important, it produces 2 to 4 times the yield of earlier varieties when grown properly (Fig. 9–14).

Today IR8 is grown in Thailand, India, Indonesia, Malaysia, Burma, Sri Lanka, and Pakistan, as well as in the Philippines. The value of the increased rice yields is said to be more than $300 million per year.

Another international institute is the International Corn and Wheat Improvement Center in Chapingo, Mexico. The research work at this institute mainly accounts for recent abundant wheat yields in India, Pakistan, Turkey, and Afghanistan. Semidwarf wheat varieties have been perfected which are short, stiff-stemmed, produce abundant grain, and resist many environmental threats. Two other international institutes are supported by the Rockefeller and Ford funds. These are the International Center for Tropical Agriculture in Cali, Colombia, and the International Institute for Tropical Agriculture at Ibadan, Nigeria.

Many scientists herald the breeding of new crops as the panacea to the world's hungry millions. The remarkable yields in many underdeveloped countries in the 1968 and 1969 growing season are cited as

Solutions to the world's food problem

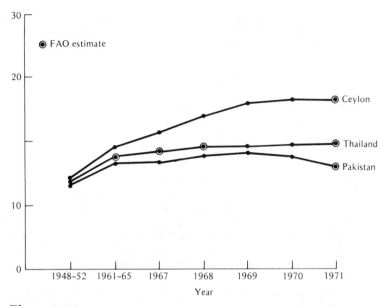

Figure 9-14
Rice yields in Sri Lanka (Ceylon), Thailand, and Pakistan. (From Food and Agricultural Organization of the United Nations, *Production Yearbook,* vol. 25, Rome, 1971.)

evidence. The increase in crop yields are impressive, especially in southern and eastern Asia. An 8.6 percent overall increase in food production in eastern Asia in 1969 over 1968 is recorded. The amount of food production increase per person in 1969 over 1968 was 5.6 percent, more than double the per capita increase of 1968 over 1967. Although eastern Asia records an overall 8.6 percent increase of food production, the increasing population size produces a lower per capita gain than the overall food production percentage increase.

Southern Asia, including India, Pakistan, and Sri Lanka, recorded an overall increase of 4.5 percent in 1969 over 1968 and an 8 percent increase in 1968 over 1967. The per capita increase was only 2 percent, for 1969 over 1968, however, less than the 6 percent increase in 1968 over 1967. Figure 9-13 illustrates the impressive yield per acre of wheat in Mexico, Pakistan, and India since 1950, and Fig. 9-14, the production of rice in Pakistan, Sri Lanka, and Thailand. Of course, the higher yields of grain crops per acre must be weighed against the rising birth rate. The increase of food per person is the significant information in the long run.

Nevertheless, the new varieties of wheat and rice are producing a new abundance of grain. Contributing factors, beside their new genetic makeup, include abundant fertilizer and plentiful water, usually through irrigation. How successfully countries can continue to supply these necessary ingredients will to a great extent determine the continuance of high grain productivity.

Some biologists are concerned that the new varieties have not had adequate time to prove themselves. The Ehrlichs are concerned that although the new wheat and rice are growing well in the countries in which they have been introduced, countries like Pakistan, where the climate is highly agreeable, not enough time has elapsed to prove that the seeds will thrive in less hospitable regions. They also maintain that the new crops are introduced into countries without adequate field testing of their resistance to insects and diseases and that the large monocultures being established around the world are ripe for devastation. Of course, increased use of pesticides would curb pests but would also increasingly pollute the environment. Another problem yet to be faced is the acceptability of the new grains. The IR8 variety of rice is reported to be less tasty than older varieties and less adaptable to milling. The poor taste and texture of Incaparina is known to have prevented its ready acceptance in Central America. Although the new varieties may help many U.D.C.s, some U.D.C.s, already too overcrowded, may still be doomed. Probably, any U.D.C. that survives will rely not only on new seeds but on other solutions enlisted concurrently. In addition, fewer children must be born.

The major problem in the U.D.C.s is their inability to feed the growing populations. The D.C.s have problems, too, but they do not usually include malnutrition and starvation. Instead, the high level of technological development, although it helps fill hungry mouths, takes its toll — pollution of all kinds. In the next chapter we shall look at air pollution, one of the most pressing pollution problems in many D.C.s.

Summary

In this chapter we have looked at one of the most serious problems of mankind today, starvation. We reviewed the food intake necessary to be fed adequately as suggested by the U.S. Department of Agriculture. We then looked at the condition in some areas of the country which already do not have adequate food. Described in detail were some of the major nutritional diseases in the world, such as goiter, nyctalopia, xeropthalmia, rickets, beri-beri, kwashiorkor, and marasmus, and the effects of malnutrition on the development of the nervous system and behavior of human beings. Finally, we reviewed some proposed solutions to the world's food problem, including food fortification, new foods such as Incaparina, the application of fertilizer to infertile regions, irrigation of arid lands, the cultivation of uncultivated areas, reduction of food loss, harvesting foods from the seas, utilizing high-nutrient algae and yeast, and perhaps most promising, the development of new varieties of grains. Possibly, the solution of man's problem of feeding his billions will come from the concurrent development of many of these potential solutions, not just one. However man's problem is solved today, one fact is clear: the history of man's development on earth was one of developing an increasingly life-sustaining relationship with his environment. As a consequence, his population grew, and today an index of his success as an organism may be the degree of difficulty in feeding his billions. In order for man to continue to live well in his global environment, or perhaps to survive at all, he must find ways to adequately feed his numbers.

Supplementary readings

Bardach, J. E., J. H. Ryther, and W. O. McLarney. 1972. *Aquaculture: The Farming and Husbandry of Freshwater and Marine Organisms.* John Wiley & Sons, Inc. (Interscience Division), New York. 868 pp.

Behar, M. 1970. "The Protein Problem." *Technology Review,* vol. 72, no. 4, pp. 25–29.

Brady, N. E., ed. 1967. *Agriculture and the Quality of Our Environment.* American Association for the Advancement of Science, Washington. 460 pp.

Brown, L. R. 1970. *Seeds of Change: The Green Revolution and Development in the 1970s.* Praeger Publishers, Inc., New York. 205 pp.

Cochrane, W. W. 1969. *The World Food Problem: A Guardedly Optimistic View.* Thomas Y. Crowell Company, New York. 331 pp.

Dumont, R., and B. Rosier. 1969. *The Hungry Future.* Praeger Publishers, Inc., New York. 271 pp.

Ehrlich, P. R., and A. H. Ehrlich. 1972. *Population, Resources, Environment,* 2nd ed. W. H. Freeman and Company, San Francisco. 509 pp.

Gullion, E. A., ed. 1968. *Use of the Seas.* Prentice-Hall, Inc., Englewood Cliffs, N.J. 202 pp.

Hardin, C. M., ed. 1969. *Overcoming World Hunger.* Prentice-Hall, Inc., Englewood Cliffs, N.J. 177 pp.

Idyll, C. P. 1970. *The Sea against Hunger.* Thomas Y. Crowell Company, New York. 221 pp.

Leisner, R. S., and E. J. Kormondy. 1971. *Pollution.* William C. Brown, Company, Publishers, Dubuque, Iowa. 85 pp.

Paddock, W. 1970. "How Green is the Green Roveloution?" *BioScience,* vol. 20, no. 16, pp. 897–902.

Paddock, W., and P. Paddock. 1967. *Famine, 1975! America's Decision: Who Will Survive?* Little, Brown and Company, Boston. 276 pp.

Pirie, N. W. 1969. *Food Resources: Conventional and Novel.* Penguin Books, Inc., Baltimore. 208 pp.

Reish, D. J. 1969. *Biology of the Oceans.* Dickenson Publishing Co., Inc., Belmont, Calif. 236 pp.

Chapter 10

This most excellent canopy, the air ... why, it appears no other thing to me but a foul and pestilent congregation of vapours!

Hamlet
Shakespeare

Air pollution

"Notice the air pollution drifting out there in case anybody thinks we don't have it." Pete Conrad, command pilot of Gemini 11, the space vehicle launched in September 1966, was discussing a photograph of the city of Houston taken from outer space. The photograph was taken during one of the rare moments when the atmosphere cleared over the city. Usually, it was covered with a thick brown smog.

In recent years smog, a mixture of fog and smoke, has produced other effects besides frustrating photographers; more often it has killed. For example, consider one of the first, and mildest, smog disasters in the United States. By midmorning of October 26, 1948, a very heavy fog settled over the industrial town of Donora, Pennsylvania. Business went on as usual, and the smoke stacks of industry continued to exhaust wastes into the atmosphere. Within 48 hours, visibility was so reduced that many inhabitants of the town reported difficulty finding their way home. As the smog cover thickened, the filth began to settle, until in some areas of the town cars cut distinct tire tracks as if moving over a thin, black snow. Soon many inhabitants of Donora complained of nausea, sore throats, breathing difficulty, coughing, running noses, and burning eyes. Pets were afflicted with similar symptoms, and many died. Of the 14,000 people in Donora, nearly 6,000 became ill in the next 4 days. Twenty people died.

Smog struck harder 4 years later in London. On December 5, 1952, a thick fog mixed with smoke from industry's stacks layered over the city for 3 days. London hospitals were soon receiving victims suffering from serious respiratory problems. The smog stayed 5 days and produced 4,000 deaths above normal. Within the next 8 weeks, another 8,000 people died. (See the insert, "Why Smog Kills.")

Smog has also taken its toll at other times. In 1956 London suffered another serious smog crisis, and in 1962 still another siege killed 400 people. Other towns around the world have experienced similar problems. For example, in 1953, 2,000 excess deaths were reported to be due to smoggy air that smothered New York City. (See the insert, "What Caused the Smog in London?")

Man's development of a technology by harnessing energy has seriously endangered his relationship with the evironment as exemplified by the effects of smog. Man, like any organism, can continue to exist only when the relationship between himself and his surroundings is life-giving; thus, the seriousness of air pollution. With this in mind, we shall examine the causes, kinds, and solutions to air pollution. Hopefully, from the discussion will come an appreciation of the significance of the problem to man's survival and insight into the importance of effecting solutions.

Why smog kills

The incidence of respiratory and cardiac illnesses increases when severe smog settles over an area. Yet to date no indisputable evidence exists that lung problems such as bronchitis or emphysema or heart disease are directly caused by air pollution. Some evidence does exist, however, that these conditions may well be aggravated by impure air. In the London disaster and the Donora disaster death came almost exclusively to those who had previous problems with bronchial pulmonary disease. Bronchitis patients in clinics in London felt a responsibility for announcing when a smog episode was due, for they would often undergo breathing difficulties up to 12 hours before the smog descended.

It is possible that hereditary disabilities and aging encourage lung and heart problems. Whatever the reasons may be, the incidence of respiration and cardiac difficulties increases every year. In the United States, *emphysema*, a lung disease to be described later, is the most rapidly growing cause of death. The incidence of this disease in males increased from 1.5/100,000 to 8/100,000 in the 10 years between 1950 and 1959. In 1962 more than 12,000 persons died of the disease. A direct relationship may exist between the increase in air pollution and the increase in lung disease. London postmen who work in the sections of the city where air pollution is high have twice the rate of incidence of bronchitis as those who work in less-polluted areas of the city. Other data from Great Britain indicate that the degree of air pollution directly influences the rate of death due to chronic bronchitis.

Another pulmonary ailment, asthma, also seems to be well-correlated with air pollution. In 1946 some American soldiers and families living in Yokohama, Japan, developed asthmatic problems. The disease resisted the treatment normally given for asthma, but a patient's symptoms were promptly relieved when he was moved from the area or flew in a plane above the polluted atmosphere. Later the disease was called "Tokyo–Yokohama asthma," because it was also common in Tokyo. Extensive studies show that the level of air pollution in the area between Yokohama and Tokyo, the heavily industrialized Kanto Plain, correlates well with the incidence of Tokyo–Yokohama asthma and the severity of the symptoms.

Minor symptoms of pulmonary distress include runny nose, a cough, and breathlessness. Serious pulmonary problems often are due to a narrowing of the bronchial tubes through which air passes on its way to the deepest parts of the lungs. This narrowing may be spasmodic or permanent. In any case, the passage of air into the lung is impeded, and the patient breathes more rapidly. Excessive secretion of mucus can accompany bronchial constriction and further decrease the amount of air passing through the tubes.

Another serious lung problem is emphysema, a disease due to the structural breakdown of the tiny sacs in the lungs, the *alveoli*, where air transfer between lung and bloodstream actually occurs. The many sacs of membranous bubbles that compose the alveolus provide a large surface area through which the gas materials diffuse. It is said that if the surface area of all the alveoli in a human being were spread out flat they would cover a tennis court. Any structural breakdown in the numerous sacs destroys the surface area available for oxygen and carbon dioxide transfer. Consequently, an emphysema victim, who has very few alveolar sacs at the ends of the bronchial tubes, suffers severe breathlessness, which frequently leads to death.

Although these pulmonary conditions are not necessarily caused by air pollution, they are severely aggravated by air pollutants. The difficulty bringing about adequate gas exchange in the lungs often causes the heart to pump more rapidly and more strenuously in order to move more blood past the alveolar sacs of the lungs. This additional cardiac stress explains the increase in rate of circulatory disease during a severe smog episode.

The sources of air pollution

Most forms of pollution occur in urban and industrialized areas. In fact, air pollution probably was first regarded as a serious problem at the advent of the Industrial Revolution, when fossil fuels were used abundantly in factories. Today, as the population grows and metropolitan and industrial areas expand, air pollution becomes increasingly widespread.

Different kinds of air pollution occur in different urban areas. Most kinds of air pollution are the results of combustion, but in London and Donora the main pollutants are particulate matter and sulfur. The exhaust from smoking chimneys which results from the combustion of high amounts of sulfur-containing fuels produces a great variety and concentration of oxides of sulfur, sulfuric acid and sulfur dioxide probably being the most common. In Los Angeles the main pollutants are hydrocarbons and carbon from combusted gas emitted in the exhausts from motor vehicles.

In fact, the prime source of air pollution in most U.S. cities is the automobile (Table 10–1). In 1966 approximately 61 percent (or 86 million tons) of the total U.S. air pollution was reported by the Department of Health, Education, and Welfare to have come from motor vehicles, whereas industry contributed only 17 percent. Consequently, the motor vehicle has become a mixed blessing in our increasingly urbanized society.

Motor vehicles contribute abundant pollutants. The main pollutant is carbon monoxide, about 66 million tons each year. Next are hydrocarbons, about 12 million tons per year. Consequently, the smog overlaying areas that have a very large motor vehicle population is mainly carbon monoxide, hydrocarbons, and some nitrogen oxides.

What caused the smog in London?

What caused the smoke and fog to settle and cause the crisis in Donora, Pennsylvania, and London? The conditions that caused these catastrophes are responsible for serious smog problems found in other places of the world. Sometimes, as with Donora and London, the dense smog settles only temporarily and then moves out after wreaking its damage. The smog layer in Los Angeles, however, is usually more permanent (Fig. 10–1). Yet the causes of these smog

Table 10-1

Estimated emissions of air pollutants by weight, nationwide, 1969 (in millions of tons per year)

Source	CO	Particulates	SO_x	HC	NO_x	Total	Percent Change 1968–1969[a]
Transportation	111.5	0.8	1.1	19.8	11.2	144.4	−1.0
Fuel combustion in stationary sources	1.8	7.2	24.4	.9	10.0	44.3	+2.5
Industrial processes	12.0	14.4	7.5	5.5	.2	39.6	+7.3
Solid waste disposal	7.9	1.4	.2	2.0	.4	11.9	−1.0
Miscellaneous	18.2	11.4	.2	9.2	2.0	41.0	+18.5
Total	151.4	35.2	33.4	37.4	23.8	281.2	+3.2
Percent change	+1.3	+10.7	+5.7	+1.1	+4.8		

[a] Computed by the 1969 method from the difference between 1969 estimates and 1968 estimates. The new method results in higher values for 1968 than those computed by EPA for 1968.

Source: P. Nobile and J. Deedy, *The Complete Ecology Fact Book,* from data gathered by the Environmental Protection Agency, as cited by the Mitre Corp., MTR-6013.

conditions are the same, an abundance of pollutant compounds spewed into the atmosphere from industry or cars, plus unique weather conditions.

Usually, despite high population density and an abundance of atmospheric pollutants, moving air overlaying a city prevents smog from settling. It dilutes air pollutants and carries them off. Sometimes, however, the air is meteorologically locked in place, a condition known as a *thermal inversion.*

Usually, warmer air occurs near the earth's surface and tends to rise, carrying away the pollutants produced near the earth's surface. Sometimes, however, as occurs in coastal areas, an influx of cold sea air will wedge itself between the earth's surface and the layer of warmer air. Then the total air mass remains motionless because the cold air does not rise. Consequently, the air pollutants in the cold air are held to the ground. In some areas geographical barriers help trap the air. For example, the border of mountains to the east of California imposes this meteorological condition on Los Angeles over 100 days each year. So far, no effective solution has been found to ease the effects of thermal inversion (Fig. 10–2).

Figure 10–1
Top: Los Angeles on a smogless day. Bottom: Los Angeles covered with smog. (Courtesy of Los Angeles Air Pollution Control District.)

The cost of air pollution

Air pollution is expensive. It injures crops, destroys metals, erodes rubber, damages stone, discolors paint, and in general, soils buildings and clothes. Anyone who lives in a polluted area knows well the amount of money that must be spent in cleaning, renovating, and repairing materials

Only the adult males and females live in the host's circulatory system. The ¹/₂- to ³/₄-inch-long females produce fertilized eggs which often clog the veins. Depending upon the species, the eggs migrate from the small blood vessels out into the bladder or into the large or small intestine and from there are passed out of the body. During their migration from the blood system the spined eggs cause great irritation, resulting in pain, bleeding, and inflamation of the migration area.

After the eggs are released from the human host they hatch into a small ciliated, free-swimming *miricidia*. Miricidia live only a few hours unless they can burrow into an aquatic snail to continue the life cycle. Within the snail the miricidia develop into nonmotile sporocysts which eventually divide asexually to produce hundreds of daughter "packages" called *rediae*.

In several weeks the nonmotile rediae produce microscopic, elongate, ovoid cells, the *cercariae*, which leave the snail. Each cercaria bears a forked tail. (The divided tail, or "split body," was the inspiration for the word "schistosomiasis.") Cercariae swim vigorously and usually within a day or so penetrate the skin of man or other host by dissolving the host tissue and energetically moving through the softened areas. Soon after the cercariae have penetrated the skin, they migrate into the host's tiny lymphatic vessels and circulatory system. Cercariae in contaminated drinking water also enter the body through the mouth. After a few weeks the cercariae develop into mature males and females, and the cycle begins again (Fig. 9-9).

The migrating fluke eggs from the small blood vessels cause most symtoms of schistosomiasis. Penetrating cercariae may produce tiny lesions, a rash, or itching. Maturing cercariae soon after entering the body may cause enlargement and a tenderness of the liver because they mature in this organ. Migrating adult worms within the blood vessels usually cause little damage, but if the lungs are invaded, pneumonia may result. A cough, fever, diarrhea, enlargement of the spleen and liver, a general wasting away of the body, and a filling of the abdomen with a fluid to produce a characteristic pot-belly are other symptoms (Fig. 9-10). Mildly infected victims usually recover, but persons who are severly undernourished or severely infected may die from the disease.

Increasing cultivated areas

According to the President's Science Advisory Committee Panel on the World's Food Problem in 1967, the amount of potentially cultivatable land is much larger than previously thought, equaling more than twice the land area that has been cultivated in the last few decades and totaling about 24 percent of the total ice-free area of the world. This is more than 3 times the size of the land area on which crops are grown and harvested

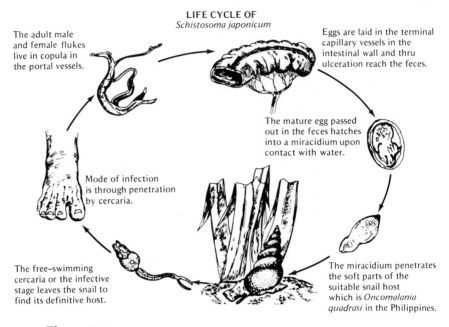

Figure 9-9
Schistosoma japonicum requires two hosts for completion of its life cycle. (Courtesy of *Santo Thomas Journal of Medicine,* Manila, Philippines.)

within a given year. More than one-half of this land lies in the tropics. Large areas of potential farmable land also lie in the temperate areas of Australia and North America.

The phrase "potentially arable land" needs definition, however, if its increase may be a factor in increasing the world's food supply. The President's panel says that it "includes soils considered to be cultivatable and acceptably productive of food crops adapted to the environment." It also says: "Some soils will need irrigation, drainage, stone removal, clearing of trees, or other measures, the cost of which would not be excessive in relation to anticipated returns." However, as Lester R. Brown of the U.S. Department of Agriculture points out: "How much land can be brought into cultivation is not a relevant question. The question becomes relevant only when we ask at what cost. Someone must pick up the tab."

Asia offers a good example. The President's panel makes clear that to increase the amount of food produced in that continent it is necessary to increase the number of crops per growing season on each acre, because there is essentially no cultivatable land in addition to that which is already used. Increasing crop production can be accomplished with irrigation. But the report also points out that, although 200 million acres could be irrigated and perhaps become more productive, the total

281 The cost of of air pollution

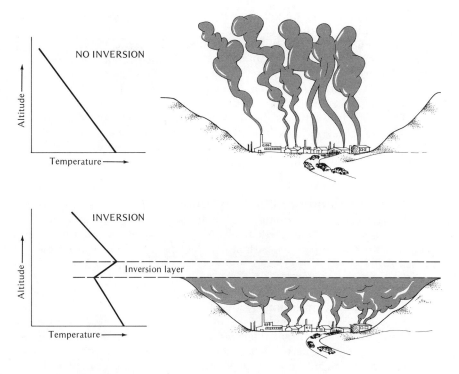

Figure 10-2
A thermal inversion in which cold air is trapped close to the earth's surface; the normal situation is shown for contrast. (From *Populations, Resources, Environment: Issues in Human Ecology*, 2nd ed., by Paul R. Ehrlich and Anne H. Ehrlich. W. H. Freeman and Company, San Francisco. Copyright © 1972.)

that have been damaged or dirtied by atmospheric pollutants. Studies show that metals such as steel, iron, copper, brass, nickel, zinc, lead, and tin corrode faster in urban industrial areas than in other areas. Aluminum and copper in a polluted atmosphere have been shown to erode 5 times faster than in cleaner air. Iron corrodes 6 times faster, brass 8 times, zinc 15 times, nickel 25 times, and amazingly enough, steel 30 times faster. As the amount of pollution goes up, the rate and cost of corrosion also rises. In New York City alone, air pollution corrodes materials at a cost of about $6 million per year. It has been estimated that approximately $2.5 million was spent each year in St. Louis for renovation of building surfaces before a campaign against air pollution was begun. As early as 1950 the director of the New York City smoke control agency said that if a program were begun to control air pollution in the city, New York City residents could reduce their expenses each year about $50 million.

Do not think that the expense comes only from the corroding effects of pollutants. The odors of air pollution in communities also reduces property value. St. Louis reported that prior to 1940, the year when this city began its anti–air pollution campaign, the values of property had been declining at an annual rate of about $25 million.

Air pollution damages crop production, too. Most of the $1.5 million loss in the United States in 1950 and 1951 attributed to air pollution was due to damage to crops. This cost averages approximately $10 per person per year. In California alone, the damage to farm crop production is said to total more than $6 million per year.

There are less obvious detrimental effects of air pollution as well. Consider eye irritation, lack of visibility and its affect on transportation, the increased need of artificial lighting on smog-darkened days, the psychological effect of the dirt and the grime, and the damage to health.

The kinds and effects of air pollutants

Many materials contribute to air pollution. Some of the more prevalent and serious include particulates, sulfur compounds, fluoride compounds, nitrogen compounds, carbon compounds, photochemical products, and lead.

Particulate air pollution

Particulate air pollutants are any materials suspended in the air which are larger than tiny molecules but smaller than 500 microns in diameter. These tiny particles are significant, because frequently they will stay suspended in the atmosphere for several months. Droplets from condensation or spraying; ash from combustion; dust from grinding, erosion, and pulverization, often with industrial machines; and salt from evaporated sea water are major particulate pollutants. No matter the size or the source, particulate air pollutants cause many problems.

One of the most significant difficulties is reduced visibility. Owing to the scattering and absorbing of light, particulate pollution reduces the amount of energy from the sun reaching the earth. Thus, the amount of sunlight that reaches the earth in urban areas in middle and high latitudes is reduced by as much as one-third in the warm months and two-thirds in the winter. The accumulation of particulate matter may have a significant effect on the earth's mean temperature. (See "Carbon Compounds," p. 285.)

Little research has been done on the effects of particulate pollutants on plants, although there is some evidence that dust from cement factories can damage trees because the stomata, pores through which gas is exchanged, become stopped.

Particulate pollutants also affect materials significantly. Besides soiling, they corrode metals, especially those used in electrical equipment, stone buildings, and statues.

Damage to the human being results from particles settling on the surfaces of the respiratory tract. The particles themselves cause damage to the cells lining this area or, if accompanied by gases, cause serious alterations in the activity of cells in other parts of the body.

More research is needed on the effects of particulates. Additional data may show that the levels of exposure considered safe must be altered in view of long-range effects.

Sulfur compounds

Compounds of sulfur, especially oxides, are among the most serious pollutants. Sulfur naturally and easily combines with oxygen to produce SO_2, sulfur dioxide. This is a colorless, pungent, heavy gas that can have seriously irritating effects on the respiratory system of man. Under usual atmospheric conditions, SO_2 can be transformed into other deleterious compounds, such as sulfurous acid (H_2SO_3) or sulfur trioxide (SO_3). *Sulfurous acid* is a colorless, mildly corrosive liquid which industry uses in bleaches. Both sulfurous acid and sulfur trioxide are eventually converted into the more irritating and corrosive sulfuric acid (H_2SO_4). *Sulfuric acid* is a strong acid and has serious corrosive effects on stone and metal as well as on living tissue.

Sulfur dioxide (SO_2) is considered to be one of the most serious pollutants of the atmosphere, in part because of its severe effect on living plants (Fig. 10–3). The gas is easily absorbed through the leaf stomata, the tiny holes on the leaf's surface through which the atmospheric gases move into and out of the inner leaf. Visible symptoms of leaf injury in plants caused by sulfur dioxide vary. Dense concentrations of SO_2 over short periods damage tissues and turn the leaves to a pale white or, in some cases, to a dark red-brown color. Tips of pine needles exposed to this

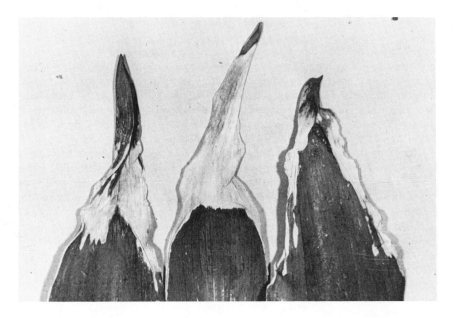

Figure 10–3
The margins of tulip leaves curl and die as a result of the presence of sulfur dioxide in the atmosphere. (Courtesy of U.S. Environmental Protection Agency. Photo by I. Leone.)

pollutant suffer a brown discoloration and then the needles fall off. Injury to a plant's leaves is due to the transformation of the absorbed sulfur dioxide into sulfuric acid and then into sulfates, which are deposited at the periphery of the leaves.

Low concentrations of SO_2 over many days or weeks slowly produce a characteristic chlorosis, or yellowing, of the leaves. Frequently, leaves then fall, and plant growth slows. Such chronic injury is caused mainly by the buildup of great amounts of sulfate in the leaves produced from absorbed SO_2. Plants vary in their susceptibility to SO_2 gas. Alfalfa, grains, cotton, grapes, and squash are all seriously affected by this pollutant. Temperature, light intensity, relative humidity, abundance of certain nutrients, and abundance of moisture in the soil all affect the degree of damage. Sulfur dioxide is also known to react synergistically with ozone and nitrogen dioxide to seriously affect some plants. Sulfuric acid droplets, falling on plant surfaces are harmful to plant tissues. The leaves of plants growing near factories that emit great quantities of sulfur have been pockmarked because of the drops of sulfuric acid. Beets and Swiss chard are especially sensitive.

The effects of SO_2 on animals and man was known early, primarily because of the Donora and London disasters. Extremely high concentrations of SO_2 and sulfuric acid seriously irritate the respiratory system. Sulfur dioxide causes the bronchial tubes to constrict in some experimental animals, a response evoked by other irritants to the respiratory system. However, the high concentrations of sulfur dioxide (from 95 to 98 percent) absorbed by the nasal passageways probably normally protects the bronchi and lungs.

Sulfur dioxide has been indicted in the erosion of many building materials, such as limestone marble, the slate used in roofing, and mortar. Unfortunately, exposed works of art, especially statues, are seriously deteriorated and discolored. Damage from SO_2 poisoning is especially prevalent in areas near the major SO_2 sources: pulp and paper mills, iron and steel mills, smelters, petroleum refineries, and chemical manufacturers such as those which make fertilizers or synthetic rubber.

Many fibers, such as cotton, rayon, and nylon, are seriously damaged with sulfuric acid. The mysterious disappearance of nylon stockings hung outside on clotheslines by housewives is attributed to the eroding effects of sulfuric acid in the atmosphere. The dyes in fabrics are also affected and fade in an atmosphere with SO_2. Leather and paper lose much of their strength and become discolored and brittle. The drying time of oil-base paints is prolonged, and their durability reduced. Metals corrode 1.5 to 5 times faster in atmospheres with SO_2 than in rural atmospheres. Metals used in overhead power lines and guy wires in really densely polluted areas have their lifetime reduced by one-third. In some especially polluted areas, electrical companies have resorted to the more resistant gold for electrical contacts.

Other sulfur compounds have also been indicted as serious pollutors. One is hydrogen sulfide gas, H_2S. This gas, as many high school chemistry students know, smells like rotten eggs. It also tarnishes silverware and copper and darkens lead. If the lead is a paint ingredient, it can seriously mar the appearance of homes. In truth, however, hydrogen sulfide is in such small amounts in the atmosphere that, as far as is known, little harm is done to plants or animals.

Fluoride compounds

Some fluoride compounds are known to be very beneficial to man, such as in the prevention of tooth decay, but some fluoride compounds in great concentration are serious air pollutants in parts of the country such as California, Florida, Idaho, Montana, and Washington.

Pollutant fluoride compounds are a result of the high-temperature treatment or acid treatment of materials that contain fluoride, such as in the manufacture of fertilizers, ceramics, and some metals. The contaminant may be gaseous or small particulate in character.

Fluorides are toxic to plants. They enter the plant through stomata and migrate to the periphery of the leaves. The leaf may absorb fluoride and eventually accumulate lethal amounts of the material at the leaf's edges. The leaves then die from the margins inward, the dried margins of the leaves delimited from the healthy inner parts by a narrow reddish line of dead tissue. Hydrofluoric acid produces small spots and turns the tips of gladioli leaves white. As with other pollutants, plants are differently affected by fluoride, even within the same species. Some plants especially susceptible to fluorides are gladioli and prune, apricot, and peach trees. Less sensitive plants include conifers, sweet potatoes, and corn. Some strains of plants most damaged by fluorides actually absorb the least amount of this pollutant. In animals, excess fluoride results in ligament calcification and the production of bony overgrowths that frequently lame the animal. More common symptoms are reduced food consumption, stiffness, emaciation, and decreased milk production in dairy cattle, and mottling of teeth.

Nitrogen compounds

The most important polluting nitrogen compounds are the nitrogen oxides, although nitrogen gas, N_2, is naturally plentiful, composing about 78 percent of the atmosphere. Nitrogen oxide (NO) is a product of combustion at extraordinarily high temperatures or under great pressure. Automobile motors and electrical power plants are its main sources. Nitrogen dioxide (NO_2), yellow brown in color and sweet smelling when dense, can be fatal at concentrations from 1 to 3 parts per million (ppm), although in the atmosphere it seldom reaches this concentration. Because nitrogen dioxide absorbs sunlight, it supplies energy for the photochemical formation of molecular compounds in the atmosphere. A third nitrogen oxide compound that pollutes is nitric acid (HNO_3). This corrosive acid is formed when nitrogen dioxide reacts with water vapor or raindrops.

Carbon compounds

Carbon is abundant in fuels; when mixed with oxygen, it yields atmospheric pollutants. One dangerous and abundant pollutant, carbon monoxide (CO), is colorless and odorless and, hence, can have serious effects before being detected. It eventually converts to carbon dioxide, but the conversion is slow. Carbon monoxide results mainly from the incomplete combustion of carbon in vehicle fuels and, consequently, is very plentiful in urban air. Concentrations of carbon monoxide vary greatly during 24 hours, usually being most abundant between 12 and 6 P.M. Concentrations can range from a few tenths per million up to 84 ppm in underground garages, loading platforms, or tunnels. Concentrations of carbon monoxide in about 30 ppm for 4 hours or more are known to convert about 5 percent of the body's hemoglobin into carboxyhemoglobin. This amount of conversion is known to impair vision and psychomotor activity, yet concentrations far more than 30 ppm are commonly found in many urban areas.

Extreme concentrations of carbon monoxide can cause serious biochemical changes in the body and ultimately death. Because carbon monoxide reacts with the hemoproteins in the blood, it reduces the capacity of the blood to carry oxygen. Long-term exposure of animals to carbon monoxide produces significant changes in the morphology of the brain and heart. Short-term exposure of rats to carbon monoxide has produced alterations in the central nervous system as detected with electroencephalographs. Brief exposures to great concentrations of carbon monoxide have produced similar effects on the nervous system in the vascular respiratory systems of human beings. Experiments that expose human beings to concentrations of carbon monoxide over a prolonged period of time seemed to indicate that the human being can produce an adaptive response, even producing more hemoglobin. There is a great need for research on the effects of carbon monoxide on the human being, especially since cigarette smoke contains significant quantities of this compound. In fact, cigarette smokers may well have levels of carboxyhemoglobin as high as 8 percent. Any additional carbon monoxide in the atmosphere might have serious effects for these people.

Carbon dioxide (CO_2) in dense concentrations can be another serious pollutant, perhaps even altering the temperature of the earth. Of course, changes in the makeup of the earth's atmosphere must be as old as the earth: the cooling of the earth 4.5 billion years ago greatly increased the concentration of water vapor in the air; the early activity of solidifying earth contributed a great amount of carbon dioxide; photosynthesis added oxygen gas; ultraviolet rays from the sun radiating onto O_2 created amounts of ozone, O_3, which filtered out the ultraviolet light and prevented the disintegration of developing life. Today, some scientists believe that the composition of the atmosphere is again undergoing a very serious change. Modern technological life depends to a very great degree on the combustion of fossil fuels which contain carbon. Therefore, a great amount of carbon dioxide is accumulating in the atmosphere. The results may be serious. Some of the sun rays that penetrate the thick layer of carbon dioxide strike the earth and are converted into heat. The heat cannot easily penetrate the CO_2 layer and cannot escape from the earth. Thus, the earth's atmosphere heats up. This phenomenon is known as the *greenhouse effect* (Fig. 10–4). Estimates of temperature increase of the atmosphere vary. Nevertheless, implications of a slight change are significant. One scientist, the physicist G. N. Plass, maintains that if the carbon dioxide content of the atmosphere were doubled, the average surface temperature of the earth would rise 6.5°F.

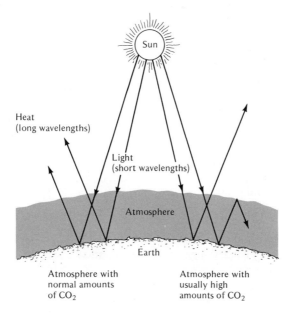

Figure 10–4
In the "greenhouse effect," a high-carbon dioxide atmosphere traps heat.

What evidence exists that atmospheric CO_2 has increased in recent years? Data gathered from 1860 through 1955 indicate that the concentration of carbon dioxide in the earth's atmosphere has increased about 15 percent in that time. Supporting data from Scripps Institute of Oceanography show that between 1958 and 1962 the carbon dioxide content of the air increased about 1.1 percent. (Remember, however, that the amount of CO_2 in the earth's atmosphere is still small. The original CO_2 percentage of the earth's atmosphere was only approximately 1 percent.) Crucially important are the recorded temperatures during the time the CO_2 content increased. The mean temperature of the earth rose 0.9°F between 1885 and 1940. Although at the moment it is impossible to indisputably correlate temperature and CO_2 concentration, perhaps the warming of the earth's atmosphere up to 1960 was caused by the increased accumulations of carbon dioxide in the atmosphere.

Although CO_2 concentration and temperature rise may be related, data gathered since 1960 prove puzzling. Since 1960 the earth's mean annual temperature has fallen. Some scientists suggest that this phenomenon does not contradict the greenhouse effect but that this phenomenon is merely affected by our new atmospheric development caused by a pollutant other than CO_2. These researchers suppose that the particulate contaminates (smoke and dust) are densely accumulating in the atmosphere. Such accumulations significantly reflect some of the sun's heat back into space, preventing it from reaching the earth. The atmosphere is deprived of the sun's heat at a rate greater than the retention of heat caused by CO_2 concentration. If CO_2 continues to accumulate, it may prevent the cooling effect of particulate pollution. The earth's mean temperature may again rise.

An increased warming of the earth has consequences. The glaciers would recede, ice caps such as those found over the Antarctic and Greenland would disappear, and ocean levels would rise. In fact, it is estimated that if all the ice on earth should melt 200 feet of water would be added to the surface of all oceans, and coastal cities would be inundated. There are other consequences. In controlled greenhouse conditions, scientists have found that in abundant carbon dioxide plants grow more rapidly, the size of lettuce and tomatoes increases, vitamin C and sugar in tomatoes rise, and resistance to virus diseases, parasitic fungi, and insects increases. What concentrations a plant can tolerate is not known, but up to 30,000 ppm have not proved lethal. Human beings, by comparison, can stand only 5,000 ppm, which, however, is 15 times the present atmospheric concentration.

Photochemical products

One of the primary ingredients of the smog that covers Los Angeles is hydrocarbons, compounds composed basically of hydrogen and carbon. The atmospheric conditions that surround this and other cities, and the abundant vehicles that burn hydrocarbon-containing fuels, account for the abundance of these materials. The smog that caused trouble in Donora and London contained a large percentage of sulfur products derived from coal; Los Angeles' smog contains abundant hydrocarbons and their derivatives.

The derivatives of the hydrocarbon exhausted from the internal combustion engine are proving to be more dangerous than the hydrocarbons themselves, but much is still unknown. The conversion process of the exhaust products is under intense investigation because of the uncertainty concerning the sequences of chemical reactions, influencing factors, and the effects on life of the compounds, both alone and synergistically. The conversion process is photochemical; that is, light energy triggers it. The main products that result from light striking the hydrocarbons from exhaust are the *olefins* (also known as *alkenes, ethylenes,* or *unsaturated hydrocarbons*), carbon compounds containing double bonds; the *aldehydes,* molecules whose terminal carbons bear an oxygen atom connected with a double bond and a hydrogen atom; and the *aromatics* (or *benzenes*), compounds structured in a ring. Some olefins and aldehydes are also exhausted directly from vehicles. The compounds produced from the photochemical products often appear to react further to produce additional compounds. In fact, the atmospheric condition over many urban areas must be a cauldron of chemical activity resulting from a plentiful and diverse supply of compounds, including oxygen compounds, and abundant energy, supplied from the sun but readied for use by the colored nitrogen dioxide.

One olefin produced directly in exhaust and produced in the atmosphere from other compounds is ethylene, $CH_2{=}CH_2$. A very small concentration of this gas, a few parts per billion, affects plants seriously. For example, it withers the sepals of orchid flowers, retards the opening of carnation flowers, and causes the petals of carnations to drop. Greater concentrations seriously retard the growth and development of tomatoes.

Two aldehydes that cause problems are *formaldehyde* and the olefin *acrolein*. Most importantly, these compounds irritate the upper respiratory system, skin, and especially the eyes. They also are suspected to be responsible for the odor of smog. A few laboratory experiments suggest that the toxicity of aldehydes increases when other contaminants are present.

The aromatics created photochemically appear to be among the most serious of all air pollutants; they also are among those about which least is known. Researchers are especially interested in one of these compounds, *benzpyrene,* because it has been proved to induce cancer. Two other photochemical pollutants appear to cause serious problems, *peroxybenzoil nitrate* and *peroxyacyl nitrate* (PAN) (Fig. 10–5). The only known effect on man of PAN in the atmosphere is tearing of the eyes, but experiments with laboratory animals indicate that in dense concentrations it is more lethal than comparable concentrations of SO_2, less lethal than ozone, and about the same in effect as NO_2. PAN also produces a silvering or bronzing of the lower leaves of plants. Lemon and grapefruit trees are especially susceptible. More experiments are needed to determine the long-range effects of the aromatic pollutants, by themselves and with other pollutants.

Figure 10–5
Peroxypropylnitrate, a common photochemical pollutant, damages vegetable crops. (Courtesy of Statewide Air Pollution Research Center, University of California, Riverside, Calif.)

Ozone

Ozone is abundantly produced from other oxygen-containing molecules. Atmospheric studies show that nitrogen dioxide, also sulfur dioxide and the aldehydes, absorb a great quantity of the sun's ultraviolet radiation that reaches the earth's surface. The resultant "excited" molecules easily react with molecular oxygen (O_2), frequently to yield atomic oxygen (O). For example, activated nitrogen dioxide (NO_2) splits to form nitric oxide (NO) and atomic oxygen (O). Atomic oxygen reacts with molecular oxygen to form ozone (O_3) (Fig. 10–6) and with nitrogen oxide to reform nitrogen dioxide. As a result of this oxidation reaction, ozone is produced abundantly.

Some effects of ozone on animals are known. Atmospheric concen-

$$CO + NO + \begin{matrix} O_2 \\ \text{or} \\ O_3 \end{matrix} \rightarrow CO_2 + NO_2 \rightarrow O^* + NO$$

$$NO + \begin{matrix} O_2 \\ \text{or} \\ O_3 \end{matrix} \rightarrow NO_2 \rightarrow O^* + NO$$

$$HC \rightarrow HC^*$$

$$O^* + O_2 \rightarrow O_3$$

$$O_3 + HC \rightarrow HCO^*$$

$$HCO^* + NO_2 \rightarrow PAN$$

$$HCO^* + O_2 \rightarrow HCO_3 \qquad \text{*Highly reactive molecules.}$$

$$HC + HC^* \rightarrow HC^* + HC^*$$

$$HCO_3^* + HC \rightarrow \text{aldehydes; ketones}$$

Figure 10–6
Interaction of molecules in a sunlit atmosphere produces many new combinations.

trations of ozone up to 0.5 ppm have been recorded, but usually the concentrations are lower. Mice, subjected in laboratory experiments to concentrations less than 0.5 ppm, showed increasing susceptibility to bacterial infection. Guinea pigs increased their breathing rate. Limited experiments with human beings yielded no clinical results, although subjects complained of chest constriction and throat irritation. Laboratory animals in denser concentrations of ozone developed chemical changes in the structural proteins of the lung and higher mortality rates due to bacterial infections. Human subjects in concentrations above 1 ppm became extremely fatigued and lacked coordination.

Effects on plants are also known. Reports indicate that ozone can depress the growth rate of plants without producing any obvious or visible lesions. However, visible damage is known and occurs as leaf blotches which disintegrate in very high concentrations. Ozone may also be responsible for the white markings on the leaves of certain grain crops and the searing of the tips of white pine seedlings, known as "tip burn." Susceptibility to ozone damage appears correlated with temperature, light, and soil moisture. Old and young leaves are more easily affected than mature leaves.

Ozone reacts easily with organic materials. Rubber is often attacked, especially when under tension (Fig. 10–6), and develops severe cracks. Antiozonant compounds have recently been added to rubber products sold in the Los Angeles area to prevent a reaction with ozone. Unfortunately, these compounds are expensive and after time migrate to the surface of the rubber material. Furthermore, compounds including gasoline and oil actually withdraw the antiozonants from the rubber product. Ozone also reacts with many fibers, especially cotton, acetate, nylon, and polyester, and with dyes. The degree of attack is determined by the amounts of light and humidity.

Lead

For a long time lead has been known to be toxic to human beings. In the past children who played with lead toys or swallowed lead-containing paint were frequent victims. Lead compounds, added to gasoline to reduce "knocking," are also emitted into the atmosphere. Because of the known danger of ingested lead products, fears arose that inhalation of atmospheric lead products would be just as dangerous. Today, lead-free gasolines are available and can be used in newer cars without an increase in knocking.

The effects of lead on the body are now known. Clinical observations indicate that ingested lead interferes with the functioning of enzyme systems in the body. Manifestations of toxicity are seen in the central and peripheral nervous system and smooth muscle. Victims of lead poisoning almost always have a hemoglobin level higher than 0.08 milligram per 100 milliliters of blood, and often much higher. These clinical observations are found in individuals who have consumed lead products. There is a lack of information about clinical lead poisoning in people who have inhaled lead pollutants.

Furthermore, lead accumulates in the automobile engine and increases the hydrocarbon emissions, whose effects have already been discussed.

Production and control of vehicle pollution

The kind of engine rather than the type of fuel is responsible for contamination of the air. Most fuels can be used in such a way as to virtually eliminate air pollution waste products. What is needed, however, is an engine that burns the fuel more efficiently than today's automobiles. Most of the pollution comes from the exhaust of the car, although 25 percent of the hydrocarbons are caused by "crankcase blow-by" and 20 percent of the hydrocarbons are due to evaporation from the fuel tank and carburetor.

Crankcase blow-by

Despite the improved construction of the piston apparatus in contemporary cars, some gas vapor still escapes between the cylinder walls and the pistons. These escaped gases enter the crankcase and are discharged into the atmosphere in vehicles not equipped with exhaust controls. Cars made in this country since 1963 have filters that capture and recycle these escaped gases into the engine. As a result, hydrocarbon emissions from many cars have been reduced by about 24 percent, from 1.5 pounds per day to 1.1 pounds. Since 1968, federal law requires that all models have filters that are 100 percent effective (Fig. 10-7). The increased cost per car is usually under $5. Research by fuel companies has produced special gasoline additives that keep these devices well-lubricated and trouble-free.

Evaporation

A second source of pollution from motor vehicles is evaporation from the fuel tank and carburetor, called "hot soak loss," which occurs mostly when the car is stopped and heat builds up. At the moment much research is underway in laboratories to control evaporation from the carburetor and fuel tanks.

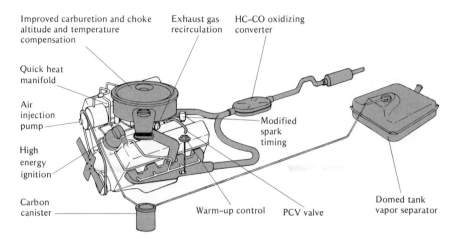

Figure 10–7
The emission control system proposed for General Motors cars should better prevent exhaust pollution. (General Motors Corp., Detroit, Mich.)

One process collects the vapors with activated charcoal when the engine is turned off. When the car is started the vapors are pulled from the charcoal and ignited in the engine. Another plan prevents the gas from evaporating by subjecting the gasoline in the tank to slight pressure. Still another suggestion involves development of a low-volatile gasoline that will not evaporate easily. Unfortunately, such a gas may not start a car easily, and that would produce more contaminants in the exhaust.

Exhaust emission

The exhaust of the car produces many atmospheric pollutants, including an unburned gasoline vapor, compounds of oxygen and nitrogen, lead, and carbon monoxide. Various controls are being investigated. Of course, an internal combustion motor that efficiently burns the hydrocarbons in the gasoline would reduce the pollutants in the exhaust. This could be accompanied by new proportions of gasoline and air, more exact timing of fuel feeding, gas additives to improve combustion, and air injections into the exhaust to convert the exhaust compounds into less-toxic materials. The most annoying contaminants from diesel engines are odors and smoke, although diesel exhaust contains basically the same kind of pollutants as the exhaust from the internal combustion engine. However, it usually contains less of them. Most could be prevented with afterburners.

The electric car

Some of the first automobile models in this country were electric. Today, in view of the great quantity and toxicity of pollutants produced by the internal combustion engine, electricity has again been suggested as a desirable source of energy for our vehicles. The electric motor converts about 90 percent of the electrical energy into mechanical energy, a higher degree of conversion than is achieved by the internal combustion engine. Furthermore, whereas the internal combustion engine is composed of gears, clutches, and other mechanical parts, the electric motor has few. It is simpler in design, smaller, and cheaply maintained. Furthermore, the electric motor is quieter than the internal combustion engine, and no exhaust is produced.

Research on the electric car continues. However, data gathered from experiments and trials with electric motors and vehicles are not very encouraging. For example, the maximum speed that an electric vehicle can reach is about 80 mph, and the best source of energy is a collection of up to 16 low-energy, expensive, and heavy batteries which occupy as much room as the trunk area in conventional cars. Furthermore, the batteries need frequent recharging, a time-consuming chore that requires an external source of electric energy, a problem now that energy of any sort is not always abundant. It is possible that recharging batteries could be done more quickly, but expensive, heavy-duty equipment would be needed. It is said that millions of dollars in a crash program or approximately 100 years would be needed to make the electric car equal in performance to the conventional internal combustion car. Nevertheless, the advantages of electrically operated cars, and a shortage of fossil fuels, may encourage their development, even though the production of additional electricity to run them may result in more environmental pollution (see Chapter 11).

Control of industrial pollutants

Particulate pollution of the atmosphere is a serious problem for industry. Many waste particles, too small to settle out easily, remain suspended in combustion exhaust. Today, schemes are being devised that will remove these pollutants from waste gases spewing into the atmosphere.

We have electrostatic precipitators that can remove about 99 percent of particulate pollution from exhaust in industrial chimneys. Within the chamber in the precipitators, the particles suspended in the waste gases become charged or ionized and, consequently, are attracted to charged electrodes. In 1965 such devices built into industrial machines removed 20 million tons of debris. Some of this collected waste material can be used commercially, as, for example, in concrete. The process works well in power stations, paper mills, cement mills, carbon block plants, and petroleum cracking plants.

Other techniques that are effective in removing particulate waste material include subjecting the waste gas and particles to centrifugal force, a process that removes about 70 percent of the particulates; capturing the particles in a liquid sprayed into the waste-gas stream; knocking particles from the gas with obstructing louvers or baffles; settling by coursing gas through large-diameter pipes to slow down its velocity; and filtering through oil or grease and paper fiber, plastic, or metal mesh. Combinations of these processes are used also. Besides particulate material, methods have been devised to remove sulphur dioxide from chimney waste. Dolomite [$CaMg(CO_3)_2$], lime (CaO), and limestone ($CaCO_3$) placed in the path of the flowing gas reacts chemically with sulfur dioxide to produce precipitation on centrifugation. The process is inexpensive and does not involve water sprays. Water in contact with sulfur dioxides produces corrosive sulfuric acid. Desulfurizing coal and oil before burning is also being tested.

A procedure used to reduce pollution of SO_2 and other contaminants includes adsorption of the chemicals on materials such as carbon. A steam bath regenerates the adsorption bed when saturated, and many pollutants can then be reclaimed. Absorbers other than charcoal can be used to pick up alcohols and benzenes. The method has been used effectively in dry cleaning plants; printing shops; paint factories; rayon, plastic, and rubber plants; food processing plants (to control odors); breweries; pharmaceutical plants; rendering plants; and sewage plants.

Combustion of the waste gas is another technique useful to some industries, such as petroleum companies. This technique is seriously considered for use on car exhaust pipes but would work well also on waste chimneys of paint and varnish manufacturers, animal rendering and fish processing plants, coffee roasting ovens, and even apartment house and office building incinerators. A catalyst might promote reactions with other chemicals and allow combustion to occur at lower temperatures.

Legal control of air pollution

In 1822 the author Chateaubriand wrote: "Soon I saw before me the black skull cap which covers the city of London. Plunging into the gulf of black mist as if into the mounts of Tartarus, I crossed the town." Smog, new product of the Industrial Revolution, was later to prove more notable. Beginning in the winters of the 1880s, smog attacks became increasingly common in London. The fog of Dickens' novels turned into the smog of medical journals.

So severe did the problem become that in 1956 Parliament passed a clean air act, 4 years after the 3-day smog disaster that killed 4,000 sufferers of bronchial and cardiac ailments in London. Today, 74 percent of London is restricted by controls that prohibit the burning of soft coal, the fuel traditionally used in fireplaces. As a result, Londoners enjoy a reduction from 1952 of the smoke emitted by homes and factories. Some of the factors that account for London's new clean air include the large rehousing program following World War II, which replaced the inadequate coal-burning heating units in old homes with devices that used other types of fuel. The government subsidized up to 70 percent of the cost of converting stoves and fireplaces to smokeless burners of coke or hard coal. Other increased legal restraints have resulted in more than 75 percent of London's area being without smoke by the end of 1970. The clean air act was toughened in 1969 to coerce local councils to define smoke control areas and to restrict the sale of smoke-producing coal.

The result is a much cleaner London, a London without a great amount of T. S. Eliot's "yellow fog that rubs its back upon the windowpanes." More important, the incidence of bronchitis is diminishing. Previously, bronchitic problems were 10 times as common in Britain as in other industrialized countries and killed up to 32,000 people per year.

The decrease in the fog over London is a result of fewer air pollutants, fewer "nuclei" around which water droplets can condense. Consequently, the sun shines more often in London now than it has for many years. In 1952 London received 8 percent less hours of sunshine per year than some towns only 25 miles from the city's center. Now an estimated 3 percent more sunshine falls on London than on surrounding areas. Visibility has also been increased. In 1958 a Londoner could see only about 1.4 miles; today he can see about 4 miles. The cleaner air also appears to have encouraged wildlife to reinhabit London and environs. In 1959 only about 60 species of birds were reported. In 1969, 138 species were identified, including house martins, not seen for 80 years, and rare birds such as the snow bunting, the hoopoe, and the great northern diver.

London may still be faced with pollution problems, however. The town is undergoing an automobile boom. Thus, hydrocarbon pollution problems may be following closely on the abatement of the sulfur dioxide pollutant problem.

The United States has also taken legal steps to purify the air. In 1955 the Department of Health, Education, and Welfare was authorized to conduct research and to provide technical support for state and local governments, which, in turn, were urged by federal legislation to take on the responsibility of prevention and control of air pollution. Nevertheless, since 1955, air pollution has become progressively more of a problem, even to critical proportions in some areas. Furthermore, because air and its pollutants move across state boundaries, effective state or local control was unrealistic. Consequently, in December 1963 the Clean Air Act was adopted. This act restated that state and local governments must be responsible for air pollution control. However, it also authorized the Department of Health, Education, and Welfare to support with grants state and local efforts in the air pollution control. Federal control has been initiated in areas where air pollution is a serious interstate problem. The act was amended in 1965 to include national standards for the control of motor vehicle pollutants.

Despite these attempts to legally control air pollution in this country, the problems worsened. Consequently, in 1967 the Air Quality Act was passed. This act coordinates even better local and state attacks on air pollution. It asks that states draw up and implement control measures for their air pollution problems and charges the Department of Health, Education, and Welfare to develop and to publish criteria indicating the effects of air pollutants on health, property, the cost, and the new techniques in air pollution control. If investigation indicates that a state has a pollution problem, the state will be expected to develop air quality standards and plans for control. States have 90 days to submit their written intent to cooperate with the standards, 180 days to define the standards, and 180 days to develop implementation of the standards. States can request a hearing of HEW's claim regarding the air pollution problem. Uncooperative states risk legal action. The Air Quality Act of 1967 also authorized creation of a 15-member Presidential air quality advisory board, the registration of additives to fuel, and a study of the possible control of pollution caused by jet and conventional aircraft.

HEW's new stipulations for the amounts and types of pollutant permitted in car exhaust illustrate its new power. Nitrogen oxide emission was reduced in 1973 model cars to 3 grams per vehicle mile (g/mile) from the average of 6 g/mile then produced from vehicles without exhaust controls. Particulate pollution was reduced from 0.3 g/mile in a car with no special adaptation to 0.1 g/mile. Hydrocarbons, previously 2.2 g/mile, are slashed to 0.5 g/mile in the 1975 model cars, and carbon monoxide is lowered from 23 g/mile to 11 g/mile in 1975 cars.

Control over the pollution of our atmosphere is under way, as the federal laws testify. State laws to control air pollution are also increasing, although some states, such as California, had legal control before the federal government. In California, local groups concerned over the stifling air pollution in cities such as Los Angeles influenced the enactment of laws controlling industrial and vehicle air pollution. In 1960 the state created the Motor Vehicle Pollution Control Board (MVPCB), which regulated certification of auto pollution control devices. In 1967 the Air Resources Board absorbed the MVPCB and controlled all air-preservation acclimation in the state.

Today, other states, with the encouragement of the federal government, are effecting increasingly stringent air pollution controls. After a persistent thermal inversion in November 1966, New York limited the sulfur content of fuel to 1 percent or less. As a result, atmospheric sulfur dioxides in New York are less than they were in 1966. Furthermore, in 1971 the permissible content of sulfur in fuel for most industries was dropped to 0.37 percent. New Jersey also has made notable advances in state-initiated air pollution control, including six continuous air monitoring stations.

Rapid progress in curbing air pollution is not to be expected, however. Bureaucratic red tape and economic pressures retard enforcement of effective controls. Yet, hopefully, laws can be put into force before worse damage is done to the environment, plants, and animals, including man.

Summary

In this chapter we have examined some aspects of air pollution. We have considered the high cost of air pollution and its primary sources. We have looked at the kinds and effects of air pollutants, including several of the more serious ones, such as particulates, sulfur compounds, fluoride compounds, nitrogen compounds, carbon compounds, photochemical products, and lead. Underlying our entire chapter was the understanding that although man evolved a life-sustaining relationship with his environment in the past, he is now producing materials which are endangering that relationship, primarily because of his increasing numbers and his utilization of energy in the development of technology. Implicit is the understanding that man, like any organism, can continue to exist only if the relationship between himself and his surroundings is life-giving. Hopefully, from this chapter comes information and insight necessary to help make right once more man's relationship with his environment.

Man is endangering his existence not only by spewing pollutants into the atmosphere but also by discharging pollutants into the water. We shall look at that kind of pollution in the next chapter.

Supplementary readings

Batton, L. J. 1966. *The Unclean Sky: A Meteorologist Looks at Air Pollution.* Doubleday & Company, Inc., Garden City, N.Y. 141 pp.

Bernarde, M. A. 1973. *Our Precarious Habitat,* rev. ed. W. W. Norton & Company, Inc., New York. 448 pp.

Cadle, R. D., and E. R. Allen. 1970. "Atmospheric Photochemistry." *Science,* vol. 167, pp. 243–249.

Crenson, M. A. 1971. *The Un-politics of Air Pollution: A Study of Non-decision-making in the Cities.* The Johns Hopkins Press, Baltimore, Md. 227 pp.

Degler, S. E., and S. C. Bloom. 1969. *Federal Pollution Control Programs: Water, Air, and Solid Wastes.* BNA Books, Washington, D.C. 111 pp.

Devos, A. 1968. *The Pollution Reader.* Harvest House, Montreal. 264 pp.

Edelson, E., and F. Warshofsky. 1966. *Poisons in the Air.* Pocket Books, New York.

Jacobson, J. S., and A. C. Hill. 1970. *Recognition of Air Pollution Injury to Vegetation: A Pictorial Atlas.* Air Pollution Control Association, Pittsburgh, Pa.

Leinwand, G. 1969. *Air and Water Pollution.* Washington Square Press, New York. 160 pp.

National Tuberculosis and Respiratory Disease Association. 1969. *Air Pollution Primer.* The Association, New York. 87 pp.

Peterson, E. K. 1969. "Carbon Dioxide Affects Global Ecology." *Environmental Science and Technology,* vol. 3, no. 11, pp. 1162–1169.

Stern, A. C., ed. 1968. *Air Pollution,* 2nd ed. Academic Press, Inc., New York

U.S. Department of Commerce. 1967. *The Automobile and Air Pollution: A Program for Progress.* Government Printing Office, Washington, D.C.

U.S. Department of Health, Education, and Welfare, National Air Pollution Control Administration. 1970. *Air Quality Criteria for Carbon Monoxide,* Publication AP-62. Government Printing Office, Washington, D.C.

U.S. Department of Health, Education, and Welfare, National Air Pollution Control Administration. 1970. *Air Quality Criteria for Hydrocarbons,* Publication AP-64. Government Printing Office, Washington, D.C.

U.S. Department of Health, Education, and Welfare, National Air Pollution Control Administration. 1970. *Air Quality Criteria for Photochemical Oxidants,* Publication AP-63. Government Printing Office, Washington, D.C.

Chapter 11

Men talk much of matter and energy, of the struggle for existence that molds the shape of life. These things exist, it is true; but more delicate, elusive, quicker than the fins in water, is that mysterious principle known as "organization," which leaves all other mysteries concerned with life stale and insignificant by comparison. For that without organization life does not persist is obvious. Yet this organization itself is not strictly the product of life, nor of selection. Like some dark and passing shadow within matter, it cups out the eyes' small windows or spaces the notes of a meadow lark's song in the interior of a mottled egg. That principle — I am beginning to suspect — was there before the living in the deeps of water.

The Flow of the River
Loren Eiseley

Water: Maintenance of quality and quantity

In the previous chapter we were concerned with one aspect of man's crisis with his environment, the production of wastes that pollute the atmosphere in such quantities that the health and possibly the very existence of man are threatened. In this chapter we shall look at the pollution of the earth's waters, another example of how man's activities are eroding the life-sustaining relationship between man and environment, a product of his evolution through the ages.

The significance of water to life will be looked at in several ways. The characteristics of water will be described in detail to show how the properties of this environmental substance helped determine the appearance and function of living things. Next, we shall look at how we are treating water, from the quantity we use now and will use in the future to the way we pollute it and, thus, endanger all life by changing the nature of this ubiquitous liquid. From the discussion in this chapter should come insight into the intimate relationship of life and environment, and awareness of the possibility of man's altering it to the degree that life, including man, may no longer be sustained.

Water and life

Four and one-half billion years ago, when the new earth was developing, vaporous materials, including water, escaped from the cracks and crevices on the solidifying crust to concentrate in a dense atmosphere, much like the giant cloud that covers Venus today. In time, as the earth's surface cooled, falling rain gathered to form the seas in the depressions in its newly solidified surface.

The seas were significant to the development of life. Perhaps 1 billion years later, as the seas grew and the falling rain eroded the earth's surface and deposited eroded materials into the seas, life arose. Significantly, the cradle for this evolving life — from simple coalesced molecules to microsphere to sophisticated cells to multicellular organisms — was the nutritionally rich sea. Millenia later life took its water origins along when it moved from its watery birthplace up to the more arid land environments. Unique structures and functions developed which discouraged water loss and permitted physiological processes to proceed as if the ocean were still home. For example, blood simulated the deeds performed by the original ocean by transferring gases and food molecules and carrying away wastes. Water affected life evolving onto land in other ways: water moving through its cycles constantly affected the cool exposed earth surface, changing, eroding, depositing, replacing (Fig. 11–1).

It is no overstatement to say that without water in its abundance in the early days of the earth life as we know it, including ourselves, could not have come about. No other molecule could have permitted the biological course of events that we believe to have occurred. Today, water is still vital to life. Let us look at the reason why, and the problems that come about when water's life-enhancing characteristics are altered by pollution.

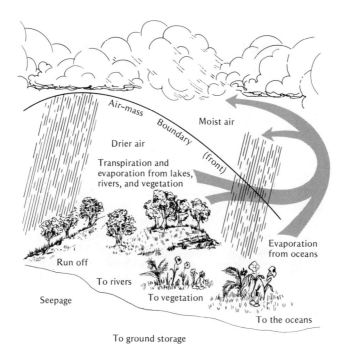

Figure 11–1
Cycling of water. (From *The Nation's Water Resources,* Water Resources Council, 1968.)

The characteristics of water

Properties of water significant to life are indirectly caused by water's atomic structure. The water molecule is composed of three atoms, two hydrogen and one oxygen. Hydrogen is composed of one positive charge, the proton, balanced with one negative charge, the electron. The single electron courses outside the proton in an orbit capable of holding a sum of two negative charges. Thus, hydrogen tends chemically to add another electron to that orbit, a property that accounts for hydrogen's extraordinary reactivity with other atoms. Oxygen has similar structural properties. However, an oxygen atom is bigger and has a positive nuclear charge of eight, surrounded by eight negative charges or electrons, two in the first energy level, or orbit, and six in a second energy level. Significantly, this second energy level is capable of containing a total of eight negatively charged electrons, so room exists to add two more. The ability of the atoms of both oxygen and hydrogen to hold more electron charges in their outer energy orbits accounts for their mutual chemical attraction. Two hydrogen atoms, each with its single planetary electron, chemically accommodates atoms by adding their orbital electrons to the outer ring of oxygen. Structurally, then, the oxygen outer level of electrons contains eight, the ring of each hydrogen contains two electrons, and each of the three atoms is chemically satisfied. Such an atomic union is a "covalent bond."

That life exists today is due to a great extent to the covalent bonding within a water molecule (H_2O). This bonding accounts for some of its remarkable physical properties, such as its molecular stability. The covalent bond is not easily broken. When it does come apart under natural conditions (in about 1 of every 554 water molecules), either one or both hydrogen atoms will break away from the oxygen atom, yielding positive hydrogen ions and negative oxygen ions.

The covalent bonding of the hydrogen and the oxygen atoms creates a polarity of the molecules, that is, a positive charge on one end of the water molecule and a negative charge on the other. This polarity results from the displacement of the charges of individual atoms of oxygen and hydrogen owing, in turn, to the manner in which they join. The two hydrogen atoms join the oxygen atom exactly 105 degrees from each other, a distance almost equal to the tilt of the Tower of Pisa. As a result, the nuclei of the hydrogen atoms are more exposed than normal and produce a positive charge to the molecule. The electron concentration around the oxygen component accounts for a negative charge on the opposite end of the molecule (Fig. 11-2).

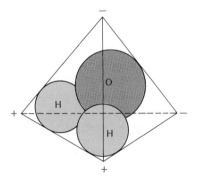

Figure 11-2
Three-dimensional sketch of the water molecule, showing polarization.

The polarity of the water molecules attracts other water molecules. Positive and negative poles link together and produce a great coherence among the water molecules (Fig. 11-3). Hence, water as a bulk of molecules tends to be quite stable. Such attraction between the negative and positive poles of a water molecule is called a *hydrogen bond,* a bond commonly found in other molecules, too.

The polarity of the water molecule and its ability to form hydrogen bonds accounts for many important properties of life. For example, the absorption of food by living forms depends greatly on the dissolution of nutrients. Dissolution of many biologically significant molecules is caused by the polar nature of the water molecule. The charges characteristic

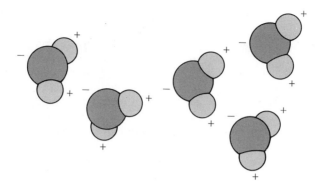

Figure 11-3
Aggregation of water molecules held together with hydrogen bonds.

of the water molecule exert a pressure on other molecules and chemically encourage them to break apart and to form ions. These ions are then surrounded by the polar water molecules and produce a solution easily ingested by life-forms.

The hydrogen bonding of water molecules accounts for other significant life phenomena. For example, *adhesion,* the tendency of water molecules to adhere to other kinds of molecules as well as to themselves mainly because of their polarity, helps explain the movement of water up a plant. The water molecules adhere to the molecules of the plant cells and, aided by the evaporation of water from the distal plant parts, helps move materials up the stem.

The hydrogen bonding of water molecule to water molecule produces other consequences of biological significance. Water becomes increasingly dense when it is cooled, but between 4 and 0°C (freezing) the water expands and becomes less dense. In fact, when it turns into ice at 0°C, water has expanded to a size $1/11$ more than the size of its liquid volume. This is because the ice molecules held in a relationship to each other by the hydrogen bond cannot move very closely together. Hence, a "lattice" of water molecules forms from 4°C to freezing, which accounts for the tendency for ice to float on the denser water.

The expanding water at these temperatures exerts great pressure and can wreak serious damage, such as bursting car radiators or breaking undrained water pipes; on the other hand, the effect of this expansion is beneficial to life-forms. They are not "frozen to death" in bodies of water in which ice forms because the low-density ice constantly floats upward, leaving the lower portions of the bodies of water warm and liquid. If the water continued to become dense as it became solid, much water now on the surface of the earth would become locked in ice. Furthermore, the seasonal turnover of the layers of water in ponds and lakes in freezing areas owing to the sinking of cold water and the rising of ice and the rising of warm water helps distribute oxygen and nutrients essential to sustain aquatic life. (This will be described in detail later.)

Hydrogen bonding of the water molecules also accounts for other biologically beneficial characteristics of water, such as its ability to absorb great quantities of heat. Heat (really the motion of molecules) may be applied to a volume of water, but the molecules are restrained from moving because of the hydrogen bonds. When an excessive amount of heat is applied, the hydrogen bonding is overcome, and the temperature begins to rise. Thus, plants and animals (many of which contain up to 75 to 80 percent water) can be subjected to a great deal of heat but not damaged, because the water will buffer the effects of the heat and prohibit destruction of the protoplasm. This buffering capacity of water is called its *high specific heat.* The overall effects of water's stabilization of temperature of protoplasm — or the sea where life began — help explain the existence of life today.

Just as water can absorb a great quantity of heat before its temperature will rise, so can it absorb a great quantity of heat before it will change from a liquid into a vapor. The general effect is a stabilization of atmospheric temperature. Great quantities of heat are taken from the environment to vaporize the water and, consequently, cool the atmosphere. One of the reasons that humid summer days are so uncomfortable is because the dense amount of water vapor in the air inhibits further vaporization, and hence, the atmosphere cannot cool. Just as liquid water must absorb a great quantity of heat before turning into vapor, so solid water, or ice, must absorb a great quantity of heat before turning into a liquid. The result, of course, is the same as before: heat is pulled from the atmosphere.

In summary, characteristics of the water molecule, notably its hydrogen bonding, account for physical properties of water in its three states: vapor, liquid, and solid. These chemical and physical characteristics explain to a major extent the development of life-forms on this earth. Without abundant water it is very likely that life as we know it could not ever have evolved.

The dependence of life on water

Survival, whether in the past or now, depends on water. Our very creation probably occurred in water. Our existence is still dependent on water, as seen in the buffering temperature, the lubrication of our bodies' moving parts, and even in the senses of taste and smell. Clearly, our future existence may depend on the availability and quality of water that surrounds us and which we incorporate into our bodies. Most threatening is the possibility that the world's supply of water, so essential to our existence, may well be depleted or so altered that biologically it is not safe. The problem of the maintenance of an adequate and pure water supply is no small issue, and concerns us next.

The human body well illustrates the acute dependency of the organism, which lives primarily on land. A 150-pound person is made up of about 100 pounds of water. If he loses only 5 percent of this normal amount of body water, his mouth and tongue will become desiccated; he will, of course, have an acute thirst, his skin will wrinkle, and he may even experience hallucinations. If he loses as much as 15 percent of his body water, he will usually die. On the other hand, an abundance of water over that which is normally needed by the human body will produce weakness and nausea. In fact, an extreme excess of water can lead to mental derangement, tremors, convulsions, and coma, and can be fatal.

The body has developed very sophisticated physiological and structural adaptations to adjust the amount of water that it contains. The water-adjustment control center is the hypothalamus, a very small part of the brain. This area ensures that there is no more variation than 1 or 2 percent in the body's water. Too much water in the brain area stimulates the kidney to excrete, the physiological device by which most water leaves the body. Too little water and the drying throat tissues stimulate the nerves and the hypothalamus, and a sensation of thirst is experienced.

Other organisms that evidence great adaptations to maintaining a careful control over volume of body water are the desert dwellers. Some plants and animals seem to be able to dwell even where water appears to be completely absent. However, these organisms do require water, and they get it by various physiological and anatomical adaptations or even behavioral adaptations to conserve their water in this intense environment. Many desert animals hunt at night when temperatures are very low. Many live underground, some have very thick hides or skin, and others actually produce their water through their metabolism when they combine oxygen with hydrogen derived from the food eaten. Plants have equal adaptations to the harsh desert environment. Their stomata normally open only in the cool night; and sometimes in a severe drought, some plants, such as the ocotillo, actually lose their leaves. Cacti have evolved thick, fleshy, water-storing stems and have reduced the leaves to merely spines. Many produce seeds that only sprout, grow, and produce a flowering plant during the rainy season.

Man's close dependence on the presence of water is well-documented. Many of the great past civilizations were born and evolved because of the presence of an adequate water supply for drinking, food preparation, commercial communication, and irrigation. The Nile in Egypt, the Tigris and Euphrates in Mesopotamia, the Yellow River in China, and the Indus in India all have been sites of significant civilizations.

Where does all the waste go?

"In the tenements of Glasgow dung was left lying in the courtyards as there were no lavatories in houses. This lack of lavatories led to the habit of house dwellers filling chamber pots with excreta and after some days, when completely full and with a shout of 'beware slops!' emptying the contents out of the window into the street below."

Man has improved on the disposal of waste since this description of Glasgow, which typified Europe's sewage disposal in the eighteenth and nineteenth centuries. Yet the problem of adequate disposal of wastes remains and in some areas is even intensified by the world's fast population growth rate and urbanization. Today, sewage is defined as the waterborne waste from home and industries and includes human excreta, soaps, detergents, paper, cloth, and numerous industrial products. The degradation of this material so that it no longer is disease-carrying or harmful, called "sanitary sewage," generally occurs in a sewage plant by a multistep process. These steps are known as *primary treatment, secondary treatment,* and *tertiary treatment.*

Primary treatment begins as the waste flows from the home or industry into the large network of subterranean sewer pipes beneath most densely populated areas. As the sewage flows or is pumped through the sewer pipes, bacteria initiate its breakdown. The bacterial decomposition is the first significant step in the treatment of the wastes. Partially decomposed waste is then filtered as it enters the sewage processing plant (Fig. 11–4). The sewage is filtered through a series of coarse and fine screens to remove the large floating objects. The waste removed by the screens is passed to a *comminutor,* or shredder, which tears the large materials into small pieces so that they will not clog the sewage treatment apparatus in later steps.

Figure 11–4
Sewage treatment plant. Three round trickling filters are in the middle, surrounded by (left to right) sludge drying beds, two sledge digestion tanks, control building, aeration and grit-removal tanks, and three round settling tanks. (Courtesy of Water Pollution Control Plant, City of Ames, Iowa.)

Next, the finely ground waste moves to a *settling chamber.* Here, during the next few hours sand and large organic pieces settle out. Unfortunately, this step, which terminates the "primary treatment" of the waste, is too often the final one before the waste is emitted into rivers and streams.

Though more than half of the materials suspended in sewage are removed by this time and processed in an organic "digester," the high organic content still supports abundant microorganisms, both benign and pathogenic, which can cause major pollution and disease problems.

Fortunately, many municipalities further treat their sewage to remove the colloidal and dissolved organic material, a process known as "secondary treatment." This includes steps to further decompose the organic material through the metabolic action of organisms accelerated with abundant oxygen. The breakdown is accomplished with a wide variety of organisms, including bacteria, protozoans, worms, snails, and flies. Often complex food relationships develop among the organisms in the filter, thus forming a kind of ecosystem. Abundant oxygen is provided with a trickling filter. This apparatus sprays sewage high in organic matter over a circular bed of crushed stone 3 to 6 feet deep (Fig. 11–5). As the watery effluent runs over the rocks, it makes contact with abundant oxygen. As a result, the nutrient molecules in the waste are decomposed. Not only is the material altered, but the oxygen deprivation that could occur in the waterway if the waste were emitted unoxidized is prevented. The amount of oxygen required to consume the nutrient molecules in this step equals the amount that would be removed from the river if the trickling filter phase were omitted.

Figure 11–5
Close-up of trickling filters, showing the spraying of sewage over beds of crushed rock. (Courtesy of Water Pollution Control Plant, City of Ames, Iowa.)

Some industries that produce a great quantity of organic matter use a different process to accomplish secondary treatment, the activated sludge process. This process treats an abundance of waste effluent, produces little odor, and is inexpensive. Organic wastes and decomposing organisms are suspended as a flocculant mass in the liquid by mechanical agitators in aeration tanks or by diffused air. The waste continuously flows through the large sludge tanks approximately 10 feet deep, 20 feet wide, and hundreds of feet long.

Despite this process, some materials low in organic material, such as phosphate-containing detergents or radioactive materials, are passed relatively untouched through the treatment. Results can be annoying at the very least. Great accumulations of detergent, for example, often pass unaltered through the sewage treatment plant and accumulate in rivers and lakes, killing plants and animals and despoiling the area for recreation (Fig. 11–6).

Figure 11–6
Foamy stream in Pennsylvania which resulted from detergent pollution. (Courtesy of U.S. Department of Agriculture. Photo by Thompson.)

Consequently, methods are sometimes used to rid the water of these and other wastes, sometimes even purifying the water to the point that it can be reused. This is known as a *tertiary,* or *advanced, treatment.* Some of the processes used include chemical coagulation, filtration, carbon adsorption, and chemical oxidation with strong oxidizing materials such as hydrogen peroxide or chlorine.

Decomposition of sewage in sparsely populated areas is often accomplished with methods other than the sewage treatment plant. Occasionally an "oxidation pond" or lagoon is used. The raw sewage is pumped into a broad pond 2 to 6 feet deep. Here, aerobic microbes degrade the organic material. In a few months the degraded effluent is usually emptied into a river or stream, and the solid sludge pumped out. Some industries that produce great quantities of high nutrient waste construct private oxidation ponds.

Still another method of waste processing is the septic tank (Fig. 11-7). Waste from homes flows into the subterranean tank, where anaerobic degradation occurs. The liquefied effluent flows out perforated pipes under the soil into a drainage field. As with the oxidation pond, the sludge from the septic tank must be pumped out occasionally.

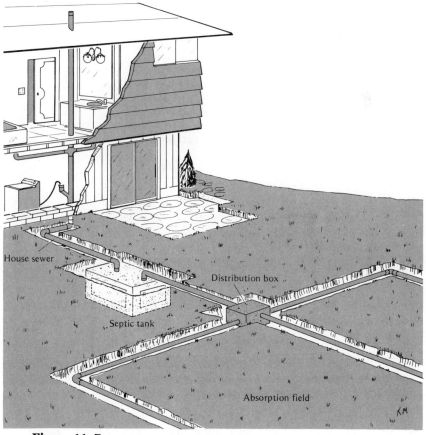

Figure 11-7
Construction of a septic-tank waste disposal system.

The demand for water

Although water covers nearly 75 percent of the earth's surface, only a very little is available to life. The oceans contain 97.2 percent of the water, but it is too salty for human consumption unless desalinized. The glaciers and ice caps, such as lie over Greenland, lock up 2 percent in ice.

Thus, only approximately 0.8 percent of all the water on earth is available (Fig. 11-8).

Man's activities require a great amount of the available water. For example, in 1960 about 320 billion gallons of water per day were used in the United States. Of this, 140 billion gallons per day were used

THE WORLD'S WATER SUPPLY

LOCATION	WATER VOLUME (Cubic miles)	PERCENTAGE OF TOTAL WATER
SURFACE WATER		
Fresh-water lakes	30,000	.009
Saline lakes and inland seas	25,000	.008
Rivers and streams	300	.0001
	55,300	.017
SUBSURFACE WATER		
Soil moisture	16,000	.005
Groundwater within depth of half a mile	1,000,000	.31
Deep-lying groundwater	1,000,000	.31
	2,016,000	.625
ICECAPS AND GLACIERS	7,000,000	2.15
ATMOSPHERE	3,100	.001
OCEANS	317,000,000	97.2
TOTALS (approximate)	326,000,000	100

Figure 11-8
Percentage of water, in various forms, on earth.

in agriculture, 98 billion in utilities, 60 billion in industries, and 22 billion by cities and towns.

It was estimated in 1944 that of the water used by inhabitants in cities, 41 percent was used in toilets, 37 percent for bathing and washing, 6 percent in the kitchen, 5 percent for drinking, 4 percent for washing clothing, 3 percent for household cleaning and watering of gardens, and 1 percent for washing cars.

In 1965, the U.S. Department of the Interior stated that 359 billion gallons of water per day, or 1,900 gallons per person per day, are used, of which 145 gallons are used domestically. (Leaking taps, faucets, and piping probably claim another 300 million gallons per day in New York City alone.) This is an increase of 200 million gallons over the previous average amount of water drawn in midsummer in normal years and about 400 million gallons more than during dry years.

Industry and agriculture use a great amount of water. Approximately 40 gallons are needed to grow the feed to produce one egg. Approximately 350,000 gallons of water are needed to grow 1 ton of corn. To refine 1 gallon of crude petroleum requires 1.7 to about 44.5 gallons; to finish

1 ton of steel takes 1,465 gallons; to make 1 pound of soap, 1.5 to 7.5 gallons; 1 pound of natural rubber, 2.5 to 6 gallons; 1 pound of artificial rubber, 13 to 305 gallons; 1 pound of aluminum, 1.24 to 36.3 gallons; an automobile, 12,000 to 16,000 gallons; a truck or bus, 15,000 to 20,000 gallons.

The quantity of water used around the world accounts for the scarcity of water experienced today in parts of the globe. Such scarcities will become more common as populations grow and urbanization and industrialization spread. Today, water shortages are not uncommon in the Sahara, the Middle East, Australia, Korea, and the United States. In the United States the underground water level in many areas keeps dropping. For example, since World War II, Nebraska has experienced a drop of more than 15 feet in its water table. It is possible that, in 50 to 100 years, water may be inaccessible. In some areas of the country a water shortage or a decrease in quality of water is already being felt. Some areas, such as Arizona, which draws 60 percent of its water from wells already overextended, have more cause to worry than others.

History tells us what happens when the demand for water is greater than the supply, a real concern when it is realized that the world's demand for water is expected to double by the year 2000. Most recently, in this country in the 1930s a 4-year drought compelled 500,000 Americans to leave what came to be known as the dust bowl states. John Steinbeck's *The Grapes of Wrath* documents this tragedy well. Droughts in 1954 and 1958 drove 1 million people from the northeastern part of Brazil to Rio de Janeiro and São Paolo, and in the summer of 1964 Tokyo experienced 41 arid days, which so reduced the supply of tap water that it was available only part of the day. In some places people must queue up for their daily water ration. The increase in the uses of water not only reduces the quantity of water in the future but also the quality. For example, municipalities use about 625,000 tons of water per day for each million citizens. Of this quantity, 500,000 tons of water result in sewage. In fact, the increase in unclean water as the world becomes more developed may threaten human life long before the water supplies are exhausted.

Types of water pollutants

The Public Health Service of the U.S. Department of Health, Education, and Welfare (HEW) defines water pollution as "the adding to water of any substance, or the changing of water's physical characteristics in any way which interferes with its use for any legitimate purpose." This definition is very broad and includes eight categories of pollutants, according to HEW: (1) sewage and other oxygen-demanding wastes, (2) infectious agents, (3) plant nutrients, (4) particulates, (5) radioactive substances, (6) mineral and chemical substances other than organic chemical exotics, (7) heat, and (8) organic chemical exotics.

Sewage and other oxygen-demanding wastes

Sewage and other oxygen-demanding wastes constitute one of the largest pollutants of our waters. (See the insert, "The Decline of Lake

Erie.") This group includes those materials usually derived from domestic sewage and animal or food processing plants which can be broken down to stable molecules when decomposed with the action of aerobic bacteria. Formerly, when quantities of organic waste deposited in natural bodies of water were low, the aerobic bacteria naturally eliminated this kind of pollution over a period of time. However, with the increase of population, the tremendous quantities of organic waste cannot be easily reduced and, hence, result in polluted water. (See the insert, "Where Does All the Waste Go?")

The aerobic bacteria, as their name indicates, require a great deal of oxygen. Consequently, any oxygen found in the body of water in which the pollutants occur is in great demand and often rapidly reduced. Such oxygen reduction is used as an indicator of the degree of pollution and is measured as the *biochemical oxygen demand*, or BOD. Along with the BOD, the quantity of oxygen in a body of water is indicated by the kind of animal and plant life which lives there. When dissolved oxygen is reduced below 4 to 5 parts in 1 million parts of water, fish are scarce; only the very durable types survive. Further reduction of the amount of oxygen results in an increase in anaerobic bacteria, that is, those which do not require oxygen to carry on their metabolism. Thus, several criteria can be used to indicate the amount of pollution in a body of water (Fig. 11-9).

Figure 11-9
Dead fish, the victims of severe water pollution. (From *Archives of Environmental Health*, vol. 22, no. 5, 1971. Courtesy of American Medical Association. Copyright © 1971.)

Infectious agents

Infectious agents, that is, disease-causing organisms, also are significant pollutants of water. These microorganisms, usually viruses or bacteria, result from the activities of tanning and slaughtering plants and also from the sewage waste disposal procedures of cities. The addition of chlorine to the water has greatly reduced the threat of this type of pollutant in many parts of the world. Diseases such as typhoid, dysentary, cholera, and tuberculosis were all caused and have produced sporadic epidemics because of the discharge of sewage containing these disease organisms into waters later bathed in or drunk. The connection between disease and water-carried microorganisms was suspected even before the bacteria that caused some of the significant diseases, such as tuberculosis or cholera, were identified. Rapid dilution of the infectious agent, exposure to sunlight, and other factors quickly affect the length of time the microorganisms will persist in the contaminated water. However, tuberculosis bacteria have been found at as much as 3.5 miles distant from the tuberculosis sanitarium where they were discharged in wastes.

Plant nutrients

Another significant group of water contaminants consists of the plant nutrients, dissolved mineral substances from various sources, such as detergents, commercial fertilizers, or animal or human wastes. Two of the main elements involved in this type of pollution are phosphorus and nitrogen, although potassium, calcium, magnesium, and sulfur can also create difficulties.

Overabundance of these elements causes many problems, including "blooms" of water plants. Blooms often withdraw great quantities of oxygen from the water, to the detriment of other organisms, and produce a bad odor when the plants decay. Some decomposing plants are known to produce a strychnine-like toxin which kills domestic animals, including cattle.

Excess nitrogen as nitrate in water can also be toxic. Domestic animals and people, especially babies, are affected because bacteria in the digestive tract converts the nitrates to nitrites. Nitrites interfere with the oxygen-carrying capacity of the blood, producing in babies *methemoglobinemia,* in which the baby turns blue and may die.

Unfortunately, in this century for the first time in the history of the earth these materials are excessively available. Previously, all these elements occurred in more balanced amounts in their role in the makeup of the biosphere. For example, until the 1930s in this country all the nitrogen needed by crop plants and forests was supplied by biological processes, mainly the breakdown of animal manure and fixation from the atmosphere by bacteria and blue-green algae. The nitrogen was removed from the soil in the protein of harvested crops or as nitrogen compounds, such as ammonia, released by microorganisms. Excess nitrogen did not contaminate runoff waters or groundwater. Today, agricultural practices such as fertilization with nitrogen or phosphorus compounds, or numerous feedlots for cattle raising with the accumulation of animal waste, supplement the natural cyclic production of these elements. Excess amounts cannot be retained by the soil and drain off into waterways.

The problem will probably worsen as the population grows and an increasing number of cattle, hogs, and fowl are raised for human consumption. The quantity is already great. In 1970 *Time* magazine reported the nation's livestock population at 50 million pigs, 38 million cattle, and 350 million chickens.

In addition, an increase in the per capita consumption of meat is occurring. For example, beef consumption in the United States rose from 80 pounds per person in 1959 to 110 pounds per person in 1968.

With more numerous animals comes abundant manure. In Minnesota it has been estimated that 14 million fowl, 5.5 million cattle, 2.5 million hogs, and 750,000 sheep produce waste equal to 60 million people. Daily manure production from 100,000 chickens is said to equal approximately 12 tons; and from 10,000 cattle, approximately 260 tons of solids plus 100 tons of liquid manure. Across the country approximately 2 billion tons of animal waste is produced each year. In terms of nitrogen or phosphorus pollution, such quantities of manure are serious. For example, 10 hogs housed for 175 days, or 1 dairy cow kept for 1 year, will excrete annually approximately 150 pounds of nitrogen, enough to adequately fertilize 1 acre of corn!

The concentration of such large quantities of manure in small areas in feedlots further complicates the problem. Although feedlots permit automated feeding and decentralized slaughtering facilities and ease the general supervision of the animals, the removal of the abundant manure is too expensive to undertake. Hence, natural runoff and seepage are allowed. Unfortunately, this practice produces abundant pollution in nearby streams and lakes. For example, it has been shown that 30 to 50 percent of the rural water supplies in Missouri contain more than 5 ppm of nitrate nitrogen, the majority of which came from livestock wastes. In Colorado studies showed that the total nitrate seeped into 20 feet of soil beneath corrals was 1,436 pounds per acre.

The large number of feedlots is not helpful either. Iowa alone has 50,000 cattle lots. Some are extraordinarily large: one-half of the 23 million cattle marketed from feedlots in 1968 came from lots with a capacity of 1,000 or more. Lots are also maintained which market up to 1,500 hogs and 20,000 chickens per year (Fig. 11–10).

Effective solutions to the problem of agricultural waste disposal are yet to come. Research is under way, but short of hauling the manure away and distributing it on fields, an old-fashioned ecologically sound form of recycling not economically sensible today, no solutions have been uncovered. A few have been suggested, however. One approach would be to reduce the amount of excreted waste. Research shows that grain-size tabs eaten by cows are retained in their rumen. There they substitute for roughage. Consequently, the cattle do not eat much hay and excrete up to 40 percent less manure than those cattle not fed tabs. Another suggestion calls for special bacteria which will quickly decompose the manure.

Figure 11–10
Large feedlots often pollute nearby waterways. (Courtesy of Iowa State University, Extension Service, Ames, Iowa.)

Chemical fertilizers also contribute to the problem. For example, in 1980 it is estimated the United States will meet its needs for food and fiber by fertilizing with 12.2 million tons of nitrogen and 2.9 million tons of phosphorus on approximately 300 million cultivated acres. Compared with the use of fertilizer in 1969, this represents an increase of 80 percent for nitrogen and 42 percent for phosphorus.

Besides animal and human wastes and chemical fertilizers, another source of pollution is detergents, especially those with phosphate. In 1969, 2.6 million pounds of the 6 billion pounds of synthetic detergents sold were eventually deposited in our streams and lakes as phosphates. The extraordinary effectiveness of phosphate detergents for removing dirt accounts for their popularity. The phosphates combine and inactivate ions such as calcium so that they cannot react with and disable the dirt-dissolving molecules in the detergent, called "surfactants."

Pollution with phosphates is usually accompanied by particulate pollution. This is because soil particles have a great affinity for phosphate molecules. In fact, the fine soil particles in a river may hold adsorbed on their surfaces up to 1,000 ppm of phosphorus, even though phosphorus in solution may be only 0.005 to 0.010 ppm. Hence, a good method for minimizing the threat of phosphorus pollution is to retard the erosion of the 4 billion tons or so of sediment that wash into U. S. waterways each year, 75 percent of which comes from agricultural and forest lands.

New detergents are now on the market which are low in phosphates. Unfortunately, they are also quite caustic and may cause infant and child poisoning if accidentally swallowed. Control of nitrogen and phosphorus pollution will depend on unproved disposal of animal wastes and probably more stringent use of fertilizers and detergents. An alerted public may be of help in pressing for solution. In the meantime, animal waste will continue to run off in rainwater, although in a few places, such as Nebraska, laws now in effect require the construction of reservoirs to retain feedlot runoff until bacterial action is well along. More controls like this are needed.

Particulates

Particulates, mainly soil and mineral particles, are serious water pollutants. Some of this pollution is natural because streams and rivers normally erode their banks. However, today, mainly because of agricultural practices, rivers are more turbid than ever. For example, from the 11,500 square miles in Maryland, Pennsylvania, West Virginia, Virginia, and the District of Columbia that lay in the Potomac River basin, approximately 50 million tons of soil is eroded per year, and of this, approximately 2.5 tons are deposited in the Potomac estuary.

The Mississippi River delivers an average of more than 500 million tons of sediment to the Gulf of Mexico each year, about one-half of the total average annual sediment carried to the oceans by all the rivers in the continental United States. The total sediment deposited in U.S. streams and rivers each year must be about 4 billion tons (Fig. 11-11). Concentrations of sediment in the river have been measured at from 200,000 to 50,000 ppm, with occasional records of 600,000 ppm.

Particulates erode power turbines and pumps, reduce populations of fish and other aquatic organisms, destroy aquatic plants through the reduction of the required amount of light, and shorten the life of dam reservoirs, severely limiting their function in water control and irrigation projects.

Radioactive substances

The increased use of power reactors and radioactive materials in industry, medicine, and research threatens water purity with new pollutants, radioactive materials. All life that comes into contact with appreciable amounts of radioactivity risks serious damage. The energy and atomic particles given off can damage parts of cells so that they malfunction. Damage to genetic material can produce mutations.

The most serious radioactive material involved in water pollution so far is radium, mainly because of its long half-life. (The half-life is the amount of time required for one-half of the element to cease being radioactive.) Approximately 5,000 grams of radium are produced as waste each year. About 2 percent is thought to escape into natural waters, resulting in a serious contamination problem.

Other radioactive materials, such as strontium and uranium, may increase in the environment in the near future as the population grows

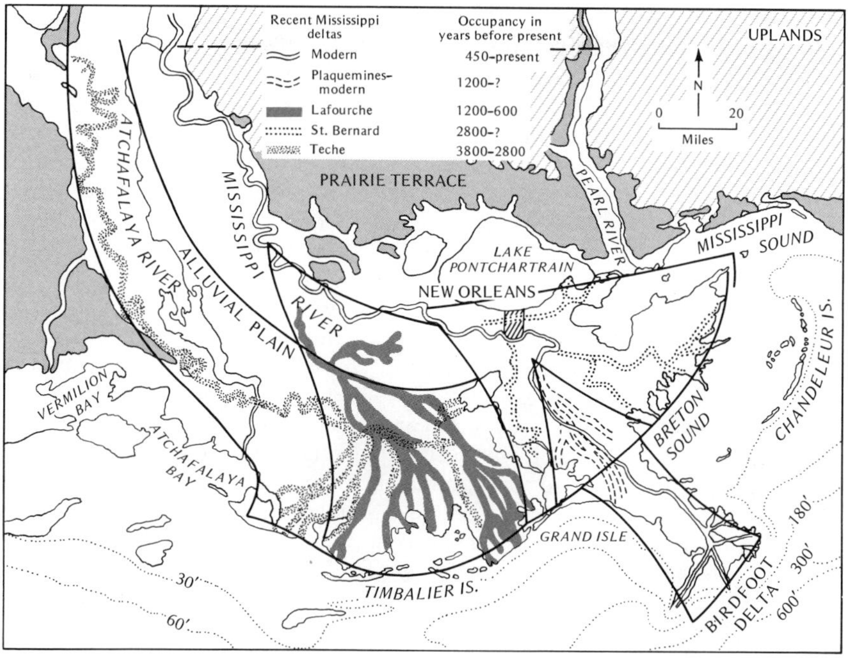

Figure 11–11
Deposition of silt from land far up the Mississippi Valley has gradually enlarged the delta plain. A, from about 450 years ago to the present; B and C, initiated approximately 1,200 years ago; D, begun approximately 2,800 years ago; E, oldest deposition, between 3,800 and 2,800 years ago. (From H. R. Gould, *The Mississippi Delta Complex in Deltaic Sedimentation*, ed. by J. P. Morgan. Special Bulletin 15. Courtesy of the Society of Economic Paleontologists and Microbiologists.)

and depends increasingly on nuclear reactors for power (see Chapter 12). Today, approximately 25 uranium mills and refineries in operation or being constructed process approximately 21,000 tons of uranium-bearing ore per day. Even today, the safe disposal of radioactive wastes is difficult. At the predicted rate of increase of use of radioactive materials, the problem may be acute in 20 to 50 years.

The effects of radiation on human beings is cumulative. Therefore, it is exceedingly important to accurately estimate tolerable doses. In the future acceptable doses will continue to depend on the amount of exposure. However, this demands that the half-life of the materials released, and the quantity of the waste, must be known. The current "maximum permissible concentration" of radium in the domestic water supply (4 micrograms per liter) and of strontium 90 (0.5 microgram per liter) may be reduced.

Mineral and chemical substances other than organic chemical exotics

Another group of serious water pollutants are the mineral and chemical wastes produced by industry or mining operations. The group includes salts, metals, acids, and oil. The water pumped from the ground in oil drilling has been a serious polluter in the past because it contains a great quantity of salt. So has the salt that leaches into rivers which course through arid regions. For example, the Colorado River at Yuma, Arizona, carries more than one ton of dissolved salt per acre-foot of water. Samples of the Pecos River water in Texas during low flows have indicated concentrations of more than 15 tons of dissolved salt per acre-foot of water. Another pollutant has been the discharge from coal mines of acidic waste such as sulphuric acid, formed when sulphur-containing coal comes in contact with moist air. It increases the hardness of water, has a serious effect on life, and corrodes concrete and steel. In 1940 damages caused by this acid were estimated at more than $2 million per year.

Perhaps the most serious pollutant in this category is oil, especially when afloat on the sea. Many minor oil pollution incidents occur every year, and occasionally a major disaster occurs which makes clear the kind of problems provoked by oil on the loose (Fig. 11–12). A typical incident occurred off Louisiana in February 1969. Fire erupted on an offshore platform of several oil wells. After one month explosions were set off to put out the blaze. Unfortunately, oil from four wells continued

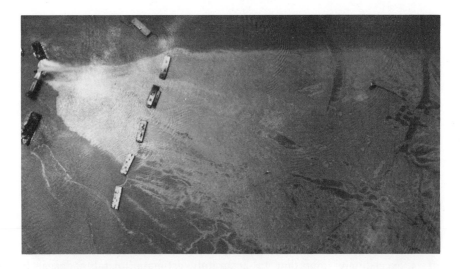

Figure 11–12
Leaking offshore oil well in the Gulf of Mexico. The fire on the drill platform has been extinguished, but oil continues to pour out. This infrared photo, taken from the air, makes possible detection of the surfacing oil, which is not otherwise easily visible when near the sea's surface. (Courtesy of U.S. Department of Interior. Geological Survey Photo.)

to spill out and soon created a huge oil slick, which threatened to coat coastal areas. The leaking wells were eventually capped. Subsequent investigations proved that the wells did not have the required "storm chokes" which automatically shut off malfunctioning wells. This incident did not lead to a serious pollution problem, but when it is realized that approximately 7,000 wells exist offshore in the Gulf of Mexico alone, and an additional 2,000 offshore wells in other places, the threat of oil pollution in the area can be better appreciated.

Another example of an offshore leaking oil well does not end so happily. On January 20, 1969, a newly drilled well "blew out." Oil escaped to cover 800 square miles of ocean off the coast of Santa Barbara, California. Eventually, much washed ashore. Eleven days later, after much damage to wildlife, notably birds, the leak was plugged with cement. By 1970, the owners of the well, the Union Oil Co., had spent up to $5 million to clean up the mess.

Perhaps the largest recent disaster involving oil pollution occurred on March 18, 1967, at 9:00 A.M., when a Union Oil Co. tanker, the Torrey Canyon, ran aground on Steven Stones Reef near the Cornish Coast. Six of her 18 storage tanks burst. Before she sank on March 30, all the oil had probably drained out, approximately 117,000 tons. The polluting effects on nearby coasts and wildlife were disastrous. By late 1969, $7.2 million in damages had been doled out.

Polluting oil from tankers and other sources will probably increasingly be a problem. Now it is estimated that 284 million gallons of oil are spilled every year into the ocean, most of it from transport tankers. This is 10 times the amount carried by the Torrey Canyon and enough to coat a beach 20 feet wide with a $1/2$-inch oil layer for 8,633 miles.

Improvements in drilling techniques and methods to locate oil will probably constantly increase oil production and, hence, oil pollution. So will shipping practices. For example, the number of oil tankers is increasing. Between 1939 and 1960 the number of oil tankers increased from 1,500 to 2,500. By 1980 the 4,000 tankers on the seas carrying more than 600 million tons of crude oil are predicted to increase 5 times. Furthermore, tankers are larger than ever. In 1970 tankers 10 times larger than those used in 1950 were common, so more oil could be transported. Enough oil is moved across the seas annually to uniformly cover the earth with a paper-thick sheet of oil. In addition, tankers are now fueled with oil, not coal, the fuel used exclusively 50 years ago.

Although tankers do not exclusively account for the 1 million tons of oil that are dumped into the sea annually in some form, they probably are the prime source, certainly contributing more to the problem than offshore drilling operations. This is because tankers do not have to be damaged to spill oil. In fact, a common practice, and a major source of pollution, is "tank flushing." Oil carried in tankers acts as ballast. After the tanks are emptied of oil in port, they are filled with water for the return trip. The oil-polluted ballast water is dumped into the open sea. Tank flushing is, in fact, probably the major source of oil pollution of the ocean. In 1962 the Shell Oil Co. developed a method to separate oil from ballast water. Shippers tacitly agreed to use the

Figure 11-13
Oil-soaked sea birds dead from fuel oil pollution of the ocean. (Courtesy of R. Furon. Photo from Photo Groupe Vert. Document de la Société Nationale de Protection de la Nature et A'Acclimatation de France.)

technique, but because the procedure proved inconvenient, dumping of tank flushings still goes on.

The effects of oil in the ocean have not been completely cataloged. Yet enough is known to incriminate oil as a serious threat to living things. Birds are frequent victims (Fig. 11-13). Some ornithologists estimate that 50,000 to 250,000 birds are killed each year by the effects of oil. The oil soaks into their feathers, displacing the air normally lodged there and interfering with bouyancy and maintenance of a constant body temperature. Once birds become oiled they seldom survive, despite efforts to clean them. Thousands of birds were cleaned after the Torrey Canyon disaster, but only a few are known to have survived. A similar situation occurred after the Santa Barbara disaster. Most deaths appeared to be due to respiratory difficulty, but little evidence has been gathered to thoroughly explain the relationship.

Deleterious effects on other forms of life are also accumulating. Studies following the wreck of the tanker Tampico Mara in March 1957 indicate that spilled oil may affect other forms of aquatic life more seriously than was previously suspected. This tanker grounded and blocked the entrance to a small unpolluted cave in the Pacific Ocean. Although the oil apparently disappeared in approximately 1 year, longer-lasting effects were noticed. The sea urchins and abalone populations remained reduced after 4 years, and by 1967, a few species of organisms still had not returned. The effect on organisms not so easily observed, such as plankton, must also have been serious.

Serious indirect effects of oil pollution on human beings are suggested by the discovery of great accumulations of the cancer-inducing hydrocarbon benzypyrene in shellfish, some, such as oysters, used as human food. If fish are also accumulating hydrocarbons, the threat may be worse. More research is needed to determine the extent of the danger. Benzpyrene and other aromatic hydrocarbons found in oil are very persistent and are suspected to be easily transferred from level to level in food chains. Since many aromatics and low-boiling saturated hydrocarbons are extremely toxic, accumulations in organisms may be fatal. Saturated straight-chain hydrocarbons of higher boiling points are apparently not as toxic, although they still affect marine organisms. Such molecules tend to resemble naturally produced hydrocarbons known to control animal behavior, such as food detection, escape, "homing," and reproduction. For example, naturally produced hydrocarbons enable starfish to detect oysters, their prey. Hydrocarbons in petroleum products resemble these naturally produced molecules and conceivably may interfere with the animal's physiology.

The conditions necessary to break down rapidly even the most persistent hydrocarbons are being studied. Apparently, oil distributed on the surface of the sea spreads into a thin layer. The lighter molecules evaporate. Heavier molecules are adsorbed onto particulate matter and eventually sink. Bottom-dwelling organisms, over a period of several months, break down the molecules into new products. Aggregations of durable hydrocarbons may persist for a longer time. More research is needed to determine to what extent different environmental conditions affect the degradation.

Some hydrocarbon aggregations float for long times. Thor Heyerdhal, who twice attempted to sail papyrus craft on the open Atlantic to prove that ancient people could have migrated in these boats, after his second voyage in 1969 remarked on the great number of "tar balls" floating around his craft as compared to his first voyage in 1947 (Fig. 11–14). His observation gives indication of the increase in oil pollution of the seas in the years between trips.

Control of oil spilled on the sea is difficult. Up to now, spreading of small spills had been done by corraling the spill with logs chained end to end. Retrieving the oil has also been tried, but this technique is costly and inefficient. Another method involves spreading some material, such as straw, which can absorb the oil. The oil-soaked straw is then collected and trucked away. This has proved most effective with oil washing up on shore. The most extensively used technique is to spray the area with detergent. The detergent emulsifies the oil and encourages its breakup. The oil is not changed chemically, however, but merely is physically broken into tiny globules which enhance bacterial action. Unfortunately, the detergents used in the past have themselves contained toxic hydrocarbons. Much damage in the wake of the Torrey Canyon was caused by the detergents used, not oil. Now "nontoxic" detergents are being tried. Any method of control of spilled oil, however, depends on a calm sea, a condition seldom found.

Several international conferences have been held to attempt to

Figure 11-14
Balls of tar collected by Thor Heyerdahl on his second ocean voyage. (From *The Ra Expeditions,* by Thor Heyerdall, translation by Patricia Crampton. English translation copyright © 1971 by George Allen & Unwin, Ltd. Reproduced by permission of Doubleday & Company, Inc.)

obtain criteria for regulation of the disposal of oil in the ocean. How severe the oil pollution of the marine environment would be without these rules will never be known. Yet it is clear today that, in spite of international agreements, oil pollution of the ocean is now a serious problem and is becoming worse as the years go on.

Heat

Another serious pollutant is heat. Water withdrawn from natural bodies of water to cool industrial apparatus is eventually returned, heated, to the natural environment. The hot water interferes with natural conditions in the lake or river. Such pollution, called *thermal pollution,* is created by chemical and metal industries, electrical power plants, notably those run with atomic energy, and coal and petroleum companies. The problem will probably worsen within the next few years, mostly because of the great increase in electrical power plants.

In the past, the production of electrical power has doubled approximately each 10 years. In 2000 A.D., the resultant thermal pollution will have increased approximately eightfold because of the increase in quantity and size of electrical power plants around the world. In 1969 the Federal Power Commission reported that 59 fossil-fuel plants, plus additions to existing plants, are scheduled for completion during 1967–1973. Forty-one more nuclear power plants plus enlargements are also scheduled.

Thermal pollution results from the generation of electricity. Water is used to reduce the temperature of the steam which passes through the turbine, generating electricity. If the steam is properly cooled, it can be recycled in the apparatus. The large quantity of heat given up in this cooling process is absorbed by water piped in from lakes and streams (Fig. 11–15).

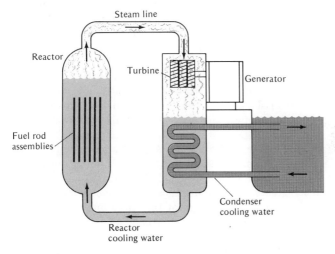

Figure 11–15
Steam heating and cooling systems in a nuclear reactor.

The heated water returned to lakes and streams appears unchanged except for temperature. Yet the elevated temperature seriously changes the physical and chemical properties of water. Some physical properties of water altered by heating are viscosity, vapor pressure, surface tension, gas solubility, density, rate of ion exchange, and evaporation rate. The impacts on living organisms, directly or indirectly, are profound. Consider solubility of gas. Heated water holds less dissolved gas than colder water. In fact, in fresh water raised from 20 to 30°C, the dissolved oxygen decreases by 17 percent. Aerobic organisms often succumb in even slightly warmer water. The almost total elimination of fish from the Mahoning River in Ohio in summer months has been traced to the high temperatures in part caused by the use of the river water for cooling by industrial and electrical power plants.

The heating of saltwater areas also has serious results. Gases, such as oxygen, are less soluble in saltwater than in fresh water. Increased solubility of salts, however, occur at higher temperatures. Furthermore, an increased rate of evaporation and greater chemical reactability may additionally increase the salt content. Hence, saltier and less oxygenated water may be produced, neither condition conducive to maintenance of life.

The effects of these changes in the nature of the water environment have noticeable and deleterious effects on wildlife. The reproductive behavior of fish is an example. In several species, falling or rising temperatures trigger the series of steps necessary for successful reproduction. Furthermore, high temperatures discourage normal development of eggs of some fish, such as the Chinook salmon. Some scientists worry that if the Columbia River should increase its temperature by approximately 5.5°F, fish species would be endangered. Carp eggs also appear not to develop at temperatures between 68 and 75°F. The banded sunfish appears unable even to produce eggs at temperatures 62°F and above. Less directly, barriers of warm water across a stream prevent some migrating fish from completing their journey up- or downstream necessary to complete their life cycle.

Other effects of heat and water on fish can be observed. Some fish are known to increase their respiration greatly — up to 4 times — and then die. The heart-beat rate and the oxygen consumed are greatly increased, probably because the solubility of oxygen in the blood is decreased.

In fact, the amount of oxygen needed by carp, generally considered one of the hardiest fish, at 95°F is at least 3 times more than that required at 33°F. Digestion is also accelerated in this fish. Food is assimilated 4 times faster at 79°F than at 50°F. These evidences of increased metabolic activity within fish suggest that they may well undergo physiological stress in heated water, frequently causing death.

In some situations the increased metabolic rate, if not too severe, may prove beneficial. In Scotland, sole have reached market size in only 2 years, instead of the usual 4 years, when grown in heated waters. Data gathered on other fish tend to indicate that cooler water may cause slower growth but that the life span is longer than if the fish are grown more rapidly in warmer waters. These data on natural populations need more study to make clear their implications.

Effects of thermal pollution on other organisms is more difficult to assess. Considering that organisms have evolved into particular environments over great periods of time, and that they today persist there because their life needs are successfully met, it is difficult to imagine that an alteration in aspects of that environment would have much positive effect. Studies show, for example, that an elevation of temperature of a few degrees can eliminate diatoms, small one-celled plants, and effect a "bloom" of blue-green algae. Besides affecting specific organisms, thermal pollution affects the community of organisms. The addition of heated water to a lake or pond produces major alterations in the natural life-sustaining thermal layering of water common to lakes, especially in temperate areas.

Normally, stratification occurs as follows: In the fall the upper layers of water cool and, as noted earlier, increase in density. Eventually, they will sink to the bottom of the lake and displace warmer water, which will rise to the surface. During the winter months the coldest water, which tends to sink because it is denser than warm water, forms ice. The decreased density of ice causes it to rise to the top, where it remains until warmer weather. When spring comes the ice melts and forms cold water, which sinks. As summer approaches, warm water layers form at the lake's surface. At the coming of autumn, the cycle begins again (Fig. 11–16). The effect of the movement of layers of water of different temperatures is highly significant to life. The abundance of nutrients and oxygen produced in the upper layer during summer is distributed throughout the entire lake in the course of a year. Life can, thus, be maintained in the lake.

Thermal pollution alters this natural turnover of the water. Usually, water is pumped from near the bottom of the lake (because it is colder there), used for cooling, and returned much warmer to the upper layers of the lake. This abundance of heated water which is lighter will prolong the conditions, normally present during the summer only, throughout the year. Consequently, the lower water layer will remain poor in oxygen and nutrients, although the upper layer will produce them in abundance. The great quantities of organic material will eventually sink to the lake bottom. The scant oxygen supply is soon depleted, and aerobic organisms die. Sometimes, the upper layer is so hot that little life can thrive there, and so the lake's demise is hastened.

In view of the deleterious effects of thermal pollution, concern is increasing for its control. Current laws require reduction of temperature of returned water at considerable cost. The Department of Interior estimated in 1968 that a total investment of $1.8 billion would be necessary over a 5-year period to restore thermally polluted waters to their original temperature.

Thermally polluted water can be cooled in several ways. Shallow artificial ponds of hundreds of square acres can cool the water from electrical plants of moderate size. The water can be pumped into the pond, and the heat dissipated through evaporation. A 1,000-megawatt power plant of the future would require a lake as large as 1,000 to 2,000 acres, that is, 1 mile wide and 3 miles long. Such a lake would be designed with a sloping bottom from a few feet deep at one end to 50 feet at the other. The water for cooling, drawn below the surface at the deep end, is pumped through the plant and discharged at the shallow end. A small stream flowing into the lake would replenish the water supply lost by evaporation.

Cooling towers are another proposed solution (Fig. 11–17). In these devices heat is transferred to the atmosphere through evaporation. The heated water is sprayed or dripped inside a 300- to 500-foot tower. Fans hasten cooling. Disadvantages accompany its use. The tower would discharge a great amount of water vapor into the air. A 1,000-megawatt power plant would eject from 20,000 to 25,000 gallons of evaporated water per minute, an amount that equals 1 inch of rain on an area of 2 square miles per day. Low temperatures would condense the vapor into fog or ice and perhaps limit its use in cold climates. It cannot be used with salt water because salt discharged into the atmosphere would destroy surrounding vegetation. The towers could also be considered unesthetic. They are expensive and would probably increase the cost of producing electrical power.

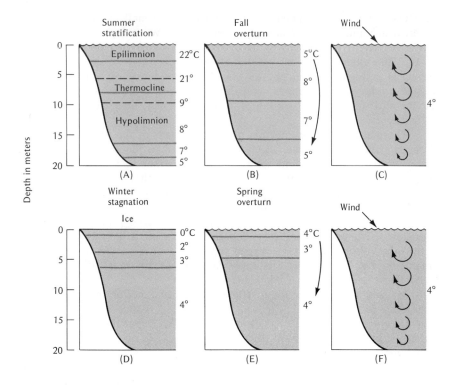

Figure 11-16
Seasonal temperature changes in a lake in the temperate zone: seasonal cycle of temperature changes in a temperate-zone lake. Water, like air, increases in density as it cools, reaching a maximum density at 4°C. In the summer, the top layer of water, called the *epilimnion,* becomes warmer than the lower layers and therefore remains on the surface. Only the water in this warm, oxygen-rich layer circulates. In the middle layer, which is called the *thermocline,* there is an abrupt drop in temperature. Since the thermocline water does not mix, it cuts off oxygen from the third layer, the *hypolimnion,* producing summer stratification (A). In the fall, the temperature of the epilimnion drops until it is the same as that of the hypolimnion. The warmer water of the thermocline then rises to the surface, producing the full overturn (B). Aided by the fall winds, all the water of the lake begins to circulate (C), and oxygen is returned to the depths. As the surface water cools below 4°C, it expands, becoming lighter, and remains on the surface; in many areas, it freezes. The result is winter stagnation (D). In spring, as ice melts and the water on the surface warms to 4°C, it sinks to the bottom, producing the spring overturn (E), after which the water again circulates freely (F). (From H. Curtis, *Biology,* Worth Publishers, Inc., New York, 1968.)

Some researchers believe that heated water can be used beneficently. Proposals include heating buildings, although in most cases the water would not be hot enough and the expense of piping would be great. The Finnish city Tapiola has used heated water since 1953, with considerable success. Other proposed uses include irrigating fields, since

327   Types of water pollutants

warm water could cause rapid germination and growth and prolong the growing season; filling ponds for aquaculture; aiding evaporation in desalinization plants; and providing recreational facilities, such as swimming areas, even in cold months.

Perhaps the control of thermal pollution must be written into law. Several federal agencies have attempted regulation. The Department of the Interior, the Atomic Energy Commission, the Federal Power Commission, and the Army Corps of Engineers are all involved in one way or another. The Atomic Energy Commission distributes operating licenses

Figure 11–17
Top: Mechanical draft cooling towers. Bottom: Natural draft cooling towers. (The Marley Co., Mission, Kan.)

for nuclear-powered plants (although no federal permission is required for electrical plants generated with fossil fuels). The Federal Water Quality Administration and the Fish and Wildlife Service both review the applicants and refer opinions to the Atomic Energy Commission. The Federal Power Commission licenses hydroelectric plants and seeks the opinions of various state and federal agencies before granting approval.

In 1970 a new law was enacted which requires that a permit be granted by the state or the interstate water pollution control agency in order to engage in activity that may result in any discharge into the waters. The water resource or water pollution control laws of 35 states require that a discharge permit be obtained to build or to operate plants that discharge municipal industrial wastes. How many of these apply to thermal pollution, however, is a question.

In 1965 the Water Quality Act forced the states to establish water quality standards for interstate streams and coastal waters. The Secretary of the Interior has the power to set these standards if the states did not comply. To help in reviewing the states' proposed water standards, the Secretary of the Interior appointed a National Technical Advisory Committee on water quality standards. The committee's recommended temperature requirements were as follows: streams should not be heated more than 5°F, the expected minimum daily flow for that month on lakes or reservoirs heated more than 3°F, and certain significant fish-spawning areas, such as the headwaters of salmon streams, should not be warmed at all. Any violators of these regulations can be acted against by the federal government. Such action was taken in 1970 against the construction of a canal by the Florida Power & Light Company to be used to drain heated water from its electric plants to Biscayne Bay, 25 miles distant.

Organic chemical exotics

Materials in this group are defined as synthetic organic chemicals, including such materials as pesticides. The impact on plants and animals owing to an accumulation of these materials is now under intensive study. In the next chapter we shall look at their effects and become familiar with the problems that surround their use.

Solid waste pollution

It is difficult to discuss water pollution without examining the problem of the disposal of solid wastes. Too frequently waterways, as well as the oceans and marshes, have been the "dump" for solid trash because other disposal methods are too expensive or too inefficient. The amount of solid trash accumulating over the country is great. The Crusade for a Cleaner Environment reported the following collection of solid trash from a 1-mile stretch of a highway in Kansas: 770 paper cups; 730 cigarette packages; 590 beer cans; 360 pop, beer, and whiskey bottles; 90 beer cartons; 90 oil cans; 50 livestock feed bags; 30 paper cartons; 26 magazines; 20 maps; 16 coffee cans; 10 skirts; 10 tires; 10 burlap bags; 4 auto bumpers; 4 shoes without mates; 2 undershirts; 2 comic books; 2 bedsprings; and 270 miscellaneous items. Such an abundance of solid trash exemplifies our national problem of prodigious solid waste production and the problem of its disposal.

The Bureau of Solid Waste Management estimated that in 1967 the average per capita production of household, commercial, and industrial wastes in the United States was 10 pounds daily, a total of 360 million tons per year. This estimate does not include the 1.5 billion tons or so per year of animal wastes, or the 550 million tons per year of agricultural wastes and crop residues. A grand total would probably be close to 3.5 billion tons of solid wastes per year.

Specifically, the yield of waste per person in 1969 included 188 pounds of paper, 250 metal cans, 135 bottles and jars, and 338 caps and covers. Add to this a 2 percent increase in solid waste each year and the addition of wastes accruing from a 2 percent national population increase each year, a total 4 percent increase in the solid waste production.

The disposal of such quantities of waste is already an expensive problem. The annual expenditure is close to $5 billion. For a public service this amount is exceeded only by school and road costs. By 2000 A.D. the expense will most likely at least double. In New York alone it costs an estimated $30 per ton to collect, transport, and dispose of the state's refuse, 3 times the cost per ton of coal mined in West Virginia and delivered in New York.

The collection and transportation of refuse is the most costly aspect of its disposal, approximately 70 to 80 percent. Annually, across the nation it probably costs $500 million to collect the roadside solid trash, such as cans, bottles, and waste paper.

After collecting the solid waste, the final stages of disposal also prove expensive, because of rising labor costs, uncertain markets for products, and problems in techniques, such as effectively separating kinds of refuse. In the future, however, recycling may be the main method for disposing of solid waste.

There is some recycling now. By 1969, 35 percent of the total paper produced in the country was recycled. A few metals were also salvaged for reuse. In fact, reclaimed iron and steel accounted for 50 percent of the total production. Approximately 15 percent of recycled old rubber was used for chemicals and fibers. As early as 1963, reused copper accounted for 40 percent of the supply in the United States, discarded lead was recovered at a greater rate (more than twice that produced from domestic mines), and scrap aluminum made up 25 percent of the total supply. Recycling methods still need improvement, however, when it is realized that the waste metals in the 180 million tons of annual municipal solid waste alone are worth more than $1 billion.

Interesting innovations are being tried for the reuse of waste. One is the construction of roadbeds with particulate glass. Federally funded research at the University of Missouri already has yielded a glass road constructed for testing in Fullerton, California. The 600-foot-long street in a heavily trafficked industrial area is composed of 24 parts broken glass, 12 parts stone dust, 2 parts asphalt, and 1 part lime. Approximately, 300 tons of reclaimed broken glass were used in the 3-inch-thick, skid-resistant "glasphalt" pavement. Glass for glasphalt should never be scarce. In 1970 it is estimated that 12 billion nonreturnable beer and pop bottles were produced, approximately 33 million per day.

Recycling of other materials is being attempted. Disposal of waste paper, one of the main ingredients of solid waste, is under experimentation. Since cattle easily digest cellulose, researchers are preparing paper trash in a manner appealing to cattle.

The Japanese have constructed large compactors to reduce waste into dense sturdy blocks that can be used in construction. The city of Cleveland is attempting the same enterprise by mixing solid waste, dried sewage, river and lake dredgings, and incinerator residue into solid blocks, hopefully to be used to reclaim submerged areas next to Lake Erie.

In a sense the use of the heat produced from burning trash is also recycling. Experiments with incineration have been tried for some time with profitable results. In Sundeberg, a suburb of Stockholm, a vacuum-powered pipeline collects the solid waste from approximately 5,000 apartments, carries it on a fast airstream to an incinerator-boiler, where it is burned. The heat generated is used to warm residential units nearly 2 miles distant.

In Milan, all streetcars and subways will be run with incinerator-generated electricity eventually. Incineration has also been tried in the United States. For example, a refuse incinerator in Hempstead, Long Island, drives both a 2,500-kilowatt electric power plant and a 420,000-gallon/day water-desalting plant; in Delaware County, Pennsylvania, 49 municipalities have formed a disposal district which includes three incinerators, each with a 550-ton capacity. The district can serve an ultimate population of 750,000. Other towns in the United States have considered the construction of incinerators but are discouraged by their expense. Furthermore, an incinerator produces considerable air pollution, a factor that soon may add more expense to the process if fines evolve as penalties for polluting.

Besides heat energy, materials are sometimes rescued from incineration ash and recycled. For example, up to 500 pounds per ton of iron and 50 pounds per ton of other significant metals have been recovered in the fly ash filtered from exhaust chimneys in some industries. Fly ash has also been under experimentation as a material useful in building blocks. Even if no materials or energy is retrieved in incineration the process is still valuable to destroy burnable solid waste. Unfortunately, until better methods of burning evolve, a great amount of the unburnable material must be separated from the destructable trash, a time-consuming and costly process.

A more efficient method is being considered which uses controlled atomic fusion energy such as occurs catastrophically in the hydrogen bomb. The process, called "the fusion torch," uses extraordinarily high temperatures to vaporize and ionize solids. Thousands of tons of solid waste could be destroyed each day in one disposal plant with this method in the future, although the vaporous product might create an air pollution problem.

Current methods of solid waste disposal

Today, trash is incinerated more conventionally or disposed of mainly in the open-pit dump, the sanitary land fill, and composting. The open-pit dump is the least sanitary, although 30 percent of the cities in the United States use it. Untreated garbage is dumped into a natural or constructed pit or depression and left to disintegrate slowly. The method has many disadvantages, including offensive odor, unsightliness, and a harbor for disease vectors such as rats, flies, and mosquitoes.

The sanitary landfill is also popular (Fig. 11–18). Trash is deposited in natural gulleys, canyons, or ravines, in artificial long, shallow trenches, or in swampy and marshy areas. Each day the trash is covered with dirt from the surrounding area and compacted. Odors and unsightliness are minimal (Fig. 11–19).

Composting is also used, although as urbanization spreads, it becomes less common. Composting is still sometimes used in rural areas for disposal of a family's garbage. Attempts to do it on a large scale, however, are more expensive than other methods. Large operations must first separate the ferrous metals (usually magnetically) and the other metals, glass, rubber, and plastics by hand. The remaining trash is shredded, subjected to short aerobic periods to increase bacterial action and hasten rotting, and then allowed to cure or incubate for several months. The final product is added to soils to increase their fertility. Some cities have practiced composting commercially. They process their sewerage products into fertilizer products which are then used on lawns and gardens. Milwaukee, Wisconsin, produces Milorganite, and Chicago barges dried sludge from its sewage treatment plants to Florida for use in the citrus groves.

New disposal methods may soon be available. In 1965 the 89th Congress enacted the Solid Waste Disposal Act to encourage basic and applied research to find new and improved methods of effective and economic solid waste management. Specifically, the act has these three goals: (1) to protect man's health and the quality of his environment, (2) to minimize the amounts of ultimate waste being generated, and (3) to maximize the salvage of useful waste material. Under the act, cooperation in waste research is encouraged between private and public institutions. As a result of such official concern, the problem of solid waste disposal may be successfully met. However, the rate of population growth and industrialization suggest that constant effort and attention is needed by an educated public to ensure success. Yet, the only alternative is a country literally covered with garbage.

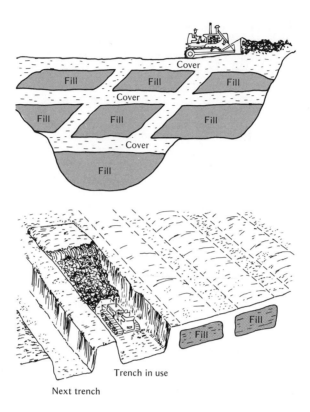

Figure 11–18
Two basic sanitary landfill methods are the area fill and the trench fill. Top: The area method is best used in abandoned quarries, strip mines, gravel pits, and rolling land. Here, a day's refuse is deposited in an area, then spread and compacted. A 6-inch layer of cover soil is placed over the fill. A 3-foot seal of soil eventually covers several fills. Bottom: The trench method is best used in dry, level sites. Often a trench is dug far enough to accommodate one day's refuse. Cover is provided by trenching for the next day's refuse. (The Caterpillar Tractor Co., Peoria, Ill.)

Used-vehicle pollution

Used vehicles are becoming a major solid waste pollutant. In 1966 New York City police estimated that 40,000 abandoned cars cluttered the city. Chicago police claim 21,000. Probably 100 million old tires in the country also add to the problem. By 2000 A.D., at the present rate of abandonment, derelict vehicles will present an acute pollution problem unless some action is taken soon (Fig. 11–20).

If inexpensive means could be found to process old car bodies to remove the valuable components, especially metals, then disposal would be facilitated. The average discarded car contains much valuable scrap: 2,500 pounds of steel, 500 pounds of cast iron, 50 pounds of zinc, 51 pounds of aluminum, 32 pounds of copper, and 20 pounds of lead. At present, not much of this material is recycled.

Figure 11–19
This was once an open burning dump. Now, a modern landfill system prevents smoke and odor. (The Caterpillar Tractor Co., Peoria, Ill.)

Some recycling efforts are proving successful, however. Shredders, of which there are only approximately 70 in the country, are used to break up the car body into fist-size pieces after the manual removal of reusable parts such as radios, air conditioners, and engines. Steel and iron are then withdrawn with a strong magnet. Already the process can purify iron to approximately 98 percent, and improvements soon should produce nearly a 100 percent pure iron scrap. Purity is important, because in the steel-making process, contamination reduces the metal's flexibility. Copper is saved by agitating the scrap in a molten calcium chloride bath. As much as 99 percent of the copper collects at the bottom of the vat and can be easily drained off. Unfortunately, aluminum, zinc, and lead are lost.

Another process is under investigation which will make possible the conversion of iron ore or scrap such as old cars and tin cans into 50,000 tons of iron powder per year by dissolving the scrap in hydrochloric acid. The ferrous chloride and hydrochloric acid produced are pumped into a vacuum crystallizer which flushes away the water and acid. The pure ferrous chloride tetrahydrate crystals that remain are dried in hot air, compacted into briquets, and pulverized into a powder when needed. (Unfortunately, the liquid waste may create another pollution problem.)

Recycling methods to retrieve valuable oils, chemicals, gas, and tar from old tires are also being perfected. In experiments involving destructive distillation in a heated reactor, 1 ton of tires has produced 1,500 cubic feet of gas. Plans are also afoot to use old tires as fuel. Actually, they contain more energy than coal and do not produce the kinds of pollutants as coal.

Thus, steps are under way to handle the 7 million discarded vehicles per year. Whether industry can reap enough profit to hasten the development of car-transmutation plants so that the increasingly acute problem of disposal is met head on is yet to be seen.

Figure 11–20
Junk automobile graveyard in a rural setting. (Courtesy of U.S. Department of Interior, Bureau of Mines.)

Summary

In this chapter we discussed the significance to life of water as an environmental factor. Examined were the characteristics of water (so as to demonstrate how the physical properties of this material were influential in determining the makeup and the activities of plants and animals), the quantity of water used, and its pollution with particulates, sewage, infectious agents, plant nutrients, sediments, radioactive substances, and mineral and chemical substances.

An underlying theme throughout the chapter is the serious deleterious effects that water pollution has on the relationship that living things share with their environment, and the serious consequences to the sustenance of life if this relationship is altered.

The decline of Lake Erie

Lake Erie was distinguished in 1812 as the site of Admiral Perry's triumphant battle with the British Fleet. His victory made the Great Lakes part of the United States of America. Today, Lake Erie is notorious because it is rapidly changing, a victim of uncontrolled water pollution. Lake Erie's decline represents clearly what can happen if the natural bodies of water are taken too much for granted.

Lake Erie, the most southern of the five Great Lakes, is only an average of 58 feet deep, and only a little over 200 feet deep at maximum. The shallow 40,000 square miles of basin contains approximately 10,000 square miles of lake and many significant industrial centers, including Cleveland and Toledo. The already densely populated area — 25 million people — may well be near 30 million by the end of the century.

Much of the dense population is due to the industries originally attracted there because of the abundant fresh water. These industries, such as steel plants, chemical and manufacturing companies, and petroleum refinery plants, every day withdraw 4.7 million gallons of water from the lake, half of which goes to Ohio alone. Thus, the decline of the lake is not only ecologically critical but also economically serious.

Lake Erie's ecological crisis is accelerated aging, *cultural eutrophication*, caused by extreme pollution from man's activities. All lakes naturally age, *eutrophy*, but do so slowly compared to Lake Erie's transformation. Some ecologists believe that Lake Erie, for example, would naturally have evolved to a condition close to its present state in 10,000 years more or less. Instead, because of pollution, the lake is estimated to have aged 10,000 years in only a few decades.

Aging, or eutrophying, lakes change from a *oligotrophic* to a *eutrophic* condition. Oligotrophic lakes are characterized by clear water, sparse algal growth, numerous species of high-oxygen-demanding fish, and very few nutrients derived from decaying organic material, such as phosphate or nitrate ions.

Eutrophic lakes are known by abundant nutrients, especially phosphate and nitrate ions, the unpleasant taste and odor of the water, and abundant algae. Dissolved oxygen is low because the bacteria, in feeding on dead algae and other organisms, deplete the oxygen reserve. Partially decomposed organic material is abundant. High-oxygen-demanding species of fish such as the lake trout or the cisco of the Great Lakes region are replaced by carp and other organisms. Clams and worms that tolerate little oxygen take over.

Most lakes appear naturally to become *eutrophic* because of the accumulation of silt from surrounding areas (Fig. 11–21). Only those lakes which are rapidly accumulating waste from mans' activities undergo cultural eutrophication, although experts disagree on what specifically controls the process. Some claim that it is the nitrates and, more importantly, the phosphates which trigger abundant algal growth. Others claim that it is the abundance of CO_2, released from the bacteria feeding on an excess of organic matter. Whatever the specific cause, Lake Erie is culturally eutrophying. The change over the past years in its fish population is testimony. Once Lake Erie supplied blue pike, walleye, yellow perch, and whitefish for a major part of the country, but today only alewives, carp, and sheepshead are abundant. In 1936, 20 million pounds of blue pike were taken from the lake; in 1963, only 1,000 pounds were harvested. A few decades ago, Sandusky, Ohio, was once one of the largest

336 Chapter 11: Water: Maintenance of quality and quantity

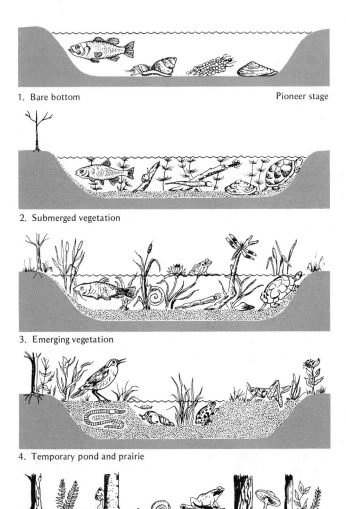

1. Bare bottom — Pioneer stage
2. Submerged vegetation
3. Emerging vegetation
4. Temporary pond and prairie
5. Beech and maple forest — Climax stage

Figure 11–21
Simplified drawings of pond succession based on events believed to have occurred in ponds at the southern end of Lake Michigan. (From *Basic Ecology*, by R. and M. Buchsbaum, reproduced by permission of The Boxwood Press, Pacific Grove, Calif.)

freshwater ports of the world. Today, it supports only a few fishing operators.

The prime cause of the lake's destruction are the abundant fertile materials which it receives from the surrounding area. The amount and kinds of pollution pouring into Lake Erie are startling. Each day 100 million pounds of sewage drains into the lake, including approximately 1.5 million pounds from the Detroit River alone. Not surprisingly, tests in 1965 showed that 89 percent of the total municipal wastes discharged into the west end of Lake Erie, the shallowest and most polluted end, have only primary treatment and no secondary treatment. Yet several millions of people are served or depend on polluted Lake Erie for their water supplies.

Phosphate, a major ingredient in agricultural fertilizers and detergents, is one of the most serious pollutants. In 1965 reports show that 87 tons of phosphates were dumped into Lake Erie each day — each pound estimated to encourage the potential maximal development of 350 tons of algal slime! As a consequence, algae mounds up at certain times of the year on the shores of the lake, where it produces unpleasant odors, clogs pipes, and interferes with navigation and fishing.

Besides phosphates, acids, oils, cyanides, and phenol from steel industries, chemical plants, and car factories in Cleveland, Detroit, and Buffalo, additional pollutants include pulp and paper wastes from Erie, Pennsylvania, and particulate wastes from cement plants and flour mills in Portland. Also pollutants are oil and untreated organic wastes from shipping vessels on the lake.

The rivers that feed into Lake Erie are also highly polluted. The Cuyahoga River was recognized for several years as a fire hazard because it contained great deposits of oil and inflammable chemicals as well as logs, rotted pilings, and old tires. In 1968 it caught fire and severely damaged two bridges which spanned it.

Recently, some steps have been taken to stop the demise of Lake Erie, although some feel that the action is too late. Others feel that, with the help of the federal government, a great deal of money, and a very strenuous enforcement of pollution control, the lake can be rescued. The rescue mission is an immense undertaking. The federal government has already identified 360 sources of industrial waste on Lake Erie, many of which do little to alter their discharged waste. In fact, in composition and effects on the lake, the amount of discharge is said to equal the raw sewage from a population of approximately 4,700,000 people.

However, in spite of the immensity of the task, some improvements have been made in Lake Erie and other polluted waters. In 1965 chlorination of waste was started. By 1970 some industries had pollution treatment facilities in operation; many others were working on them. By 1970, in New York, for example, 14 of the 41 sources of pollution had been controlled, and 18 were under the state commissioner's order to abate their pollution. Eleven were still working on the problem and would be installing pollution devices shortly.

Some federal and state funds have been appropriated in the fight. However, it has been estimated that approximately $1 billion dollars will be needed for pollution control projects in the next 20 years to prevent Lake Erie's death.

Other Great Lakes are following the same route. The depletion of certain game fish in Lake Michigan is noted as a serious sign. Hopefully, the industries and municipalities on the Great Lakes will alert themselves to the symptoms and halt the process incurred by human neglect and abuse.

Supplementary readings

Bigger, J. W., and R. B. Corey. 1969. *Eutrophication: Causes, Consequences, Correctives.* National Academy of Sciences, Washington, D.C. 661 pp.

Bloom, C. 1970. "Heat: A Growing Water Pollution Problem." *Environment Reporter,* vol. 1, no. 1, pp. 1–21.

Coan, G. 1971. "Oil Pollution." *Sierra Club Bulletin,* March, pp. 13–16.

Coker, R. E. 1954. *Streams, Lakes, and Ponds.* University of North Carolina Press, Chapel Hill, N.C. 327 pp.

Cole, L. C. 1969. "Thermal Pollution." *BioScience,* vol. 19, no. 11, pp. 989–999.

Degler, S. E. 1971. *Federal Pollution Control Programs: Water, Air, and Solid Wastes,* rev. ed. BNA Books, Washington, D.C. 176 pp.

Deglar, S. E. 1971. *Oil Pollution: Problems and Policies.* BNA Books, Washington, D.C. 142 pp.

Degler, S. E., and S. C. Bloom. 1969 *Federal Pollution Control Programs: Water, Air, and Solid Wastes.* BNA Books, Washington, D.C. 111 pp.

D'Itri, F. M. 1972. *The Environmental Mercury Problem.* CRC Press, Cleveland, Ohio. 124 pp.

Engdahl, R. B. 1969. *Solid Waste Processing.* Government Printing Office, Washington, D.C. 72 pp.

Fay, J. A. 1969. "Oil Spills: The Need for Law and Science." *Technology Review,* vol. 72, no. 3, pp. 32–35.

Furon, R. 1967. *The Problem of Water: A World Study.* American Elsevier Publishing Co., Inc., New York. 207 pp.

Gilluly, R. H. 1970. "Finding a Place to Put the Heat." *Science News,* vol. 98, no. 5, pp. 98–99.

Golueke, C. G. 1970. *Solid Waste Management.* Government Printing Office, Washington, D.C.

Holcomb, R. W. 1969. "Oil in the Ecosystem." *Science,* vol. 166, pp. 204–206.

Kneese, A. V. 1964. *Water Pollution: Economic Aspects and Research Needs.*
Resources for the Future, Washington, D.C. 107 pp.

Krendel, P. A., and F. L. Parker, eds. 1969. *Biological Aspects of Thermal Pollution.*
Vanderbilt University Press, Nashville, Tenn. 407 pp.

Maass, A. 1951. *Muddy Waters: The Army Engineers and the Nation's Rivers.*
Harvard University Press, Cambridge, Mass. 306 pp.

Macan, T. T. 1963. *Fresh Water Ecology.*
John Wiley & Sons, Inc., New York. 338 pp.

Macan, T. T., and E. B. Worthington. 1968. *Life in Lakes and Rivers,* 2nd ed.
William Collins & Co., Ltd., London. 272 pp.

MacKenthun, K. M. 1969. *The Practice of Water Pollution Biology.*
Government Printing Office, Washington, D.C. 281 pp.

Marx, W. 1967. *The Frail Ocean.*
Coward, McCann & Geoghegan, Inc., New York. 248 pp.

National Academy of Science. 1969. *Eutrophication: Causes, Consequences, and Corrections.*
National Academy of Sciences, Washington, D.C. 666 pp.

National Conference on Solid Wastes Management, University of California. 1966. *Solid Waste Management: Proceedings.*
University of California, Davis, Calif. 214 pp.

Petrow, R. 1968. *In the Wake of Torrey Canyon.*
David McKay Co., Inc., New York. 256 pp.

Popham, E. J. 1961. *Some Aspects of Life in Fresh Water.*
Harvard University Press, Cambridge, Mass. 127 pp.

Smith, J. E., ed. 1968. *Torrey Canyon Pollution and Marine Life: A Report by the Plymouth Laboratory of the Marine Biological Association of the United Kingdom.*
Cambridge University Press, New York. 196 pp.

U.S. Public Health Service. 1960. *Pollution Abatement.*
Government Printing Office, Washington, D.C. 38 pp.

U.S. Water Resources Council. 1968. *The Nation's Water Resources.*
Government Printing Office, Washington, D.C.

Viets, F. G., Jr. 1971. "Water Quality in Relation to Farm Use of Fertilizers." *BioScience,* vol. 21, no. 10, pp. 460–467.

Wadleigh, C. H. 1968. *Wastes in Relation to Agriculture and Industry,* Miscellaneous Publication 1065.
Government Printing Office, Washington, D.C.

Willrich, T. L., and G. E. Smith, eds. 1970. *Agricultural Practices and Water Quality.*
Iowa State University Press, Ames, Iowa. 415 pp.

Wolman, A. 1969. *Water, Health, and Society: Selected Papers.*
Indiana University Press, Bloomington, Ind. 400 pp.

Chapter 12

Technology has but one justification: to serve man's needs for food, shelter, and clothing so that he can be free to develop his unique assets — mind and spirit. Technology whose end result is an impoverished setting for the human mind — let alone technology that kills people — is worthless, a total failure.

Pesticide Pollution
Clarence Cottam

Biocides

Since the beginning of history man constantly has striven to eliminate those organisms which threatened him, either as predators or as destroyers of his health or food. Only recently, through the use of biocides has man come near his goal. Biocides (pesticides) are chemicals intended to kill man's competitors for food, clothing, and shelter. The word "biocide," rather than "pesticide," is used here to emphasize that, because of the unity of life processes, what is intended to kill "pests" probably has effects on all life-forms. The early biocides derived from natural materials and, especially the potent synthesized materials produced in the last few decades, have proved astonishingly effective. Today, however, despite success, these materials have also been proved to have serious disadvantages. Yet whatever the advantages or disadvantages of biocides, one statement is true: the definition of man's relationship with his environment had been forever changed once they were used in abundance.

A study of biocides brings into focus once again the nature of the relationship between environment and living things, including man. The evolution of species and ecosystems has been millenia long, and their relationship is a sophisticated and delicate one. To alter this relationship unthinkingly as with biocides could bring about disaster for many species, despite apparent benefits. Thus, a study of biocides is important, because these chemicals may be one of the most illustrative examples of how man is affecting his natural world and altering the life-sustaining relationship between plants and animals and environment.

Biocides in the past

The first biocides, the "botanicals," were derived from plant materials. Powdered tobacco leaves dusted on plants were discovered in 1763 to effectively control sucking insect pests. Another botanical, pyretherum, derived from the chrysanthemum plant, and rotenone, derived from the roots of some tropical plants, were used in the early 1800s. Pyretherum and rotenone are still used today.

Later, other materials were used. In the nineteenth century sulfur was commonly sprinkled on leaves and stems to control chewing insects, and is still used today. Arsenic, in the form of Paris green, was used as early as 1865 against the Colorado potato beetle. By 1900 arsenic and lead were realized to be effective against insects, notably the larva of the gypsy moth. So were soaps, detergents, kerosene, oils, and other petroleum ingredients.

The effectiveness of the early biocides was discovered accidentally, but deliberate research more recently has produced new and more potent materials. Analysis of lead arsenate and Paris green implicated arsenic as the effective ingredient. By 1912 experimentation showed calcium arsenate effective against the boll weevil, a notorious parasite of the cotton plantations in the southern United States, and its subsequent widespread use changed the cotton industry. By the 1920s commercial biocide production in this country was well under way. The number of biocide manufacturers in the United States totaled 35. Research during World War II gave additional impetus to the development of biocides, and following the war, notably with the appearance of DDT, a chemical discovered in 1874, the age of biocides was upon us. Abundant quantities of biocides were produced and used in the next decades. For example, 78 million pounds of DDT were used in the United States in 1958–1959, the peak years, and although less was used in subsequent years, 40.3 million pounds was applied in this country in 1966–1967. DDT began to be abundantly exported, 113.8 million pounds in 1962–1963 and 80.3 million pounds in 1966–1967.

Today, biocide production is still a big business, despite the outlawing of the use of some chemicals. In 1967 the United States, which manufactures 50 to 75 percent of the amount used globally, produced more than 1 billion pounds, including 103.4 million pounds of DDT and 348.2 million pounds of herbicides. Until very recently the manufacturing of DDT alone was probably a $20-million-per-year business.

The kinds and effects of biocides

Biocides can be classified according to the organisms which they affect. Robert L. Rudd lists rodenticides (chemicals that control mammals, most frequently rodents), fungicides (chemicals effective on fungi), herbicides (materials effective on leafy seed plants), and insecticides and acaricides (materials that destroy insects and spiders and their relatives) (Fig. 12–1).

Rodenticides

The rodenticides include a diverse group of chemicals; therefore, the effect on organisms varies chemically and physiologically. One of the common rodenticides, Warfarin, lowers the ability of the blood's clotting factor, prothrombin, to induce clotting. Effective use of the chemical depends on repeated ingestion. Mammals killed by this chemical evidence hemorrhage internally. Other rodenticides affect other parts of the mammal's body. For example, strychnine affects the nervous system and produces stiffening of the body, convulsions, and paralysis. It produces cardiac problems, particularly fibrillation of the heart. Zinc phosphide, a compound less stable than strychnine or sodium fluoroacetate, induces coma and severe damage to the liver. Thallium also affects the nervous system, producing depression, difficult breathing, and difficult coordination of the muscles, the digestive organs, and the kidneys.

Fungicides

The fungicides are frequently used to destroy the rusts of some food grains, mildews, and other fungal parasites that prey on animals and plants. Most of the compounds contain sulfur, copper, or mercury. The accumulation of mercury in living tissue and its known serious toxicity possibly indicts mercury compounds as a new major threat to life.

Mercury can easily be fatal. Between 1953 and 1960 more than 60 Japanese fishermen are known to have died after eating fish and shellfish which contained high mercury deposits. The fishermen hauled their catch from Minimata Bay and River, where a plastics factory deposited its mercury by-products. Symptoms leading to death included deafness, blindness, convulsions, coma, and mental retardation.

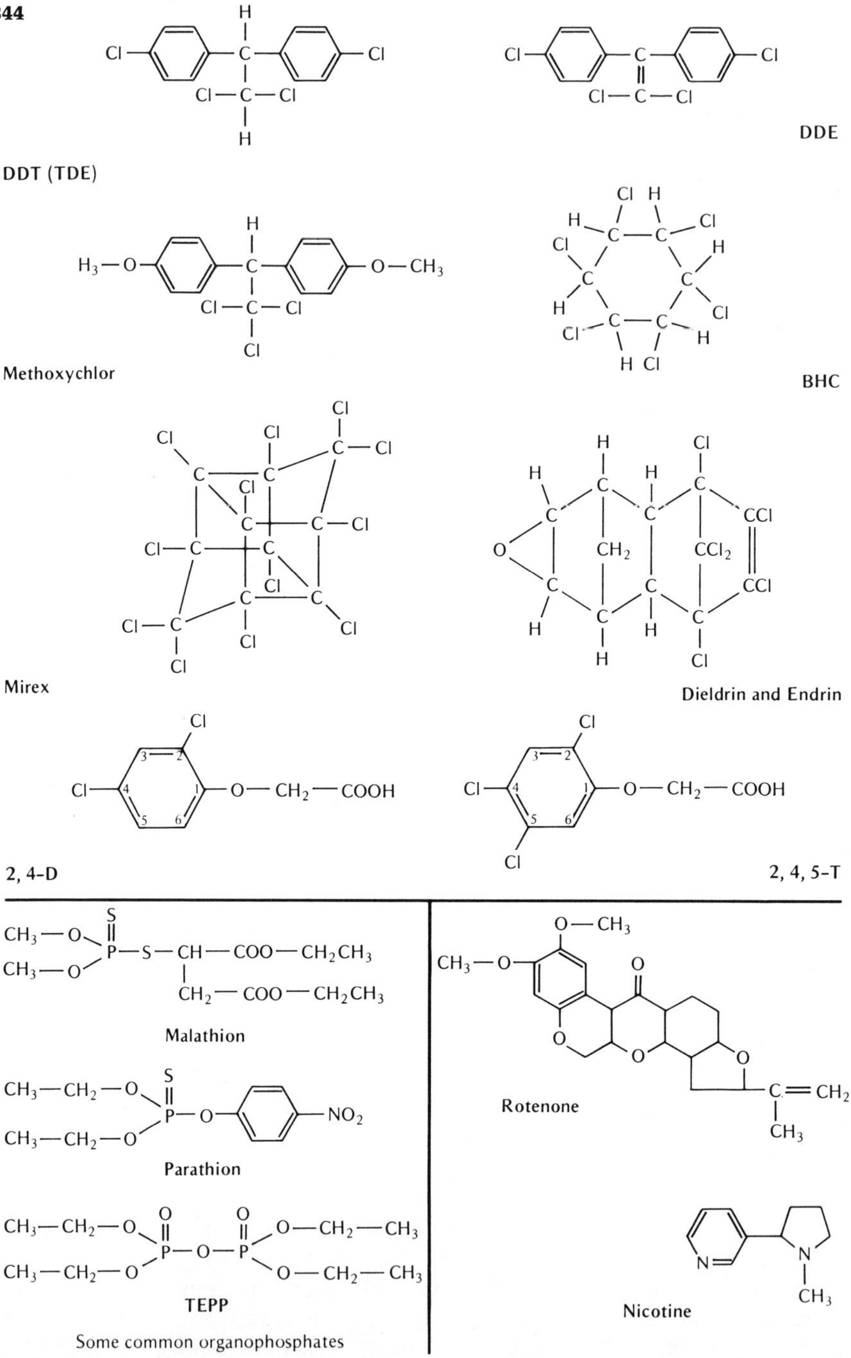

Figure 12–1
The molecular structure of many biocides shows similarities. Top: Some chlorinated hydrocarbons: mirex resembles a distorted box; dieldrin and endrin have different three-dimensional configurations. Bottom right: Rotenone and nicotine. Bottom left: Some common organophosphates.

Nonfatal symptoms of mercury poisoning are serious. In large amounts, damage occurs to the liver, the eyes, and the chromosomes. Deformation of the unborn is not uncommon. Severe damage to the brain can occur when mercury reaches 10 ppm, leading to behavioral aberrance, including irritability, timidity, and tremors, symptoms of the Mad Hatter of *Alice in Wonderland*. These symptoms are known to plague those who worked in factories that manufacture hats, a process that involves mercury compounds. Behavioral symptoms of mercury poisoning in the elderly — headaches, fatigue, insomnia, anxiety, lethargy, loss of appetite — are often dismissed as senility. No antidote for mercury poisoning is known. Mercury as a fungicide in the form of methyl mercury is used to control smut on seeds such as oats. The chemical is also used to control apple scab and diseases of potatoes and tomatoes. Unfortunately, mercury compounds are easily translocated and so can occur easily in plant parts eaten by man.

Mercury compounds are frequently used for other industrial purposes, too. Phenyl mercury acetate (PMA) is used to discourage mold in commercial launderies, especially diaper services. Mercury is also used in paint and paper manufacture. In fact, in 1946, 46,000 pounds of mercury were used in paper and pulp production in the United States to keep the machinery clean of fungi, which grow easily in the warm rich pulp. Large quantities of mercury are also used in the manufacture of plastics, sodium, and chlorine.

Mercury tends to accumulate in organisms. For example, mercury in chicken eggs reached average values of 0.029 ppm when the birds fed on mercury-treated seeds. The accepted maximum quantity in food is 0.5 ppm. Canned tuna has been removed from the market because it tested at 0.7 to 0.9 ppm, and Canada has banned fishing on its side of the St. Clair River, in Lake St. Clair, and the Detroit River after discovering high levels of mercury in fish there. Some experts say that it may take hundreds of years before fish can safely be taken from these areas. Fish from Lake Erie also contain large amounts of mercury, approximately 3.5 ppm.

Tuna, as an example, probably contain a large amount of mercury because the element accumulates in organisms and is, thus, more concentrated in organisms high in the food chain. This food chain often begins with bacteria, which have the ability to convert inorganic mercury into lipid-soluble and highly toxic methyl mercury. Ingestion of the mercury-containing bacteria eventually results in intensive concentrations in the predators. Tuna have a high metabolic rate and so ingest more mercury than other fish.

Herbicides

Herbicides are used to control plants either by directly poisoning their photosynthetic systems or by inducing rampant growth of some plant parts. Many of the herbicides are called "selective herbicides" because they affect significantly one group of plants. However, other herbicides sterilize the soil so that all plants cannot grow. Others destroy only a part of a plant. For example, sodium arsenate in contact with the bark of trees causes the bark to fall off and, thus, girdles unwanted trees.

Monuron and simazin are herbicides that interfere with photosynthesis. 2,4-D and 2,4,5-T are herbicides that stimulate excessive growth of some plant parts such as the phloem cells and, thus, interfere with the normal movement of nutrients. These chemicals also appear to cause defoliation, by affecting the abscission layer at the base of the leaf petiole. 2,4-D and 2,4,5-T are very selective, stimulating the growth of broad-leaved but not narrow-leaved plants (Fig. 12–2). Both were developed in the 1940s for use in chemical warfare. Since then they have been used to control brush on forest lands and roadsides, to clean weeds from lawns, and to defoliate as a defensive measure cultivated areas and forests of Vietnam at the rate of approximately 1.5 million acres per year (Fig. 12–3).

These chemicals are known to affect animals and man. In the 1950s and 1960s workers in herbicide plants were noticed to develop the symptoms of the disease *chloracne* — aberrant electrocephalograms, liver damage, nervous and mental disorders, depression, loss of appetite and weight, and reduced sex drive. Recovery requires a long time.

Human beings are not the only ones affected. In the 1950s a new disease, called *chick edema,* was observed in chickens. Diseased chickens breathe with difficulty, appear droopy with ruffled feathers, and stop laying eggs; many die. Autopsies show fluid accumulation in the pericardial sac and abdomen and under the skin. The liver and the kidneys are damaged. Nonviable chick embryos also show symptoms of the disease. Hatched chicks show eye defects and cleft palate, edema, and

Figure 12–2
Herbicide applied to a field of plants to control weeds. Left: Untreated peanuts heavily infested with nutsedge and other weeds. Right: Weedless field resulting from herbicide applied to the field before planting. (Courtesy of U.S. Department of Agriculture.)

Figure 12-3
Forest in Vietnam defoliated with herbicides (upper left) contrasts with untreated forest (lower right). (Photo by A. H. Westing, Windham College, Vt.)

short twisted feet due to the slippage of tendons, which often causes them to walk on their knees (Fig. 12-4). Laboratory tests show that gross abnormalities are produced in mouse embryos. Kidney abnormalities and cleft palates are observed also. Birth defects of Vietnamese children now being observed may possibly relate to the accidental spraying of people during the 30,000 defoliation sorties carried on in Vietnam.

Insecticides

The most widely used biocides, and most significant to man, are the insecticides and acaricides. These diverse chemicals are often classified into groups according to their chemical nature, that is, as organic compounds, natural organic compounds, or synthetic organic compounds.

Major inorganic compounds are the sulfur compounds and arsenicals, such as lead arsenate. Human poisoning with arsenicals produces damage to the liver, kidneys, and vascular system, gastral enteritis, severe diarrhea, violent convulsions, and then death.

The natural organic pesticides are the "botanicals" described above. Recent research has produced materials that are concentrations of these active compounds and more potent than the botanicals from which they were derived.

348   Chapter 12: Biocides

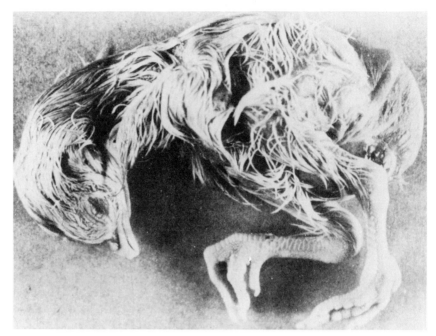

Figure 12-4
Developing chick subjected to "Agent orange," a mixture of 2,4-D and 2,4,5-T used to defoliate Vietnam forests, shows deformation of the rump. (Courtesy of E. W. Pheiffer, University of Montana. Photo by J. Verrett, U.S. Food and Drug Administration.)

The synthetic organic compounds perhaps have proved to be the most effective of all the insecticides. Many are extraordinarily toxic, affect diverse groups of organisms, persist over long periods of time, and accumulate in the tissue of nontarget organisms. Included in this group are the carbamates and some phosphorus compounds, such as parathion. Both chemicals inhibit the function of cholinesterase, the enzyme responsible for blocking the nerve impulse from nerve cell to nerve cell. Inhibition of this enzyme stops the inhibition of the transmission of nerve impulses. Consequently, mammals that have been poisoned with these compounds show general weakness, vertigo or dizziness, excessive tearing, and tremors. Interestingly, aquatic organisms appear less affected by these chemicals than terrestrial animals.

Probably the best-known group of organic synthetic biocides is that comprising the chlorinated hydrocarbons. Public concern over these compounds has reached such a pitch that when the word "pesticides" is used, DDT, dieldrin, or lindane come to mind. These chemicals, as well as benzenehexachloride, chlordane, heptachlor, methoxychlor, toxaphene, aldrin, endrin, and the polychlorinated biphenyls (PCBs) are chlorinated hydrocarbons. The effect of these chemicals on organisms was little known until recently. Today, data about chlorinated hydrocarbons increasingly implicates them as seriously affecting organisms not intended as targets. Research and the abundant use of the chemicals has convinced many that, from this point on, biocides must be used cautiously.

One chlorinated hydrocarbon, DDT, has been researched extensively. This chemical is known to affect the central nervous system. Organisms sensitive to DDT exhibit hyperactivity, severe tremors, and convulsions.

The advantages of biocides

The Bible records ancient man's many confrontations with pests. Recall that among the difficulties which Moses and Aaron faced in removing the Hebrews from Egypt were three insect plagues: gnats, flies, and finally locusts. "And the locusts came up over all the land of Egypt, and settled on the whole country of Egypt, such a dense swarm of locusts as had never before, nor ever shall be again. For they covered the face of the whole land, so that the land was dark, and they ate all the plants in the land and all the fruit on the trees which the hail had left; not a green thing remained, neither tree nor plant in the field, through all the land of Egypt."

Losses due to pests, particularly insects, and diseases run as high as 40 to 50 percent in some countries. In parts of Pakistan, locusts, caterpillars, and crickets still destroy as much as 80 percent of the food. India has a similar problem. In South and Central America the cattle tick and the warble or torsalo fly incur very large losses to beef products and milk production. Calf mortality may run as high as 70 percent in some of the heavily infected parts of these countries. Recently, in Argentina 50,000 calves succumbed to screwworm infection, a parasite that can now be very effectively controlled, as we shall describe later (Fig. 12–5).

In the United States the loss of materials, especially food materials, to pests is great. It is estimated that damage caused by rodents comes to about $2 billion per year; loss to insects, perhaps $4 billion; and the total loss to weeds, about $11 billion annually. Worldwide losses to pests are most likely $70 to $90 billion, about one-third of the world's agricultural production, enough food for 1 million people.

Studies of crop destruction by specific insects reveal alarming waste. For example, cotton insects account for approximately a $200 million loss to the cotton crop every year. The worst offender, the boll weevil, accounts for about 10 percent of the annual crop loss. Between 1940 and 1944 the coddling moth caused a yearly loss to the apple crop in the United States of about 15 percent, approximately $25 million. In 1948 the U.S. Forest Service estimated that the loss of saw timber caused by insects was about 5 billion board feet.

Because of the tremendous destruction by pests, the invention and perfection of biocides has been a great boon to mankind. Much of the destruction of foodstuffs is now only history because of the systematic and careful application of the proper biocide. In fact, Robert Metcalf of the University of Illinois says in *Organic Pesticides: Their Chemistry and Mode of Action* (John Wiley & Sons, Inc., New York, 1955):

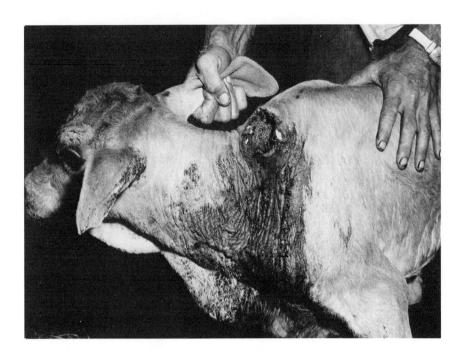

Figure 12–5
Severe infection of cattle by screwworm larvae can be a serious threat to cattle raisers. (Courtesy of U.S. Department of Agriculture.)

Pesticides are a major reason why a farmer is able to feed and clothe forty-six persons today, compared with twenty-five persons in 1960 and only four persons in 1850. In our present era of the managed ecology of monocultures, farm mechanization, and complex system of food harvesting, processing, distribution, and storage, the use of pesticides often represents the slender margin between production and crop failure and between economic profit and economic loss. In developing countries, where food supplies are marginal, pesticides used may represent the difference between survival and starvation.

Increased apple yields illustrate the benefits of biocides to farmers. The coddling moth larva produces holes in apples. Unsprayed apple trees in West Virginia produced an average of 87 worms per 100 apples. Trees sprayed with an insecticide (Guthion) produced an average of only 3 or 4 worms per 100 apples. Other fruit-attacking pests, such as the apple maggot, the plum culio, the orchard mite, and the San Jose scale, have been curtailed similarly. Production of corn also increases after the application of insecticides. In a Florida test in 1957 and 1958, an untreated field of sweet corn produced an average of only 1.6 worm-free ears in every 100 ears. However, in a similar field treated with Sevin, a carbamate insecticide, the quantity of worm-free ears per 100 was 86.2.

Another insect, the alfalfa weevil, a European import, has proved to be a serious pest for alfalfa growers during this century. In one test an insecticide applied 18 days previous to harvest of the crop increased the first and second cutting of an alfalfa crop from 1.6 to 2 tons per acre. Furthermore, the protein content increased from 16.9 to 20 percent, and the price of the hay went up $5 per acre, to $20.

DDT has been used to increase potato yields. The average potato yield in New York State for 1936 to 1945 under the best growing practices and treatment with arsenical insecticides was 110 bushels per acre. In 1946 and 1947, when DDT was first used extensively to control potato pests, the average yield was 172 bushels per acre, an increase of 56 percent.

Citrus crop yields benefit. Fumigation of soil in California citrus orchards with dichlorobromopropene to control nematodes increased lemon yields by 20 percent and orange yields by 33 percent.

Besides insecticides, herbicides and fungicides have also had a role in increasing crop yields. Black rot of grapes has been significantly controlled with a carbamate poison called "ferric dimethyldithiocarbamate." Application of this chemical increases grape yields from 1,000 to 8,000 pounds per acre. It has proved effective against other diseases, such as peach leaf curl.

In one experiment, the application of the herbicide 2,4-D increased the yield of forage plants per acre from 1,100 pounds to 2,800 pounds, 254 percent. In a very significant experiment, the chemical Dalapon applied to a field of bird's-foot trefoil, a significant forage legume, so decreased weeds that the yield of trefoil rose 4,825 percent, from 80 pounds to 3,860 pounds per acre.

Besides these uses, biocides have significantly reduced human disease. We know that at least 27 diseases, including some of the world's deadliest, such as typhus and malaria, are effectively kept in check with the use of biocides. Early in 1944, 60 people a day in Allied-occupied Naples died of typhus. DDT was used to delouse 1,300,000 of the inhabitants, and within 3 weeks, the typhus epidemic was over. Similar incidents have since occurred elsewhere.

Malaria, transmitted by the mosquito, may threaten one-half of the world's population. Without biocides, it has been said, 3 million people would die each year from this disease. For example, in India use of DDT reduced malarial cases from an estimated 75 million per year in 1952 to approximately 0.1 million in 1964 (Fig. 12–6). Life expectancy has increased from 32 years in 1953 to 47 years in 1967. Such effects for the first time in many countries are mainly due to the use of DDT since World War II.

Other diseases are diminishing, such as yellow fever, dengue or breakbone fever, and encephalitis, all carried by mosquitoes; typhoid fever, cholera, yaws, tuberculosis, and trachoma, all carried by the housefly; and African sleeping sickness and river blindness, caused by the black fly.

In view of all the advantages to man from the use of biocides, you may ask: What could ever be criticized about the use of biocides? Until 1962 most biocides appeared to be the salvation for man in many

Figure 12-6
Malaria in India has been controlled by spraying walls and ceilings with DDT solutions to kill resting mosquitoes before they have time to deposit their larvae. (Courtesy of WHO.)

parts of the world. However, in 1962 Rachel Carson published a book, *Silent Spring*, in which she described the threat of biocides to many life-forms, including man. Although originally dismissed by many critics, most of Carson's warnings have proved to be true. Today, we know well that many biocides unwisely used have serious effects on life and its environment.

The disadvantages of biocides

One significant reason why biocides have been severely attacked as threats to life is because many persist in the environment. (Some do not, such as parathion, TEPP, diazinon, schradan, systox, or malathion.) Persistent biocides cling tenaciously to particles of soil and do not vaporize, wash away, or decompose quickly. Soon after application of a biocide, a residue often accumulates containing a great amount of viable biocide. In a short time, the residue may also contain chemical derivatives of the biocide, as well as molecules of the solvent in which the biocide was carried.

The persistence of some residues may be as brief as a day or two; then again, with other biocides, such as DDT, chlordane, dieldrin, or heptachlor, their persistence may be much longer, even years, although the duration of persistency appears to depend on soil moisture, soil type, soil temperature, and degree of cultivation. Areas of ground spread with 100 pounds of DDT per acre in 1947 yielded a residue of 28.2 pounds of DDT per acre in 1951. This indicates only an 18 percent loss per year of the total amount of biocide applied. Other studies show the rate of destruction of DDT per year to be even less. Heptachlor, for another example, appears to decompose more quickly than DDT but apparently still persists actively in small amounts 5 years after application.

Biocide residues may be absorbed by plants growing on the area, but this does not mean that the biocides are lost. If the plant material is plowed under or allowed to decay, the biocide material once again is found in the soil. Such a problem is common in corn and potato fields.

High amounts of biocides occur in fields where biocides are regularly used. DDT concentrations of 113 pounds per acre have been detected under apple trees in orchards; 61 pounds per acre have been found between the trees. In soil in which corn had been grown, the quantity of DDT averaged approximately 11 pounds per acre; in soils in which potatoes were raised, about 7 pounds per acre.

A further complication may be the toxicity of the breakdown products from the biocides. Dieldrin and heptachlor epoxide are two toxic materials derived from biocides. Dieldrin has been shown to be 6 to 12 times more prevalent several years after application to the soil than its parent molecule, aldrin. Heptachlor epoxide has been found 2.5 times more plentiful in the soil 6 months after the treatment with the less-toxic heptachlor.

The threat of biocides increases when the toxic parent materials and their residues migrate from the region of application to new areas. This happens when animals eat plants that have absorbed biocides. Dissolution or suspension of the biocides in surface water, such as rain runoff or irrigation water, also distributes the chemicals. Lake Michigan, for example, contains approximately 2 ppt of DDT primarily as a result of surface runoff.

The distribution of DDT and other biocides across the globe, even into areas where they are not directly applied, is believed to be occurring. Measurements indicate that Alaskan Eskimos store 0.8 ppm DDT in their body fat. Most of this DDT is ingested in foods imported from areas where the insecticide is abundantly used. DDT is carried in other ways, too, such as by migratory fish and birds, or ocean currents and wind. The DDT gets into the wind on spraying. Rain or snow often brings the biocide down to the ground, where animals can eventually ingest it. Although no extensive sampling of DDT distribution has been done, evidence suggests that its distribution pattern is similar to those of some radioactive materials.

Some biocides are a threat because they accumulate in biological tissue. Organisms in any ecosystem characteristically depend on each other for energy (Chapter 2). Consequently, in any natural community some species, the predators, prey on others. Many organisms destined to be eaten by predators concentrate unaltered biocides in their bodies. When eaten, the biocides pass into the predator's body. If the predator is eaten by another predaceous species, the biocide is once again transferred. Eventually, the organism at the end of the food chain harbors great quantities of biocide. An example of an animal in several food chains which concentrates extreme amounts of biocides is the oyster. Experiments show that this animal in a 40-day exposure of 0.1 ppt DDT can concentrate 70,000 times that amount of DDT in the body. It also concentrates endrin and 2,4-D. Presumably, the great amount of water passing through the body, 6 gallons per hour, accounts for the extraordinary storage. Such an accumulation of toxic material can have serious effects not only for the oyster, but also, presumably, for those predators which eat oysters.

Delayed effects on some predators in the food chain have been well-substantiated. Fish succumb to the ingestion of DDT-killed insects; turtles die because of eating DDT-killed fish. Scientists know of a few other examples of the delayed effects of biocides in the food pyramid, to be finally expressed in an organism far removed from the intended victim of the biocide application. R. L. Rudd cites the death of birds that fed on fish from Clear Lake, California, after the area was sprayed with DDT to control the Clear Lake gnat. Clear Lake, a popular recreation area, was plagued for many years with gnats, despite numerous attempts to eradicate them. In 1948 DDT, concentrated 1 part to 70 million parts of water, was applied. It killed the gnat larvae and other aquatic invertebrates. In September of 1949, 1954, and 1957, additional applications of DDT were made but at higher rates than previously.

(The use of increasing amounts of DDT to eradicate populations of insects is common. Mainly, this is necessary because after every biocide application a few insects survive which are especially resistant to the applied concentrations of insecticide. These insects live to breed a new population, with most of the new individuals inheriting the resistance of the parents. Consequently, when the population builds up it has a higher resistance to the chemical. In short, a more resistant variety of the pest has evolved, an event thoroughly consistent with Darwin's ideas of natural selection, as explained in Chapter 1. To eradicate the new population, a greater concentration of the biocide must be used. This can go on and on and clearly illustrates the foolhardiness of relying exclusively on biocides to eradicate a species or variety of organism.)

After the first summer of DDT application in Clear Lake, grebes, fish-eating birds normally common to the area, were affected. Most left; a few returned periodically. Many died from suffocation, a symptom of DDT poisoning. An incidental investigation of the visceral fat of the dead grebes exposed abundant DDT, 1,600 to 2,000 ppm, a concentration of 80,000 times greater than that applied to the lake. Immediately, other lake inhabitants were analyzed for DDT. Alarming quantities of the biocide were found in fish, other birds, frogs, and plankton. Plant-eating fish concentrated between 40 and 300 ppm. Carnivorous fish contained up to 2,500 ppm in their fat, the greatest concentration of the chemical. Presumably, the grebes ate the fish and ingested the toxin. People living in the area expressed alarm that they, too, were in danger, but fears were somewhat allayed by the realization that people mostly eat the fleshy parts of the fish and discard the fat. Consequently, their DDT intake should be substantially lower.

Apparently, the organisms that initially absorbed the DDT passed on through the food chain were the plankton, composed of algae and other small organisms. Plankton showed an average of about 5.5 ppm, a 265-fold increase over the initial application of DDT, and about one-half the quantity found in the small fish eaten by carnivorous fish.

A study of a marsh in Long Island also illustrates how DDT accumulates in the food chain. The marsh had been sprayed for 20 years with DDT to control the mosquitoes. DDT residues up to 32 pounds per acre were found in the upper layer of marsh mud. Carnivorous animals in the area contained up to 1,000 times more DDT than the marsh plankton (0.04 ppm). The fish that lived on the plankton contained 1 ppm; the birds that ate the fish concentrated 75 ppm.

DDT is not the only culprit in food chains. Others include aldrin, dieldrin, and strychnine. A study in California indicates another chlorinated hydrocarbon, toxaphene. Toxaphene accumulated in marshy areas in runoff from agricultural areas and from there entered food chains. Pond mud and invertebrates contained approximately 0.2 ppm. Many fish contained approximately 8 ppm. Fish-eating birds carried 650 ppm in the fat of their muscle tissue. Such concentrations proved lethal to some of the bird species. White pelicans, common egrets, great blue herons, western grebes, and ring-billed gulls died in great numbers between 1960 and 1964.

Rudd, in *Pesticides and the Living Landscape*, illustrates the delayed effects of DDT used to control the bark beetle, the insect that carries Dutch elm disease. Dutch elm disease was discovered only a few decades ago in the United States. Today, the American elm in the United States is nearly wiped out (Fig. 12—7). Control programs up until recently involved DDT spraying. In the 1950s, shortly after spraying programs were begun in earnest, certain birds were observed dying in great numbers. At first it was suspected that the birds, notably robins, were drinking water that contained great amounts of DDT. Investigations showed, however, only small quantities of dissolved DDT in rainwater puddles. Examinations of the food of robins, notably earthworms, showed a great abundance of DDT, from 4 ppm in the central nerve cord to over 400 ppm in the earthworm's craw and gizzard. All organs except the central nerve cord concentrated DDT greater than 48 ppm. Analyses of the robins indicated that they, too, stored DDT, in concentrations at least 10 times greater than that found in the soil where the spraying occurred. The death of the robins presumably were due to a greater susceptibility of the birds' central nervous system to the chemical than that of the worm. One report from Michigan State University in 1961 reported a decline in the robin population on that campus from 370 to 4 birds in just 3 years. Estimates of robin mortality after DDT spraying are as high as 90 percent. Many other species of birds are also known to be susceptible to this chemical.

Figure 12-7
Two views of the same street show the beauty lost when stands of American elms die from the Dutch elm disease. Top: Before the disease struck. Bottom: After the removal of stricken trees. (Courtesy of Elm Research Institute, Harrisville, N.H.)

A final example of the effects of insecticides in the food chain occurred in Borneo. DDT was used to kill flies. The lizards that ate the flies also died — from DDT poisoning. Finally, the cats that ate the lizards began to die. At this point, the World Health Organization, alarmed at the great number of rats, potential carriers of the dreaded bubonic plague, dropped cats by parachute to reestablish the feline population and restore ecological balance.

The effects of DDT on fish

In the spring of 1966, conservation officials in Michigan, hoping to initiate a new era for sportsmen and commercial fishing companies, released 4- to 6-inch Coho salmon into the tributaries of Lake Michigan (Fig. 12–8). Hopefully, only a few years would be needed for the fry to mature into 15- to 20-inch meaty game fish. The extermination from Lake Michigan of many game fish because of pollution and the invasion of lamprey eels and alewives had given rise to this ingenious plan. By 1969 all hopes appeared shattered. The Food and Drug Administration seized 14 tons of frozen Coho. The fish showed contamination with DDT of 19 ppm. The state loss in recreation dependent on fishing probably was small compared to the loss to the fishing industry, $2.5 million. The FDA promptly limited marketable fish to those with a DDT content no greater than 5 ppm, a parameter that probably excludes 80 percent of the commercially caught fish in Lake Michigan from being sold, including lake trout, white fish, lake herring, and chubs, according to the U.S. Bureau of Commercial Fisheries.

Since these events, research on the effects of biocides in fish has begun. Much is yet to be learned. It is already known, however, that young fish are seriously affected by DDT. DDT is passed into the eggs produced by mature females that have DDT in their body tissues. As the young fish develop, the yolk sac containing stored food — and stored DDT — is absorbed. The effect is usually fatal. In experiments, 1 milligram of DDT per kilogram of fish did not kill adult brook trout, for example, but did prove lethal to many fish fry. This concentration of insecticide is close to that known to occur in eggs of fish collected in many freshwater lakes, including those of Coho salmon taken from Lake Michigan, where fry mortality is increasingly a problem.

Other effects of insecticides occur to adult fish. The research is in an early phase. Consequently, many questions remain unanswered. However, it is known that adult brook trout have a higher rate of mortality when exposed to concentrations of DDT and then undergo environmental

Figure 12–8
Stocking Lake Michigan waters with Coho salmon was an attempt to revitalize the fishing industry of the Great Lakes. (Courtesy of Michigan Department of Natural Resources.)

shocks of some sort. Ninety-six percent of the trout fed 3 milligrams of DDT per kilogram of fish for 6 months died in the following 3 months when suffering abrupt termperature changes and food shortages, whereas 99 percent of the brook trout not on the DDT diet survived. Similar experiments have been done with salmon.

The effects of DDT on birds

The sport of falconing deserves the credit for exposing the effects of insecticides on the reproductive ability of many birds. In Britain, where the sport is serious business, falconers make annual pilgrimages to the birds' nests, so as to capture the birds when young and trainable. In the late 1940s collectors began to complain of a decreasing number of birds. One researcher suggested that DDT might be at fault, as it was becoming well-known that insecticides accumulate in the body tissues of predatory birds. He studied the nests visited by the falconers and found that between 1960 and 1966, in the 68 nests studied, 44 broken eggs were found, a great increase of the number of broken eggs found in these nests in years previous to 1960. Aware that sparrow hawks and eagles also suffered from broken egg damage, he arranged to visit museums to weigh and measure the wall thickness of 1,600 "blown" eggs. He found that both egg weight and shell thickness of eggs collected decreased after 1946, the year DDT was introduced.

The British research led to similar investigations in the United States on the eggs of some of the 12 or so species of American birds known at present to be declining in the wild, as well as producing a high proportion of cracked or broken eggs. Bird species included in the study were the peregrine falcon, the bald eagle, and the osprey, among others. Studies proved that the eggshells of some of this group were about 18 percent thinner now than before the late 1940s. These flesh eaters, along with others, have notably declined in the wild. Peregrine falcons within the last 40 years are known to have nested on the tops of skyscrapers in New York. Until recently, it was said that no breeding pair existed east of the Rocky Mountains. Studies of these birds along the Colville River in Alaska indicate that the average number of young per pair in 1962 was 2.5; in 1968, it was 1. Furthermore, the unhatched, often broken eggs contained a high insecticide content. In many areas bald eagles were faring no better. In 1945, 44 pairs of eagles nested around Lake Michigan. In 1969, 1 pair existed, and it had not raised young successfully since 1964. In the early 1960s counts of immature birds in the eagle population in the entire United States showed a noticeable decline: 26.5 percent (1961), 23.7 percent (1962), and 21.6 percent (1963).

The Bermuda petrel is also endangered. This bird was thought to have been extinct for 300 years. Approximately 30 years ago, a few were located near Bermuda. Ornithologists were delighted with the find. Today, the bird may truly be on the road of extinction. The survival rate of chicks dropped from 67 percent in 1958 to 36 percent in 1967. High DDT residues are found in the birds' eggs. A similar fate may await the brown pelican off California. In 1969, hundreds of pairs of these birds attempted to breed, but to no avail. Their eggs squashed beneath them, and almost no young were produced. Abundant DDT was found in the eggs.

In the 1940s, although DDT appeared incriminated in the demise of certain populations of birds, no laboratory experiment had yet been done. Soon experiments were set up to produce valid data indicating cause and effect. At the Patuxent Wildlife Research Center, Bureau of Sport Fisheries and Wildlife of the U.S. Department of Interior, experiments have been performed with the effects of DDT on several kinds of birds. Three groups, each of 12 pairs of sparrow hawks in flight pens, were fed differentially. The diet of one group contained no insecticide, the diet of another contained moderate doses of the DDT and dieldrin, and the final group was fed large amounts of the chemicals. Results showed that the hawks on the insecticide-free diet hatched 84 percent of their eggs, whereas those on other diets hatched approximately 60 percent of their eggs.

Kestrels, relatives of the peregrine falcon, were subjected to similar experiments. On a diet of 2 ppm DDT and 0.33 ppm dieldrin, hawks produced eggs with 15 percent thinner shells than those eggs laid by hawks on insecticide-free diet. Mallard ducks behaved similarly. The ducks fed 3 ppm DDT produced eggs that broke 6 times more frequently than the eggs from ducks on pesticide-free diets. Also, the shells were 13.5 percent thinner. The mallard is the only herbivorous bird known that is affected by DDT in this way.

Experiments with pheasants and quail, even when fed 200 ppm DDT, do not normally produce such results, but these birds and others may be affected by the insecticides when subjected to stress, such as low temperature, starvation, or mating. Presumably, under these conditions the stored fats are mobilized, and consequently, so is the DDT. For example, starved adult herring gulls from Lake Michigan had high amounts of DDT in their bodies — 21 ppm in the brain, 99 ppm in the breast muscle, 2,441 ppm in the body fat. The dying birds showed symptoms known to have been characteristic of DDT poisoning. It is also known that a large number of male pheasants succumb during the mating season if they have been fed large quantities of DDT. Strangely, during mating and the egg-laying season, no noticeable mortality occurs among the females. However, within a few weeks of this period, those females on DDT diets of 100 to 250 ppm frequently die. The eggs produced by adults on a 100-ppm-DDT diet produced only a 25 percent hatch. Explanation for this unusual sequence of events has not yet been made, but whatever the details it probably revolves around the actions of DDT within the bird's body. Not everything is now known, but research proceeds apace.

It is known that DDT apparently induces the production of enzymes from the liver, at least in the female bird. The enzymes, in turn, depress the activity of estrogen, the female sex hormone, by hydroxylation. Similar effects of DDT are known to occur on other steroid sex hormones, such as progesterone, and the male hormone, testosterone. Normal estrogens function in production of the egg shell. Altered estrogens account for the thin egg shells and cracked eggs. Why DDT affects the estrogen of some bird species and not others is yet to be discovered.

The effects of DDT on man

A news magazine headline stating that "Babies receive more DDT in breast milk than cow's milk" exemplifies a major concern for many people, concern about the possible threat to human life posed by biocides. Is the 0.1 to 0.2 ppm of DDT in breast milk, 2 to 4 times higher than the 0.05-ppm tolerance level allowed by the Food and Drug Administration in cow's milk sold interstate, sufficient to really worry about?

The implications to man of the nature and behavior of biocides has concerned the public, as well as researchers and doctors, since their first use in the late 1940s. Yet little research was done in this area until recently. Now, investigation progresses rapidly on animals with the hope that experimental results gathered with experimental animals will give insight to the threat to man.

DDT and some related products are known to be extraordinarily lipid-soluble. This means they are stored easily in the fat of living organisms, man included. At present, the amount of DDT measured in human beings is approximately 12 ppm, a level higher than that allowed in commercial meat products, a sobering thought even if human flesh is not part of your diet.

Some individuals, such as workers in DDT plants or insecticide sprayers, are known to harbor more. DDT sprayers working in malarial control programs in the Far East and the Caribbean area have picked up quantities many times the usual amount. It is not uncommon to find workers who have several hundred ppm of DDT in their body fat. Some information is known on individuals who have taken in certain amounts of DDT over a period of time. Employees in plants where DDT is manufactured absorb up to 40 milligrams of DDT per day. In one experiment volunteers were fed 35 milligrams of DDT per day for 18 months. In neither case did any short-term adverse symptoms occur.

In view of the effect of DDT on other organisms and its persistence in the body, what is known of its effect of large concentrations in the human body? In brief, no serious deleterious effects have been proved, although some medical facts regarding the pesticide must be mentioned. Human cancer victims appear to carry 2 to 3 times more DDT in their bodies than accident victims. Work with rats, mice, and trout appear to support a possible connection between DDT and cancer induction. Other observed incidental effects of DDT include enlargement of the uterus in laboratory rats; increase of the activity of enzymes in the liver of the rats; depression of the effect of testosterone; visible changes in cancer cells *in vitro,* including vacuolation, elongation, and granulation; apparent deterioration of memory time; and the movement of DDT through the placenta into the fetus in small mammals. Incidentally, experiments also indicate that at least two anticonvulsant drugs, phenytoin and phenobarbital, appear effective in reducing DDT stored in the body. Phenobarbitol reduces concentrations of aldrin and dieldrin in meat and dairy cows, too.

Solutions to the biocide problem

The appearance of Rachel Carson's book *Silent Spring* in 1962 triggered public concern about the use of biocides, which produced several effects, including the encouragement of research to produce data to rebut or prove Carson's remarks. As information became more plentiful, many outraged groups of citizens around the country demanded action, usually legislative, to curtail the use of biocides. State laws resulted, and national laws were being proposed soon thereafter.

As a result the action of citizens in the courts, Michigan was one of the first states to ban DDT — when the newly developed game fish, the Coho salmon, proved to contain 3 times the maximum amount of DDT allowed in beef sold for human consumption. Fish prepared for commercial consumption were withdrawn from the market. Now, California has banned the use of DDT in households and gardens. Maryland has outlawed it.

Federal action has been slower. In 1969 the Department of Agriculture canceled registrations of DDT uses on shade tree pests, aquatic areas, house and garden pests, and tobacco pests, and announced plans to cancel all nonessential uses of DDT by December 30, 1970. In August 1970, additional registrations were cancelled for other uses on livestock, food crops, flowers and ornamental plants, and lawns, but the December 1970 deadline was not met, mainly because pesticide companies appealed the decision. DDT was finally phased out nationally on December 30, 1972, except for uses deemed essential to public health. In the United States, 2,4,5-T was successfully banned nationally in April 1970.

Other countries have also taken action against biocides. Since January 1, 1970, Canada has reduced its use of DDT by 90 percent. The biocide can be used for controlling insects in forest parks and other outdoor areas under emergency conditions only. Sweden also announced restrictions on pesticides in house or garden in 1970. All other DDT uses are banned for a 2-year test period. Several countries are ready to ban aldrin and dieldrin as seed dressings or for any use. Britain has greatly restricted use of aldrin, dieldrin, and heptachlor.

Natural pest control

Besides laws to control the quantity of biocides applied to the environment, methods are being studied to hold pests in check without applying synthetic material. Several new techniques of pest control have resulted. The artificial introduction of a predator to an area with subsequent reduction of its prey, the pest, has been tried successfully several times (Fig. 12–9). The introduction of the ladybug into California citrus groves to control scale insects in 1880 is an example (Fig. 12–10).

One approach, "sterilization," has been used on the screwworm fly of southern and western United States. This parasite until recently frequently devastated herds of cattle, sheep, goats, and wildlife. The several life stages of the parasite include the adult flies, approximately 3 times bigger than houseflies, which lay eggs on the animals, most often in skin wounds such as those caused by barbed wire, but even in the nostrils of sleeping people. Maggots hatching from the eggs feed on the animal's flesh. Secondary infection often sets in. In people the maggots have been known to burrow into the sinuses and cause infection and even death.

The insect was first controlled in the 1950s. Pupae, the developmental stage between larvae and adults, were treated with cobalt radiation. As a consequence, hatched males were sterile. A mating between the sterile males and females produced no fertile eggs, and since the females mate only once, proved effective as a birth control measure. Now, quantities of sterile males are released regularly. The U.S. government maintains a

Figure 12-9
Alligator weed in Florida has been controlled by its natural predator, the flea beetle. Top: Channel clogged with alligator weed. Bottom: The same channel after release of the flea beetle, which feeds exclusively on the alligator weed. (Courtesy of U.S. Department of Agriculture.)

300-mile buffer zone in Mexico into which is released 175 million sterile males per week. To further the attack on this pest, Mexico is progressing with this treatment toward the bottleneck of the Panama Canal area. Hopefully, some day the screwworm will be eliminated in this country and in Mexico, although the migration of fertile males from Mexico is still a big problem.

Other insects on which this technique shows promise of control include the boll weevil, which costs the cotton industry $300 million per year; the corn earworm, which wreaks $500 million per year damage; and tropical fruit flies.

Besides male sterilization, several additional techniques may prove effective. The deliberate infestation of pests with parasites is now considered possible. For example, the bacterium that can fatally infect the cabbage looper insect can be grown in quantity in the laboratory. Viruses may also be effective pest controllers. Viral control of the gypsy moth in the Northeast is under consideration. Natural chemicals may be used to lure insects, such as the cockroach, into traps. Manipulation of temperature, suspected of controlling in some insects major biological events such as migration and hatching, also may work.

"Ecological diversity" may also be significant as a pest control device. For example, in California the grapevines are parasitized by a leaf hopper. Fortunately, this grape leaf hopper is kept under control in the warm months by an attacking wasp. In the winter the leaf hopper dies from the cold. Its death would appear to threaten the overwintering of a wasp population large enough to control the leaf hoppers in spring. However, control is effected because a second species of leaf hopper, which lives on wild blackberries, survives the winter and carries the wasp population over until spring. Consequently, the wasp population is large enough, even in early spring, to control the grape leaf hopper. Thus, leaf hopper control can be effected if wild blackberries are strategically planted. Such an example makes an eloquent plea for the preservation of diverse species in every habitat, ecological diversity. More research may unearth other insect controls arising from ecological diversity.

The present and future of biocides

The most significant biocides are the synthetic organic compounds. Not only have they proved to be the most effective of the chemicals used, but some are extraordinarily poisonous, affect many kinds of organisms, and are not easily disintegrated in the environment. In the past, because of their remarkable advantages in killing certain life-forms, notably insects, they have been used in excessive quantity and often without significant long-range concern.

Today, experience with these biocides proves that some have many deleterious effects, which suggests that they should be used far more carefully and wisely. Some laws are already in existence which recognize the need for limiting the use of these poisons. Still more control appears to be needed, however. Considerable research on the effects of these biocides is in progress, but more intense efforts seem necessary before we can knowledgeably use these materials with the proper controls. At the same time it appears sensible to research carefully other

364　Chapter 12: Biocides

Figure 12–10
A common and effective natural predator, the ladybird beetle (ladybug), feeding on mealy bugs, which seriously infect citrus trees. (Courtesy of I. M. Newell, University of California, Riverside, Calif.)

ways in which pest populations can be contained. Some of the most promising areas seem to be in biological control, as well-illustrated by the control of the screwworm.

Possibly, the most significant need in the future in regard to biocides is the need for a careful examination of the reasons for the use of excessive quantities of these chemicals. To develop greater harvests may be a justifiable reason for their use in view of the rapid population growth of the world. However, the use of excessive biocides to grow produce that is totally insect free may, in fact, be no longer justifiable. Perhaps in countries such as the United States we need to readjust the standards and values by which we live — standards that make it intolerable to produce apples without an occasional larva. Such emphasis on purity of product may no longer be fitting when the production of products of such quality takes immense quantities of biocides that we know to be of potential danger to many forms of life besides the larvae.

In the next chapter we shall examine the use of nonrenewable resources, again attempting to show that man is often foolishly and ignorantly exploiting his environment. As a result, he appears to be endangering the life-giving relationship that he maintains with his environment.

Summary

In this chapter we have investigated the types of biocides commonly used, including rodenticides, fungicides, herbicides, and insecticides. We have looked at the effects of each of these kinds of chemicals, noting that the synthetic organic compounds appear to be the most dangerous. We compared the advantages of the use of biocides with the disadvantages, and concluded that the use of these materials exacts a price, such as affectation of food chains or debilitation of fish and birds. We examined the legal controls to biocide use and the advantages of natural pest control. Throughout the chapter one major theme prevails: man and other organisms continue to exist because they have evolved life-giving relationships with their environment. Alteration of themselves or their environment, as, for example, by adding large amounts of biocidal chemicals, threatens that life-giving relationship. As a consequence, the future well-being and maybe even the existence of life-forms, including man, may be in jeopardy.

Supplementary readings

Boffey, P. M. 1971. "Herbicides in Vietnam: AAAS Study Finds Widespread Devastation." *Science,* vol. 171, pp. 43–47.

Carson, R. L. 1962. *Silent Spring.* Houghton Mifflin Company, Boston. 368 pp.

Cox, J. L. 1970. "DDT Residues in Marine Phytoplankton: Increase from 1955 to 1969." *Science,* vol. 170, pp. 71–73.

Graham, F. 1970. *Since Silent Spring.* Houghton Mifflin Company, Boston. 333 pp.

Guftafson, C. G. 1970. "PCBs: Prevalent and Persistent." *Environmental Science and Technology,* vol. 4, no. 10, pp. 814–819.

Huffaker, C. B., ed. 1971. *Biological Control.* Plenum Publishing Corporation, New York. 511 pp.

Kilgore, W. W., and R. L. Doutt, eds. 1967. *Pest Control: Biological, Physical, and Selected Chemical Methods.* Academic Press, Inc., New York. 477 pp.

Kramer, J. R. 1969. "Pesticide Research: Industry, USDA Pursue Different Paths." *Science,* vol. 166, pp. 1383–1386.

Miller, M. W., and G. C. Berg, eds. 1969. *Chemical Fallout.* Charles C Thomas, Publisher, Springfield, Ill. 531 pp.

O'Brien, R. D. 1967. *Insecticides: Action and Metabolism.* Academic Press, Inc., New York. 332 pp.

Rudd, R. L. 1964. *Pesticides and the Living Landscape.* University of Wisconsin Press, Madison, Wis. 320 pp.

U.S. Department of the Interior. 1966. *Fish, Wildlife, and Pesticides.* Government Printing Office, Washington, D.C.

Whiteside, T. 1970. *Defoliation.* Ballantine Books, Inc., New York. 168 pp.

Woodwell, G. M. 1970. "Effects of Pollution on the Structure and Physiology of Ecosystems." *Science,* vol. 168, pp. 429–433.

Woodwell, G. M., P. P. Craig, and H. A. Johnson. 1971. "DDT in the Biosphere: Where Does It Go?" *Science,* vol. 174, 1101–1107.

Chapter 13

Conservation means the wise use of the earth and its resources for the lasting good of men.

Breaking New Ground
Gifford Pinchot

The use of energy, minerals, and soil

Chapter 13: The use of energy, minerals, and soil

Man's history has been a taming of nature. His survival has always been tied to how well he could manipulate the natural materials and energy to his own needs. Until recently, his triumphs over the earth were almost always to his benefit. However, today, because of the rapidly increasing population and a technology ravenous for natural materials, man may be using the earth in a manner that will eventually be suicidal. The symptoms of an exploitation of the earth's energy and matter to a point that threatens survival itself is the subject of this chapter.

Energy sources

Early man experienced several significant revolutions during his slow development. One was the use of fire; another the cultivation of animals and plants for food. A third must have been the first use of animal energy to accomplish tasks never before possible or done only over long periods of time and with many human beings. A fourth important revolution, and a step in the development of culture as we know it today, must have been the use of wind and water.

Yet it probably was not until the twelfth or thirteenth century, when certain black rocks along the northeastern coast of England called "sea coales" were burned, that man's abilities to dominate and manipulate nature took a giant step. The use of coal led in the early nineteenth century to the Industrial Revolution, which allowed the expansion of man's potential through the smelting of metals, followed by the inventions of the locomotive, the steamship, steam electric power, and domestic heating. With the coming of petroleum and natural gas midway through the century, man's abilities to manipulate the environment increased with the internal combustion engine in the automobile and airplane and diesel electric power. Finally, man reached a previously unimagined influence over his environment with the development of atomic power in the 1940s.

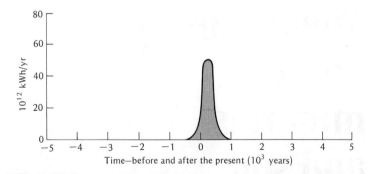

Figure 13–1
Exploitation of fossil fuels in historical perspective from minus to plus 5,000 years from present ("0" represents the present; each interval on the time scale represents 1,000 years). (From M. King Hubbert, "Energy Resources," in *Resources and Man*, publication 1703, P. Cloud, ed., Committee on Resources and Man, National Academy of Sciences–National Research Council, W. H. Freeman and Company, San Francisco, 1969.)

Today's astonishingly rapid use of fossil fuels to maintain environmental dominance confronts mankind with a serious problem. The fossil-fuel reserves are being exhausted (Fig. 13–1). This is primarily because the meteorological conditions necessary for the production of coal that were present 250 million years ago, which favored the luxuriant growth of huge tropical seed ferns and giant nonflowering trees in vast swamps, no longer exist. Also, the peculiar decay and weathering that the plant material underwent before burial and consolidation depended on unique geological conditions involving pressure, temperature, and time no longer extant on a large scale. The coal-making process is no longer going on, so any coal mined reduces the global store.

The unique geological conditions that helped produce petroleum are not now common either. Probably, petroleum was formed from the bodies of tiny ancient plants and animals which lived in seas that once covered contemporary landmasses. When the innumerable organisms died, they sank into the mud and sand at the sea's bottom. There, under great pressure as sedimentation continued above, the mud and sand became rock. Eventually, after the seas withdrew, the earth's crust, including the sedimentary layers, buckled under great pressure, burying the rock organisms. Such heat and pressure and decomposition of the organisms is believed to have formed oil from the dead microorganisms. Today, this oil held between rock strata in vast deposits is petroluem. Petroleum, a fossil fuel formed under ancient conditions, is apparently not being formed in plentiful quantities today.

Consequently, in time man's survival may be threatened unless he can successfully exploit types of energy other than fossil fuels just to maintain, let alone further expand, a culture developed with abundant coal and petroleum.

Coal

Experts estimate that close to 6,000 billion tons of coal originally lay under the earth and that by 1960 less than 100 billion tons had been mined. The rate of use increased each year, however. Palmer Putnam, in his book *Energy in the Future* published in the early 1950s, states that in the United States one-half of all the coal ever used was burned between 1920 and 1950, a 30-year period. Annually, this meant a coal harvest of nearly 2 billion tons per year by the 1960s. Consequently, a look to the future suggests that enough coal is available to last several centuries, after which other fuel types must be exploited.

Some countries are already experiencing shortages because of the unequal global distribution of coal deposits. Ninety-five percent of all coal lies in the Northern Hemisphere. Asia, the USSR, and North America possess nearly equal amounts of approximately four-fifths of the world's coal. Europe owns less than 13 percent; Africa, 4 percent; Australia, 1 percent; and South America, less than 1 percent.

The abundance of coal in the United States — two-thirds of the states have deposits — and Canada helps account for their prosperity, of course. So does the belt of coal stretching from Britain through northern

France, Belgium, Holland, Poland, Germany, Czechoslovakia, and Russia account for the developed nature of these countries. China, too, has abundant coal and appears to be rapidly developing an economy to match it.

In addition to the exhaustibility of the unevenly deposited coal reserves, the retrieval of the coal on a large scale presents another ecological problem. A popular method of mining coal today is *strip mining*. The economy made possible by the development of large earth-moving machinery developed since the 1920s accounts for its popularity. Unfortunately, the coal is reached by removing the overlying earth. Enormous ugly scars are left, devoid of any vegetation and usually subject to erosion (Fig. 13–2).

Figure 13–2
The Anaconda Company's Berkeley Pit copper mine at Butte, Montana, described as the "richest hill on earth," is an example of mining in which the upper layers of earth are removed to reach the lower layers of ore or coal. (Courtesy of U.S. Department of Interior, Bureau of Reclamation.)

Petroleum

Modern drilling for petroleum began as recently as 1859, the year Darwin's *Origin of Species* was published (Fig. 13–3). Yet because of intensive use of this fossil fuel, the world's production will probably peak near 2000 A.D. and then decline, mainly because the deposits of petroleum will have been exhausted. Some countries will reach a peak even earlier.

In the United States, for example, production of petroleum (exclusive of Alaska, whose deposits are bountiful but not yet accurately quantified) is estimated to reach its peak by the early 1970s (Fig. 13–4).

The high rate of use in this country of approximately 900 gallons per year for every man, woman, and child (or 8 times that used in the rest of the free world per person) explains the decline in U.S. reserves. Much of this petroleum is used as fuel, and as distillation processes become refined, more and more is used as lubricating oil, wax, asphalt, and petrochemicals, such as synthetic rubber, fertilizers, and plastics.

Originally, it is estimated that close to 1,250 billion barrels of oil were deposited in the earth, and that of this, approximately 100 billion barrels were extracted by the 1960s. In fact, it was said in 1968 that one-half of the world's cumulative production of petroleum had occurred in the 12 years since 1956. As with coal, man may well have a serious problem as the petroleum is exhausted, and no other adequate energy source is developed. The search for new reserves of petroleum continues at an accelerating pace. In 1947, approximately 7,000 wells were dug, whereas in 1953, approximately 13,000 wells were dug. As the search ac-

Figure 13-3
Modern oil drilling rig. (W. F. Hale, Texaco.)

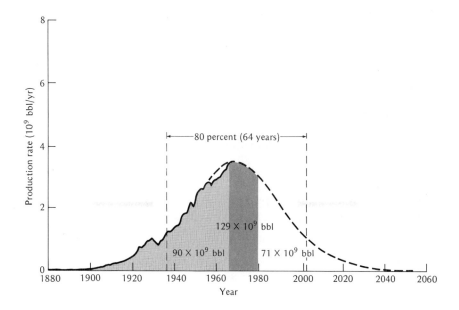

Figure 13-4
Rate of production of petroleum liquid in the contiguous United States and adjacent continental shelves. (Data from M. King Hubbert, "Energy Resources," in *Resources and Man,* publication 1703, P. Cloud, ed., Committee on Resources and Man, National Academy of Sciences–National Research Council. W. H. Freeman and Company, San Francisco, 1969.)

celerates, the scarcity of petroleum becomes increasingly apparent. For example, average depths of wells in 1947 were 3,547 feet; in 1953 they reached 4,069 feet.

Perhaps new oil fields will open up to offset the great demand. Alaska, for example, is a relatively untapped source of petroleum. The field near Prudhoe Bay on Alaska's north slope may prove to be equal to or larger than the east Texas field or the largest fields in the Middle East. The quantity of petroleum in the presently drilled fields is fairly well-calculated, however, so the oil resources in South America (Venezuela, Colombia, Ecuador, Peru, Bolivia, and Uruguay), in Iran, Iraq, and Arabia, and in Burma and Egypt are not expected to reveal any large hidden deposits that would postpone man's dilemma.

Natural gas

Natural gas is believed to have been formed in a manner similar to petroleum. Consequently, exploration for oil has revealed vast quantities of natural gas, mostly methane, as well as oil. Natural gas contains components heavier than methane, but distillation converts the gas into a most desirable fuel for domestic use. In the United States, in fact, thousands of miles of pipelines carry the gas to all parts of the country. As with petroleum and coal, natural gas is an irreplaceable commodity. Natural gas and petroleum are said today to supply approximately 60 percent of the world's energy, and 61 percent of U.S. energy for industrial purposes. Furthermore, the demand for natural gas has grown about 6 percent annually, and promises to continue growing at a 3 to 4 percent rate per year. Once used up at this incredible rate, it cannot be replaced. Natural gas supplies may last only a few more decades unless abundant new resources are found (Fig. 13–5).

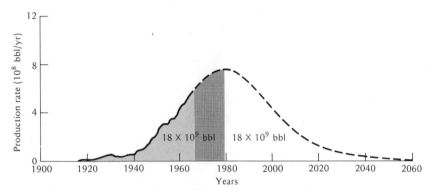

Figure 13–5
Production of natural gas liquids in the contiguous United States and adjacent continental shelves. (Data from M. King Hubbert, "Energy Resources," in *Resources and Man*, publication 1703, P. Cloud, ed., Committee on Resources and Man, National Academy of Sciences–National Research Council. W. H. Freeman and Company, San Francisco, 1969.)

Thus, the main sources of energy in the industrialized world today are all fossil fuels — coal of various kinds, petroleum and its products, and natural gas. Several generalizations apply to the use of all three. The geological conditions under which each was formed no longer exist; the fuels are a capital resource subjected to a "robber economy" — that is, once used, they are forever gone — the world's supply of the fuels, especially oil, is often in a land other than that which needs it; and some countries with abundant fuel reserves have very few other commodities, The political and economic consequences of the last two points have considerable significance.

Other fossil fuels

Other fossil sources of fuel, such as shales and tar sands, have been investigated in preparation for the era when coal, petroleum, and natural gas supplies are exhausted. Certain shales (laminated rocks formed at the bottoms of ancient seas from mud, sand, and silt) are found in Scotland, Canada, Austria, and France, which contain considerable quantities of oil, varying between 10 and 130 gallons per ton of shale. Shale in the Colorado Plateau yields between 10 and 60 gallons per ton of rock. Unfortunately, contaminating materials such as uranium are found in the oil, and 1.5 tons of waste rock are produced for each barrel of oil, enough to cover Colorado 10 feet deep if all the oil-rich shale in the state were processed. Oil is produced from shale by feeding crushed rock into "retorts," which break down, or "crack," the organic material (kerogen) in the shale at 350 to 500°C. The oil produced is similar to petroleum but contains more sulfur, nitrogen, and oxygen.

Tar sands, deposits of sand or sandstone impregnated with the heavier components of petroleum, are also looked to as potential energy sources. At present little is known of their global distribution, although in the United States only Utah and California appear to have any important quantities. They are also difficult and expensive to process because, unlike oil shales, simple distillation does not work. Hot water and solvents are needed to release the oil from the sand.

Solar energy

Scientists who realize the impending calamity of exhausted fossil fuels look for other sources of energy to power the globe. Research is under way to tap the sun's energy. Already functioning in Israel is a simple system of heating homes and the water supply with solar energy (Figs. 13-6 and 13-7). Commercial solar hot-water heaters are available in Florida, Arizona, and California. A solar cooker consisting of an alumi-

Figure 13-6
Painting of a solar farm developed by scientists at the University of Minnesota and Honeywell, Inc. The structures in rows harvest the sunshine, then convert it to thermal power and eventually to electrical energy. (Shirlee Houglum, Honeywell, Inc., Minneapolis, Minn.)

Figure 13-7
Solar home under construction in Delaware. Roof top "solar cells" capture sunlight, convert it to electricity, and store it in batteries. Heat from solar radiation is pumped to insulated tubes of "eutectic salts," common salts that absorb heat by melting and resolidify when the heat is used. (Courtesy of The Institute of Energy Conversion, University of Delaware, Newark, Del.)

num reflector 10 square feet in area has been tried in India. The use of solar energy for heat is extremely costly, however, so it is not likely that it will be extensively used in the near future. Probably, it will first be exploited in parts of Africa, Asia, and Australia where the percentage of sunshiny days is high, fuel is scarce, and construction costs are low.

The use of solar energy to generate electricity appears feasible. Required are semiconductors, which will be charged by the sun's radiation. The commercial Bell solar battery can be used to generate electricity for a variety of small enterprises. Attempts to generate electricity with the sun's energy on a large scale are proving to be complex and expensive, but further research and development may overcome these obstacles.

Water power

The power of moving water has been used since Roman times. Grist mills, textile mills, and saw mills helped build the economy of many countries in the eighteenth and nineteenth centuries. The more recent use of dammed water has brought electrification to many places otherwise undeveloped.

Some researchers believe if the numerous waterways in the world were today harnessed as in times past, enough energy might be available to maintain the world's present level of technology and development. Immense technical problems – for example, preventing silt from filling in dammed water reservoirs — would have to be worked out before such a plan could be instigated, however.

Perhaps a more likely use of water for energy is tidal power, or the harnessing of the energy produced from the rising and falling of tides. Advantages include its independence of weather, a condition not true with the capture of solar energy. Tidal mills to grind grain have been in use since the twelfth century, but any use depends on favorable sites. Probably, only 50 highly desirable places exist in the world for capitalizing on the tides. These are high latitude bays with long irregular shorelines with partially enclosed smaller bays. The Bay of Fundy is an excellent example. In some desirable sites the water level ranges 50 feet between high and low tides.

The most important application of tidal power is the generation of electricity. France in 1966 constructed the first major tidal electrical plant. Russia built one in 1968 in the Kislaya Inlet on the Barents Sea

Figure 13-8
The first major tidal electrical plant also functions as a bridge. (Courtesy of French Embassy Press and Information Division, New York.)

coast and is planning others. All plants function by directing the water flowing into or out of the dammed coastal basins through hydraulic turbines attached to generators (Fig. 13-8).

The world's potential tidal power fully developed would probably provide only a fraction of the world's power needs. Yet in some areas it would be very beneficial, because it minimally disturbs the ecology and esthetics of an area, produces no noxious pollution, and exhausts no energy resource.

Geothermal energy

Some electrical power plants have been constructed that use the steam produced underground by volcanic heat and directed to the earth's surface through cracks and crevices or drilled wells. The earliest naturally produced steam-generated plant was constructed in 1904 in Tuscany Province, Italy. The largest, in New Zealand, took 8 years to build. The third largest plant is in northern California; Japan, Mexico, and the USSR also have geothermal power plants. Geothermal plants probably will not be the main suppliers of electrical energy in the future. The demand for electricity will be too great, and the number of natural steam-yielding sites too few.

Atomic energy

Atomic reactions will probably become the principal source of energy as the fossil fuels are used up. Atomic energy has many advantages, but the most important are an abundant source of radioactive material from which energy is emitted and the incredible quantity of energy which the radioactivity yields. One gram of uranium^{235}, an important radioactive element, yields energy to equal the combustion heat of 2.7 metric tons of coal!

In the future almost any country may have radioactive ore available once extraction techniques are improved. That is because the common igneous rock granite contains 4 ppm of uranium and 12 ppm of thorium, another important radioactive element. Although the quantities seem low, in 1 ton of rock they are equivalent in energy to 50 tons of coal.

The channeling of atomic energy, usually into steam energy for the generation of electricity, requires a nuclear reactor. The core of the reactor, or critical assembly, is the site of activity. Here is placed a radioactive, or fissionable, material such as ^{235}U (Fig. 13-9).

The nuclear reactor operates through fission, the ability of the radioactive materials to naturally split apart. This occurs when a slow, or "thermal," neutron naturally occurring in the environment strikes the uranium nucleus, knocking it apart. Among the reaction products are heat and fast-moving neutrons. The neutrons are slowed by collision with a "moderator" substance. Then they may collide with another uranium nucleus in a repeat performance. A chain reaction ensues, with the production of a great amount of heat. Unfortunately, in this type of reactor, called a "burner" reactor, all the ^{235}U is soon exhausted. In view of the

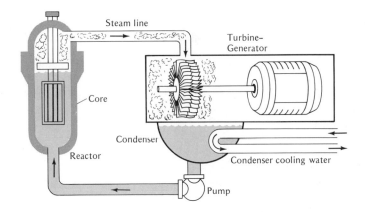

Figure 13-9
Basic construction of a fission power reactor.

relative rareness of the element, ^{235}U, the burner reactor is not a promising source of energy for the future (Fig. 13–10).

$$^{235}_{92}U + neutron \longrightarrow \text{fisson products} + \text{neutrons} + \text{energy}$$

Figure 13-10
Reaction in a burner reaction.

The "converter" reactor may have promise, however, because it can convert fissionable material from nonfissionable elements, all the while producing heat. The convertor reactor will continue operation as long as the proper nonradioactive "fuel" is constantly added.

The nonfissionable materials used usually include another isotope of uranium, ^{238}U, or thorium, ^{232}Th. Uranium-238, upon the absorbance of neutrons, can convert to plutonium (^{239}Pu) through the sequence of steps outlined in Fig. 13–11. Plutonium is radioactive, and its subsequent breakdown yields heat and radioactive particles. Thorium, through the capture of a neutron, also produces a radioactive uranium isotope, ^{238}U. Thus, nonfissionable ^{238}U and ^{232}Th can convert to fissionable isotopes if struck with neutrons to yield fissionable isotopes ^{239}Pu and ^{233}U.

$$^{238}_{92}U + neutron \longrightarrow {}^{239}_{92}U \xrightarrow{\beta-} {}^{239}_{93}Np \xrightarrow{\beta-} {}^{239}_{94}Pu$$

$$^{232}_{90}Th + neutron \longrightarrow {}^{233}_{90}Th \xrightarrow{\beta-} {}^{233}_{91}Pa \xrightarrow{\beta-} {}^{233}_{92}U$$

Figure 13-11
Reaction in a breeder reaction.

In the converter reactor an initial quantity of ^{235}U-yielding neutrons is mixed with plutonium or thorium to start the reaction. Then neutrons discharged from these two elements in their process of decay are recycled to begin decay in other atoms of these elements. Furthermore, in a fast breeder reaction the creation of fissionable ^{239}Pu occurs at a rate even greater than the consumption of the initiating ^{235}U. Hence, a convertor, or breeder, reaction can continue indefinitely as long as a nonfissionable fuel is constantly added, always incidentally radiating great amounts of heat.

The heat is the important product of the nuclear reactions. In fact, the upper limit of heat production does not depend on the fission process, or sequence or reactions, but rather on the rate of removal of the heat produced. The temperature will continue to rise if the heat is not removed, possibly eventually destroying the reactor. The abundant heat is used to make steam, which is channeled through a turbine connected to a generator. Nonbreeder, nuclear-powered electric generators are also now in operation, the first begun in Shippingport, Pennsylvania, in 1957.

Radioactive energy will probably be used for other purposes, too. Fuel for marine vessels is one application, as, for example, in the nuclear-powered submarine USS *Nautilus,* built in 1955. Power stations in remote regions when prolonged unattended projects are under way; industries, such as chemical plants or food processing plants, when great quantities of continuous heat are needed over a long period; or long-voyage spaceships traveling to distant planets will probably provide other uses for radioactive energy.

The great amount of energy available from a small quantity of the nuclear material is the main attribute of nuclear-produced energy, but it has at least one serious disadvantage, the disposal of the radioactive wastes emitted from the reactors, especially when they are cleaned. One common method of disposal is to dilute the material and then disperse it. This is commonly done by pouring the wastes in lakes, rivers, or the sea. Greatly concentrated wastes cannot be treated this way, however. Rather, they are stored in underground tanks, embedded in concrete blocks sunk in the sea, buried in a remote area, or stored in underground geological formations such as salt beds. At Oak Ridge National Laboratory the wastes as a slurry are injected into hydraulically induced fractures 700 to 1,000 feet deep in shale formations. Since the deposits are between horizontal shale strata, little threat exists that the radioactive materials will reach the surface and contaminate living organisms. Such a possibility haunts researchers using radioactive materials because of the serious deleterious effect of radioactivity on life.

Fusion reactions, such as occur in the sun, are other forms of nuclear reaction that have been proposed as an energy source. In these reactions atoms are fused instead of fissioned as described above (Fig. 13–12). In either case energy is given off, but fusion emits vastly greater quantities. Hence, scientists have experimented with the possibility of harnessing the fusion process. Control seems a long way off at this point. The energy emission is too sudden and too concentrated to manage with any equipment or technique now known. The most successful taming

$$^2_1H + ^3_1H \longrightarrow ^4_2He + neutrons + energy$$

Figure 13-12
Fusion reaction.

of fusion has produced the hydrogen bomb, a long way from generating electricity or powering a spaceship.

Minerals

The exhaustion of fossil fuels in the next centuries will present man with a problem never before confronted. Atomic energy may maintain the human population, although technology must develop a great deal before the problem can be resolved.

The exhaustion of mineral resources may even be as great a crisis, and one that man must face sooner or later. In fact, the amount of time before some vital mineral resources are exhausted can be reckoned in only a few decades. To better understand man's dilemma so as to possibly avoid disaster when the mineral deposits run out, several generalizations should be understood.

The deposition of mineral ores around the world was accomplished by unique geological processes over long periods of time. For example, many mineral deposits probably began in the slow migration of molten rock, or magma, toward the surface. As it slowly cooled, it allowed for a differential settling out of various crystals. Solidifying from the most viscous magma were the oxides of many common elements, such as silicon, aluminum, iron, and magnesium. Less viscous magma composed of minerals containing rare elements, such as uranium, tungsten, and beryllium, probably formed "dikes," offshoots into cracks and fissures of surrounding rock.

Finally, supposedly ionically complexed materials in aqueous solution containing many metals, such as tin, gold, copper, mercury, and zinc, were selectively deposited under varying conditions of temperature and pressure into very permeable and porous rock.

After solidification, weathering occurred to further act on the deposits. Mineral crystals dislodged by wind or water accumulated, because of their high specific gravity, along stream bottoms or beaches. Iron and aluminum deposits in some tropical areas have been deposited this way. Dissolution of some compounds has left rich concentrations of compounds, too. Carbamates of copper, zinc, and lead are known to be formed from primary deposits by CO_2-containing groundwater.

Numerous formation processes have occurred to create the diverse kinds of mineral ore deposits found around the world, some of which are not well-understood today. What is appreciated, though, are the great periods of time needed for the ore depositions. Today, the demand for most mineral ore is so great that it is impossible to expect any mineral to be replaced at the same or greater rate than it is used, although examples of contemporary mineral depositions are known.

Minerals are nonrenewable and exhaustible. Unlike energy sources that once used are forever gone, many lost as heat, mineral resources may be utilized only in a percent concentration or in a particular, economically feasible combination with other elements. At a later time as the demand for the mineral becomes greater or extraction or processing methods improve, ores once discarded as waste may be utilizable. Such was the case with the 25 percent taconite iron ore of Minnesota, which was discarded until recently when a new magnetic process allowed its conversion into 63 percent iron pellets. Increasing demand for iron prompted development of new and inexpensive technology. Unfortunately, increased water pollution results from the new process.

Furthermore, estimates of time needed to exhaust natural mineral supplies do not include deposits of minerals which may be undiscovered, of course. Certainly, some areas are not well-exploited, for various reasons. For examples: costly transportation prevents thorough mining in Australia; Chile has an inadequate continuous supply of water; and in Alaska, labor is unavailable. When the ores are more thoroughly worked, a more accurate concept of their extent will be known. Estimates of the size of some mineral deposits will be revised.

Nevertheless, new technology and new deposits will probably not meet the increasing demands. For example, although many metals may be available as waste for recycling, such as tin cans, economical methods of preparing for reuse must keep pace with the demand.

The political and economic implications of the exhaustion of a country's resources are serious. History is ripe with examples of the influence of the discovery or exhaustion of ore reserves on the internal economy and politics of countries, as well as on their influence with other nations. For examples: the west of the United States was opened by the gold rush; the discovery of gold in Alaska and Australia quickly sprouted very large towns in both areas which turned into ghost towns when the mineral ran out; silver from Spain provided mercenaries for Carthage to fight the Romans; and the bountiful supply of metals available to victorious Rome allowed development into new areas — many of which supplied additional silver, gold, copper, and tin. Numerous other examples could be quoted from the past.

Today, as an example, 70 percent of Chile's national income comes from the sale of copper to foreign powers. The political stance within and without the country is obviously linked to the economy based on copper sales. Communist countries are other examples of nations whose political influence internationally in the future may well be linked to their mineral reserves of great importance, such as tungsten, paladium, platinum, and antimony.

Iron

Iron has been used in tools since antiquity (Fig. 13–13). The fusion of iron and carbon to produce steel was mentioned by Homer. The durability of the alloy has made iron the most useful metal for civilized man. Certainly, the Industrial Revolution could not have occurred without it. As technology develops, the utilization of iron increases. In the 1960s industry produced over 250 million net tons of steel, an increase over the previous decade of 25 million tons.

Iron occurs in several ores, all of which are common, easily accessible, and very pure. The main areas of occurrence are the Lake Superior region in the United States and Canada, the Franco-German border region extending into Luxembourg and Belgium, and Great Britain. Other notable, but unexploited deposits occur in Brazil, Cuba, the USSR, China, India, the Philippines, and Indonesia.

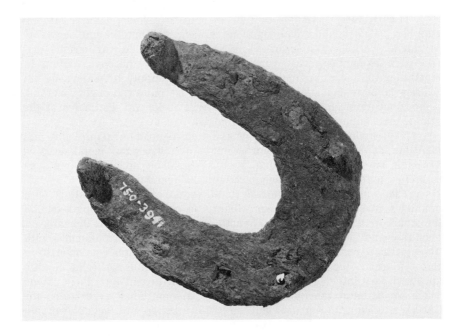

Figure 13–13
The forging of this ancient horseshoe found near Jutland, Denmark, must have been one of the first uses of iron. (Courtesy of American Museum of Natural History, New York.)

The great demand for iron ore is draining the mined deposits of high quality and forcing industry to use lower grade. For example, the average iron content of lake ores in the United States dropped from 55.5 percent in 1902 to 51.75 percent in 1910, and is now maintained at 51.5 percent. Also, the original rich ore deposits in Great Britain are essentially depleted now, forcing exploitation of ore beds containing only 27.5 percent iron (Fig. 13–14).

How long will the iron ore deposits last? Precise estimates are impossible, but at the present rate of consumption a conservative guess is approximately 200+ years (Fig. 13–14).

383 Minerals

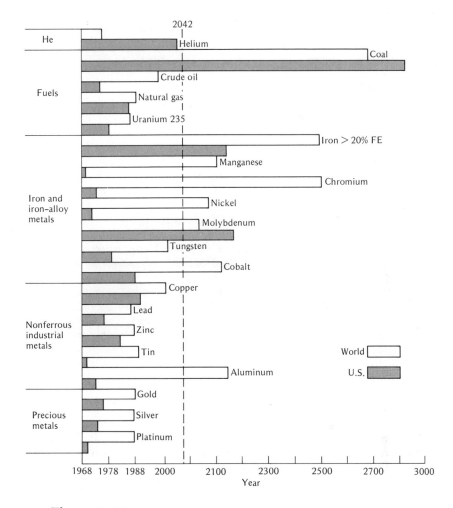

Figure 13–14
Predicated lifetimes of various fuels and metals. (From P. Cloud, "Mineral Resources in Fact and Fancy," in *Environment*, W. W. Murdock, ed., Sinauer Associates, Inc., Stamford, Conn., 1971.)

The validity of the estimate is dependent on at least two factors. One, the fairly accurate assessment of the earth's store of iron (a difficult feat in view of the inaccessibility of China and the USSR, for example) and two, the improbability of any major improvements in extracting and processing the ore, an unlikely condition.

New processes have been developed to make economically feasible the extraction of the metal from inferior Swedish and Minnesotan ores. Taconite ore from Minnesota containing 25 percent iron is collected and powdered. The iron oxide is separated with magnetism and pelleted with a clay binder to yield balls containing 63 percent iron. This new process has revitalized areas previously abandoned. Other processes may be developed as the demand for iron increases.

Ferroalloy metals

Several metals are frequently used in alloys with iron. Included in this group are chromium, cobalt, manganese, molybdenum, nickel, tungsten, and vanadium.

Chromium deposits are rare, especially in the Western Hemisphere. Ninety-eight percent of the world's reserve occurs in southern Africa. This metal lends strength and hardness to steel alloys. The total world reserve is estimated to last 500 years; in the United States, only 5–10 years (Fig. 13–14).

Cobalt occurs with other metals. The largest deposits are in Cuba and Africa. Cobalt is in great demand, mainly because it helps form noncorrosive alloys. The reserves in the Free World are estimated to last approximately 150 years; in the United States, only about 25 years (Fig. 13–14).

Manganese as an oxide is common in nature, but concentrated ore is rare. Most manganese is found in Brazil, Africa, India, Mexico, and the USSR. Unfortunately, the United States has none. The metal is used in great amounts, primarily in metallurgy to remove impurities from chemical industries. Thirty-five pounds of manganese ore is needed, for example, to make 1 ton of steel. The manganese reserves are estimated to last approximately 140 years (Fig. 13–14).

Zinc

Zinc occurs primarily in nature as ZnS, or sphalorite, although several other ores are also found. The USSR, Australia, Peru, Mexico, and Japan mine zinc ore, but the main producers of zinc ore are Canada and the United States. Zinc is used in coating steel to prevent corrosion, in alloys for car parts, and in dry-battery cans. The demand for zinc increases 2 percent per year. Consequently, reserves are said to probably last 20 years or so, although new discoveries of the metal are likely (Fig. 13–14).

Aluminum

Aluminum is perhaps the most recently exploited metal. In our modern technology new applications for the metal seem to be discovered every day. Mainly, this is because of its light weight, ability to combine with other metals to produce durable alloys, high thermal and electrical conductivity, and resistance to corrosion.

Aluminum stands little chance of being exhausted from the natural environment. The metal makes up approximately 8 percent by weight of the earth's crust. Most of the aluminum from the original igneous rocks has weathered into soil as aluminum silicate, or clay, and technology that allows its exploitation is well-advanced. In severely weathered rock the silica has been dissolved and deposited in dense ores such as bauxite, containing as much as 60 to 70 percent aluminum oxide. Today, Jamaica, Surinam, and Guyana supply most of the world's bauxite.

Copper

Copper was the metal first used by man, probably about 8000 B.C. (Fig. 13–15). Its malleability and ease in forming alloys accounted for its popularity initially, but as in the Iron Age, copper's continued use arises from its beauty and durability. By the nineteenth century its unusual ability to conduct electricity accounted for an intense exploitation of natural ores.

Great deposits of ores occur in every continent, but the major mines are found in the Rocky Mountain ore of the United States, the Republic of the Congo, northern Rhodesia, central Canada, northern Michigan, and the western Andes, in Chile and Peru. The quantity of the ore suggests that the metal will probably be exhausted soon, probably in 50 years or so (Fig. 13–14). However, demand goes up continuously, as seen in the annual production of copper in the world between World War I and 1960. Previous to World War I the annual production was near 1 million metric tons. In 1939 production reached 2 million tons; by 1943, 3 million tons; and by 1960, 4 million tons. Furthermore, the quality of ore decreased significantly. Today, the average percent of copper in ore is one-sixth that found in ore in 1900.

Tin

Tin has long been used, primarily in bronze, an alloy made with copper. Bronze artifacts dating 3500 B.C. have been found. Tin occurs in lean veins in granitic rock as in Bolivia or, more commonly, in alluvial deposits containing cassiterite, an ore with 70 to 77 percent tin. The

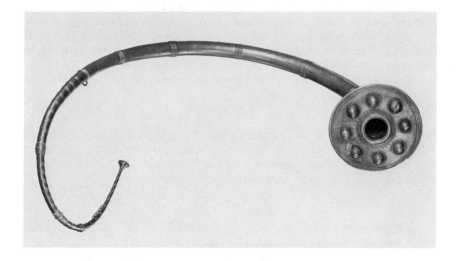

Figure 13–15
A trumpet, relic of the Bronze Age, is evidence of the early use of copper and tin in alloys. (Courtesy of American Museum of Natural History, New York.)

metal is used in small quantities in alloys in tin plate containers, solder, artwork, pipes and valves, and car parts.

Almost no tin is found in the United States and Canada, and less than 2 percent of the world's output comes from Europe. Hence, the biggest users of tin have the fewest natural deposits. About 60 percent comes from southeastern Asia; specifically, Malaysia, Thailand, lower Burma, and two small Indonesian islands. The rest of the material comes from Bolivia, the Congo, Nigeria, and China. Notably, underdeveloped countries supply most of the world's tin. The expected expiration of the tin reserves is approximately 25 years, excluding China's reserves (Fig. 13–14).

Molybdenum

Molybdenum occurs abundantly in the United States, even exported at about 25 million pounds per year. Most is used in the iron and steel industry. Very likely, this metal will last another 100 to 150 years (Fig. 13–14). It is used in industry because of its strength and resistance to shock. In the United States, the reserves will last even longer.

Nickel

Nickel is not distributed evenly around the world; the United States, for example, has little. Most metal comes from Canada and Norway for use in hard, heat-resistant alloys and electroplating anodes. Nickel reserves may last about 140 years with the present 2 percent per year increase in demand (Fig. 13–14).

Tungsten

Little tungsten occurs anywhere in the world except Asia. Yet the developed countries demand large quantities for use in carbide manufacture and in the high-speed-tool steel industry because of tungsten's durability and resistance to heat in alloys. Significantly, China probably contains 77 percent of the metal. The world reserves are estimated to last 40 years or so. The deposits in the United States are nearly exhausted now, although plentiful reserves may be uncovered in California brine deposits (Fig. 13–14).

Vanadium

Vanadium is widely distributed, but not in high-grade ore deposits. Major reserves occur in the United States, Africa, Finland, China, and the USSR. Most is used in hard alloys and in the production of chemicals. Vanadium reserves presumably will last hundreds of years.

Soil

The dust bowl: An ecological lesson·

In *The Grapes of Wrath* (Modern Library, New York, 1939), John Steinbeck described one result of severe mismanagement of the soil, the development of the dust bowl in the 1930s: "During a night the winds raced fast over the land, dug cunningly among the rootlets of the corn, and the corn fought the wind with its weakened leaves until the roots were freed by the prying wind and then each stalk settled wearily sideways toward the earth and pointed the direction of the wind.

"In the morning the dust hung like fog and the sun was as red as ripe new blood. All day the dust sifted down from the sky, and the next day it sifted down. An even blanket covered the earth. It settled on the corn, piled up on the tops of the fence posts, piled up on the wires; it settled on roofs, blanketed the weeds and trees" (Fig. 13–16).

The land described by Steinbeck was not always dust. Initially, most of the American Great Plains was covered by grass and provided an excellent environment for grazing animals, such as the bison. Throughout history, however, the area has been subjected to droughts, often long and severe. When the first settlers reached the area in the 1880s, they came during a decade of plentiful rainfall. However, the following decade was much drier, and many of the original farmers who plowed the land and planted the crops moved on. By 1890 the rains came again and attracted more migrating Americans. But the problem of alternate droughts and moist years was to persist. By 1910 lack of rain and a soil broken by cultivation accounted for considerable wind erosion. Yet cultivation continued, especially with the coming of World War I, when food was in great demand and prices were high.

Prosperity reigned until 1931. Then the drought returned. Because of extensive farming in the area the effects of the lack of rain were devastating. In 1933 a new and violent series of dust storms alarmed the nation and awakened it to the disaster being born at its center. Millions of acres were destroyed because the topsoil, 2 to 12 inches, was picked up by the wind, then dropped to bury homes and roads in the area. Some dust clouds were carried eastward to the sea, past the nation's capital.

Lawmakers in Washington who witnessed effects of the dust storms resolved to take action. In 1935 the Soil Conservation Service was

Figure 13–16
Desolation of the dust bowl. (Courtesy of U.S. Department of Interior, Bureau of Reclamation. Photo by L. Axthelm.)

established. Soon states formed soil conservation districts to locally handle soil problems. This legislation was eventually to help preserve millions of acres of soil from erosion and destruction, but it did not immediately solve the problem of dust bowls.

The drought of the 1950s, after a moist decade in the 1940s, further devastated the dust bowl area, so severely that the federal government declared an emergency. The influx of government money and the ability of many farmers in the area to support themselves in cities saved the unfortunates in the area from total disaster.

The story of the dust bowl is an example of soil mismanagement. Other examples abound. The Mississippi River is said to carry as much as 730 million tons of soil per year. In the Southeast, estimates of a 75 percent loss of the original soil from the Piedmont Plateau, the Upper Coastal Plain, and other areas are believed to be valid. The total load of eroded soil supposedly carried into the seas by rivers probably exceeds 1 billion tons per year. And, in the West close to 600 million acres of 800 million acres of western range land are so seriously eroded that a 50 percent reduction of carrying capacity is common!

In these days of increasing demand for resources, such tragedies of soil abuse cannot be afforded. The characteristics of the soil and its uses for man, which make it a precious commodity to be cherished, illustrate its important role in nature and why it needs to be protected.

Soil structure

In earlier chapters we discussed evolution in order to emphasize the great amount of time needed to produce the kinds of relationships seen today between organisms and their environment, and the threat to life when these relationships are carelessly altered. A discussion of the evolution of soil and the dependence of life on it helps make the same point.

In the beginning, the earth had no soil. Slowly, soil developed through the milennia, out of the natural environmental conditions impinging on the original rock, such as frequent rains and running water, and the alternate freezing and thawing of moisture. Soil came from the small eroded pieces of rock, often too small to be seen. Some, in fact, are aggregates of atoms often less than 1/1,000 millimeter and so are very mobile. Frequently, they are transported in suspension thousands of miles from their point of origin. Their deposition in quantity, along with water, gases, and organic materials derived from plant animals, is soil.

Usually, soil is characterized by *layers*, or *horizons* (Fig. 13-17), products of weathering, type of parent rock, and type of accumulated organic material. Topmost is the topsoil (horizon A), often a fertile soil with much organic material, gases, and small organisms, including nematodes, algae, amebae, bacteria, actinomycetes, and fungi. The bacteria alone are said to be so numerous as to weigh 1 ton per acre or more. Beneath this is the subsoil, a well-weathered small-particled soil (horizon B). Next is a layer of partly weathered and broken-down parent rock (horizon C). Beneath this is the parent material, or original rock (horizon D).

Although most soils generally show the horizons, the thickness of layers, the kind of rock particles composing them, the density of the layers, and other characteristics vary greatly. Mainly, it is meteorological conditions that account for the variations. For example, warm climates enhance the weathering of soils, so soil samples taken across the country from north to south show an increase in clay content. Also, low temperatures and abundant rainfall increase the amount of soil nitrogen, and abundant rainfall carries dissolved materials deeper into the soil. Calcium carbonate ($CaCO_3$), for example, often forms a band in the soil, but the depth varies with mean rainfall. The acidity or alkalinity of soils is also a result of rainfall. In general, in arid regions such as some grasslands, the soil is alkaline or neutral because the lack of rain has not leached away the materials. On the other hand, wetter areas with more leaching often have acidic soils.

The kinds and sizes of particles resulting from weathering make up the soil and account for many of its characteristics. These characteristics include its permeability to water, water-retaining capacity, seration, ability to stand repeated cultivation, ability to supply nutrients to plants, and resistance to erosion.

Clay, for example, is composed of the tiniest particles, often complex aluminum from silicates, and makes up about 30 percent of the soil. The clay particles are charged, or ionized, and, hence, attract other charged particles. These positively charged particles are not easily disassociated with pure water, although other positively charged particles can replace them. Common absorbed ions include phosphate, borate, and molybdate. Negatively charged ions are not absorbed, such as nitrate,

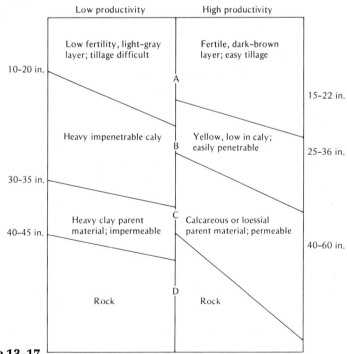

Figure 13–17
Profile of soils with low productivity and high productivity. (Data from *Encyclopedia Britannica*.)

sulfate, and chloride, and consequently, they occur abundantly in waterways.

Clay aggregates also have the ability to swell with water. This is because the structure of clay particles resembles a crystalline lattice. The lattice expands when it picks up water molecules. Thus, clay contributes a great deal to the texture, properties, and uses of soils.

Raymond F. Dasmann, in his book *Environmental Conservation,* cites an illustration of the way structure and permeability of soil is affected by cultivation. The Seabrook Farms in New Jersey were faced with the problem of disposing of large quantities of waste water from the processing and freezing of vegetables. The water was highly polluted with organic matter and dirt, so it could not be discharged directly into nearby streams. Attempts to dispose of the waste water by spraying it on agricultural lands were frustrated because the pollution plugs the pores in the soil so that little water seeps down. The waste was then sprayed on previously uncultivated forest land. These undisturbed soils could absorb 5 inches of water per 10-hour period compared to 1 inch on the cultivated lands. As a result, the equivalent of 600 inches of rainfall per year was sprayed on the forest soils, and the vegetation grew incredibly.

Soils of value agriculturally, for example, have a particular mixture of clay (particle size 0.002 meter) and silt (above 0.002 millimeter). Soil with too much clay compacts easily, and plant roots cannot penetrate it. Soil with too little clay does not retain water. Cultivation often affects the clay content of a soil and changes its properties. For example, unplowed soils have a very open and loose structure, because erosion has not occurred to carry off the clay. On the other hand, frequently plowed soil can be very dense because the clay particles will pack, and then plowing and penetration by plant roots is difficult, and percolation of water with dissolved materials is less efficient.

A soil mixture suitable for farming includes other ingredients besides clay. Abundant humus is important to help hold moisture and supply nutrients, and abundant nitrogen is vital. The legume-nodule bacteria (*Rhizobium*) and others, such as *Clostridium, Azotobacter,* and *Beyerinckia,* and the blue-green algae help add nitrogen to the soil. The bacteria are astonishingly efficient in the conversion of atmospheric nitrogen. The nodule bacteria alone are calculated to fix as much as 250 pounds of nitrogen per acre per year, although the average is closer to 80 pounds. Thus, organisms play vital roles in the maintenance of the nitrogen cycle.

Soil preservation

Soil has never stayed in one place. Running water and wind have moved soils since they were first formed from the early rocks. Thus, soil erosion is not new, nor is man exclusively the cause of it. Nevertheless, today man's agricultural practices certainly aggravate the problem in a major way.

For example, in 1882 land classified as desert or wasteland equaled 9.4 percent of the earth's land area; in 1952 it grew to 23.3 percent. Today, some farmed areas on earth are known to lose inches of topsoil per year through erosion by wind or water. In fact, it is estimated that soil erosion is so rampant in India that one-half of the farmland is threatened with total removal of topsoil.

Soil erosion occurs easily when trees are cut or the soil is plowed. Plant roots bind the soil, and their destruction allows the soil to be readily moved by wind or flowing water. Such destruction, aided by drought, accounted for the formation of the dust bowl in this country. Secondary causes of soil instability include the disturbance of the cycles involving humus and nitrogen. In undisturbed soils, these materials are normally replaced at least at the rate at which they are removed. In disturbed soils, the lack of recycling of organic matter demands that fertilizers be used, often rich in nitrogen, which greatly increases the nitrogen content of the soil and adjacent waterways. If fertilizers are not used, soil fertility continues to decline, and in time this is reflected in the crop. For example, the protein quality of wheat grown in Kansas declined greatly between 1940 and 1951. In some counties it was as great as 8 percent.

Several techniques have been attempted to impede the degree of soil erosion. Included are *minimum tillage, terracing, contour plowing,* and *strip cropping*. Minimum tillage attempts to cultivate only to the degree necessary to ensure quick seed germination and satisfactory growth of row crops. The seed bed is only roughened, so water sinks easily into the soil. Thus, little flows off to erode. Contour tillage is plowing along the contours of the fields rather than up and down the hills (Fig. 13–18). Consequently, water cannot run off as easily. It is claimed that 50 percent more erosion occurs with up and down plowing than with contour plowing (Fig. 13–19). Another method, strip cropping (Fig. 13–20), involves the planting of strips of grass or small grains. Such a technique is effective on hilly land, especially if coupled with contour plowing. It is known to reduce soil losses about 75 percent. Terracing is widely used on long slopes. The slope is converted into a series of broad, level areas. Water draining down the incline is diverted into the flat areas. It slows down and reduces its erosion ability. Furthermore, the water is retained for watering crops grown in the terraces. Terrace cultivation is used very extensively in China, Japan, the Philippines, around the Mediterranian, and in the Andes of South America. Other erosion controls include grass waterways for water runoff and the construction of concrete or masonry structures to prevent the overflow from cutting a gulley back into the waterway.

Probably, one of the best types of water erosion control is the maintenance of a cover vegetation on land. Of course, such a technique cannot be practical on land destined for farming. Yet most likely, a great amount of unplowed farmland should or could be reforested or regrassed. If so, many reservoir dams used to catch and retain runoff water during rains would probably not be needed. As it is, these reservoirs often have a life span of only 10 to 20 years because of the abundant silt brought into them in floodwater. Other techniques used to reduce the eroding force of rushing water include the straightening of old streams, banking streams with concrete walls, and constructing bypass channels.

392 Chapter 13: The use of energy, minerals, and soil

Figure 13-18
Contour plowing and strip cropping on an Iowa farm. (Courtesy of Iowa State University, Extension Service, Ames, Iowa.)

Wind erosion is also a major problem in agricultural areas. Methods for its control include mulch tillage, or stubble mulching, in which crop remains are left on top of the soil to hold the soil; rough, or coarse, tillage, which provides a rough surface to trap blown soil; barriers, such as crop strips or windbreaks to break the force of the wind; and plowing in the spring to avoid exposure of loose soil to intense winter winds.

Often water, as well as wind erosion, threaten land in a particular area. Methods of control depend greatly on the peculiarities of the area, because no single set of controls works everywhere. The amount and intensity of precipitation, the nature of the soil, the velocity of the winds, and the steepness of slopes vary profoundly.

In conclusion, one theme underlies this discussion of soil erosion. Man's existence in the past and today is to a great extent a product of a life-sustaining relationship which he has maintained with the soil. To abuse the soil is to endanger that relationship and, thus, to seriously threaten man's very existence.

The soil conservation movement

When man first settled the Americas, his primary intent was to cut the trees, clean the land, and expose the soil for efficient farming. As soon as the first trees were cut, the soil began to erode significantly. Even before the Revolutionary War, the soil was known to turn our eastern rivers dark, and records show that farmland under cultivation even then was being abandoned because of a decrease in its fertility. In the early 1900s, President Theodore Roosevelt, alarmed at the destruction of the American natural environment, called a conference of governors to discuss natural resource conservation. Soon the first national forest was denoted, and the U.S. Forest Service was created. In the 1920s, Congress approved the establishment of conservation experiment stations which would quantify erosion loss and develop erosion control measures. In the wake of the tragedy of the dust bowl in 1935, Congress passed

Figure 13-19
Steps to prevent rapid runoff of water from this severely burned hill in Idaho included contour trenching. (Courtesy of U.S. Department of Interior, Bureau of Land Management.)

Figure 13-20
Strip plowing and terracing on an Iowa farm prevent severe erosion. (Courtesy of U.S. Department of Agriculture.)

the first Soil Conservation Act in the history of any country. This act established the Soil Conservation Service, an agency directed to cooperate with farmers and to develop a program for proper land use. By 1948, all 48 states followed the national example and enacted soil conservation laws, and the Soil Conservation Service adjusted its program to work with these districts. Now its responsibilities include rural recreation, resource development, and conservation in general.

Summary

In this chapter we have examined man's use of energy, minerals, and soil. Specifically, we have discussed the significant sources of energy, including fossil fuels, solar energy, water power, geothermal energy, and atomic energy. It was pointed out that the fossil fuels are becoming exhausted, with no hope of their renewal, and that atomic energy may hold the greatest hope for meeting man's future energy needs. The supply of many minerals is also becoming depleted, and by the year 2000 several will no longer be available for commercial use. We discussed the use and abuse of soil, emphasizing the dependence of life on it, the time and processes required for its formation, and the specific ways it can be preserved. Underlying the discussion in this chapter has been one general theme: that man and organisms exist today because they have evolved a life-sustaining relationship with their environment. To foolishly exploit that environment to the point that in the future we are deprived of some of the essential materials we need means that we are endangering that relationship and, hence, openly threatening our own existence.

Supplementary readings

Averitt, P. 1969. *Coal Resources of the U.S.,* U.S. Geological Survey Bulletin 1275.
Government Printing Office, Washington, D.C.

Claudill, H. M. 1971. *My Land Is Dying.*
E. P. Dutton & Co., Inc., New York. 144 pp.

Ehrenfeld, D. W. 1970. *Biological Conservation.*
Holt, Rinehart and Winston, Inc., New York. 226 pp.

Eyre, S. R. 1968. *Vegetation and Soils: A World Picture,* 2nd ed.
Aldine Publishing Company, Chicago. 328 pp.

Farb, P. 1959. *The Living Earth.*
Harper & Row, Publishers, Inc., New York. 178 pp.

Fenner, D., and J. Klarman. 1971. "Power from the Earth." *Environment,* vol. 13, no. 10, pp. 19–26, 31–34.

Flawn, P. T. 1966. *Mineral Resources: Geology, Engineering, Economics, Politics, Law.*
Rand McNally & Company, Chicago. 406 pp.

Flawn, P. T. 1970. *Environmental Geology: Conservation, Land-use, Planning, and Resource Management.*
Harper & Row, Publishers, Inc., New York. 313 pp.

Garvey, G. 1972. *Energy, Ecology, Economy.* W. W. Norton & Company, Inc., New York. 235 pp.

Hammond, A., W. Metz, and T. Maugh, II. 1973. *Energy and the Future.* American Association for the Advancement of Science, Washington, D.C.

Holdren, J. P., and P. Herrera. 1971. *Energy: A Crisis in Power.* Sierra Club, San Francisco. 252 pp.

Huberty, M. R., and W. L. Flock. 1959. *Natural Resources,* 2nd ed. McGraw-Hill Book Company, New York. 556 pp.

Jones, C. F. 1968. "Energy Resources." *Texas Quarterly,* vol. 51, no. 2, pp. 84–89.

Lincoln, G. A. 1973. "Energy Conservation." *Science,* vol. 180, pp. 155–162.

Lovering, T. S. 1968. "Future Metal Supplies: The Problem of Capability." *Texas Quarterly,* vol. 51, no. 2, pp. 127–147.

Mouzon, O. T. 1966. *Resources and Industries of the United States.* Appleton-Century-Crofts, New York. 486 pp.

National Academy of Sciences, Committee on Resources and Man. 1969. *Resources and Man.* W. H. Freeman and Company, San Francisco. 259 pp.

Owen, O. S. 1971. *Natural Resource Conservation: An Ecological Approach.* Macmillan Publishing Co., Inc., New York. 593 pp.

Park, C. F. 1968. *Affluence in Jeopardy: Minerals and the Political Economy.* Freeman, Cooper & Company, San Francisco. 368 pp.

Rocks, L., and R. P. Runyon. 1972. *The Energy Crisis.* Crown Publishers, Inc., New York. 189 pp.

Skinner, B. J. 1969. *Earth Resources.* Prentice-Hall, Inc., Englewood Cliffs, N.J. 149 pp.

Stamp, L. D. 1969. *Land for Tomorrow: Our Developing World,* rev. ed. Indiana University Press, Bloomington. 200 pp.

U.S. Bureau of Mines. 1965. *Mineral Facts and Problems.* Bureau of Mines, Bulletin 630. Government Printing Office, Washington, D.C. 1,118 pp.

U.S. Department of Agriculture. 1957. *Soils: Yearbook of Agriculture.* Government Printing Office, Washington, D.C.

Weinberg, A. M. 1968. "Raw Materials Unlimited." *Texas Quarterly,* vol. 51, no. 2, pp. 90–102.

Weinberg, A. M., and G. Young. 1966. "The Nuclear Energy Revolution." *Proceedings of the National Academy of Sciences of the United States,* vol. 57, pp. 1–15.

Chapter 14

For that which befalleth the sons of men befalleth beasts; even one thing befalleth them: as the one dieth, so dieth the other; yea, they have all one breath; so that man hath no preeminence above the beasts; for all is vanity.

Ecclesiastes 3:19

Man and living resources

Chapter 14: Man and living resources

Perhaps even more obvious than man's exploitation of earth's matter and energy is his exploitation of the life that has evolved out of the matter and energy. This chapter tells of man's thoughtless destruction of plants and animals. Such destruction has serious implications for man's future because what he has done to other forms of life suggests that he lacks reverence for life and environment in general, and for himself specifically. Ultimately, this attitude could destroy him.

Forests and their use

Now they were down under that ocean of leaves. A red-tailed hawk screeching high over the treetops would hardly reckon there was a road down here. You had to be a porcupine rooting under the branches to find it or the black cat of the forest that could see in the dark and that some called the fisher fox.

This place ... must be the grand-daddy of all the forests. Here the trees had been old men with beards when the woods in Pennsylvania were still whips. All day ... eyes watched daily ahead for some sign that they might be coming out under a bit of sky.

Down in Pennsylvania you could tell by the light. When a faint white drifted through the dark forest wall ahead, you knew you were getting to the top of a hill or an open place. You might come out in a meadow or clearing, perhaps even in an open field with the corn making tassels and smelling sweet in the sun. But away back here across the Ohio, it had no fields. You tramped day long and when you looked ahead, the woods were dark as an hour or a day ago.

This description in The Trees by Conrad Richter (A. A. Knopf, New York, 1940) of travel in the pioneer era of our country before agriculture was a major industry accurately represents the vast forest cover over much of the land at that time. Today, this vast resource is greatly depleted because of the need for soil to farm and trees for lumber. The history of the forests in this country, their uses, their management, and their future is the concern in this section.

When colonists reached America, the new land was most likely one of the most forested areas of the world, probably 800 to 1,000 million acres, containing up to 8,000 billion board feet of potential lumber. Today, the total available forested areas have shrunk to about 760 million acres, or one-third of the 2.3 billion acres in the 50 states (Fig. 14–1). Of this, according to the U.S. Forest Service, only about 509 million acres are commercially available. S. W. Allen and J. W. Leonard, in Conserving Natural Resources, point out that this reduction in acreage occurred in only 350 years of settlement. Many early settlers viewed the forest resources as a nuisance and strove to eliminate them from acres when settlement seemed promising. As the population increased, the need for forest products further hastened their destruction. The great demand for lumber to build ships and dwellings rapidly destroyed the great white pine

Forests and their use

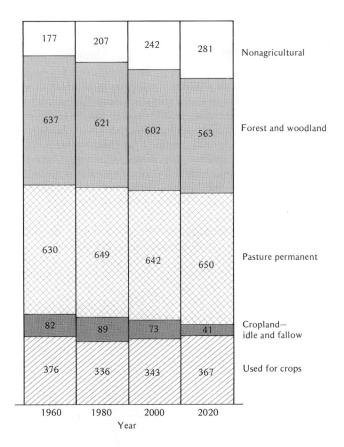

Figure 14–1
Projected use of land in the contiguous United States in 1980, 2000, and 2020 (in millions of acres). (From *The Nation's Water Resources,* Water Research Council, 1968.)

forests stretching inland from the east coast. Little thought was given to the future, so conservation practices were nil. Finally, by the beginning of the century when the white pine in Wisconsin and Minnesota was finished, the logging industry sought other sources of trees.

Between 1900 and 1910 the lumber companies advanced on the southern forests, where southern pines grew on abandoned cotton and tobacco fields. The rapid growth of the trees, which made them an excellent source of lumber then, still accounts for the thriving pine lumber industry in the South.

Lumbermen not only looked to the South for new lumber sources, but also to the West. The impressive and abundant Douglas fir held as much economic promise as the white pine once did back East. Fortunately, as the lumbering wave hit the western states, the federal government recognized the threat of extinction of the forest as a lumber and recreational resource, and laws preserving sections of forests for future generations were put into effect.

The distribution of forests

The distribution of kinds of trees account for much of the social and political history in our country. Much of the Great Lakes area, for example, was settled originally because of the accessibility of the region provided by logging camps and roads. The distribution of the kinds of trees around the country, indeed the world, is, in turn, due to environmental conditions over great periods of time. For example, the great white pine forests are reliably said to have originated as secondary growth in the wake of a major natural catastrophe, such as a hurricane or fire.

Other environmental factors, mainly temperature, moisture, and soil conditions, determined areas of forests distinctly different. Although the

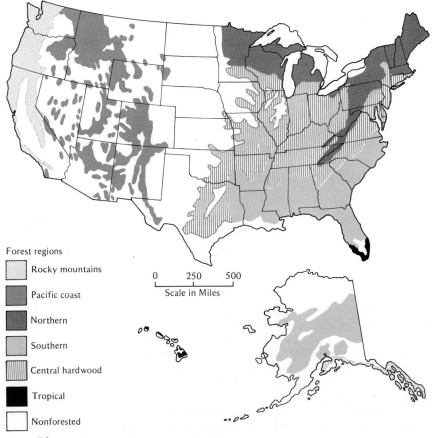

Figure 14-2
Forested regions in the United States.

classification varies somewhat, generally the following types of forested areas are recognized (Fig. 14–2).

The Northern Forest

This area is characterized by white, red, and jack pine; red, black, and yellow spruce; birch; beech; elm; ash; aspen; and several other hard and soft woods. The growing season is short, although 30 to 40 inches of annual precipitation occur, and the soils do not favor farming. Consequently, forestry in these areas is still a major enterprise, especially since the terrain allows easy logging. Sometimes the lower portion of the Northern Forest is known as the "birch–beech–maple–hemlock forest" and separates this area from the central hardwood forests to the south and the conifer growth in the more northern Northern Forest.

Central hardwood forests

This area lies below the Northern Forest and above the Southern Forest. It contains many oaks and hickories instead of conifers mainly because of the fertile soil, 30 to 60 inches of annual precipitation, and a generally warm climate. Today, the area is mainly used for agriculture — forests occupy only 20 percent of the land. Many pockets of trees in woodlots and farms supply quantities of significant lumber trees, such as hickory. Major stands occur in the Arkansas Ozarks and the Appalachians.

The Southern Forest

In the Southern Forest lies two-fifths of the nation's total commercial forest, an area that also produces two-fifths of the nation's annual lumber harvest. The abundant precipitation of 35 to 70 inches annually, high humidity, and high temperatures favor rapid tree growth. Hence, the southern pines and hardwoods, such as pecan, gum, oak, and hickory, grow rapidly and harvest of stands of trees in 20- to 25-year cycles is the rule. The terrain is not excessively sloped, and markets are near, so lumbering is a vital industry in this region. Today, 60 percent of the area is commercial forest.

Rocky Mountain forests

Sparse moisture, low temperatures, and high elevations favor the growth of conifers such as Douglas fir and Englemann spruce in this region. These conditions, as well as the rough terrain and distance of markets, account for a low 6 percent annual harvest of trees from the area.

Tropical forests

Florida and Hawaii are the only areas in the United States that bear tropical forests. Furthermore, the palm and mangrove trees in Florida, and other moisture- and warmth-loving trees, are of little commercial use, except possibly as erosion control. The forests have suffered extensively from abusive mismanagement. Native Hawaii forests are being altered by the importation of exotics.

Pacific forests

Pacific forests produce almost one-half of softwoods harvested. Most of the trees are in federal hands, a result of legal action taken in the early part of this century when the forests appeared threatened by lumber companies. The region has an annual precipitation of 40 to 150 inches and relatively fertile ground. The trees, including the Douglas fir, western hemlock, and western red cedar, grow exceptionally fast and large, often hundreds of feet tall and up to 8 to 12 feet in diameter. Subregions of pine occur in Oregon, Washington, and California.

The uses of wood
Fuel and lumber

One of the most significant uses of wood in the world is as fuel. In fact, one-half of the volume of wood cut is for fuel, and much of the leftover waste from wood used for other purposes is also burned. In the underdeveloped countries, wood is a major fuel. In Latin America, for example, 80 percent of all wood cut is used for heat. Even in Europe, 40 percent of all wood cut is used for fuel.

In the United States, however, the use of wood for fuel is declining. Between 1900 and the 1950s the amount of fuel wood used dropped about 70 percent; in 1952 fuel wood composed only 16 percent of the consumption of all timber products. The U.S. Forest Service predicts that it will continue to decrease in the near future.

A major use of wood cut from our forests is for lumber. In fact, of the 68 billion board feet of forest products in 1962, 37.3 billion feet was lumber alone. The rest was pulpwood, fuel wood, and other wood products. Allen and Leonard in their book state an equivalent value to better understand the amount of lumber produced that year — enough to cover 860,000 football fields with boards.

The great amount of lumber used is a new high, and the prediction of the Forest Service is that it will grow higher. Previously, the amount of lumber used was less, although a rapid increase in the amounts used was seen in the nineteenth century. Between 1900 and now, however, the amount of lumber used has vacillated, spiking in 1906 and 1907, again in 1923 and 1925, and finally just before World War II. Since World War II the demand for lumber has remained high. The irregular demand for lumber is tied to the fluctuations in the economy. For example, little lumber was produced during the Great Depression of the 1930s.

The economy of lumbering also accounts for the increasing prevalence of small lumber companies, or the division of large ones into small units. The cost of transporting lumber long distances and the increasing distances between forested areas account for this division into small operations. Beside watching expenses, another concern in lumbering is the protection of particular stands of lumber. When timber was plentiful, few companies worried about the exhaustion of the resource that fed thin operations. Today, unless a company practices reforestation measures such as those outlined below, it could easily put itself out of business. This problem also explains the small lumbering company today. If the stand runs out, little is lost when the lumbering operation is demantled and moved.

Pulp and paper industry

Over the recent decades the use of paper and cardboard in the United States has increased greatly. For example: in 1909, 4.1 million tons were used; in 1929, 13.4 million tons; in 1941, 20.4 million tons, and in 1954, 31.1 million tons. The demand is even greater today, and the United States must import trees to meet the demand. Canada supplies most of the imported pulp. To illustrate the demand, it is estimated that the Sunday edition of the *New York Times* for 1 year requires 125,000 tons of newsprint. R. F. Dasmann points out that this, in turn, requires the annual growth of 1,250 square miles of Canadian forest!

Better technology in producing paper from pulp has expanded the kinds of trees that can be used. New developments may further expand the industry technology in attempts to meet the rapidly increasing demand for paper and paper products.

Other uses of forests

Forests are also utilized in other major ways. Their value in erosion control and recreation is inestimable. Forest products such as plywood, charcoal, poles, railroad ties, and naval stores (rosin, turpentine) are increasingly in demand as the population grows.

Forest management

The great needs now and in the future for wood demand that the lumber companies and governmental agencies manage their trees in some way so that the forest resource is not exhausted. Federal action initiated preservation early. In 1891 President Benjamin Harrison set aside the Yellowstone Timberland Reserve surrounding Yellowstone Park. Later, President Theodore Roosevelt, attuned to the advice given by ardent conservationists, added areas until the national reserves totaled 148 million acres. These areas were transferred to the new U.S. Forest Service directed by Gifford Pinchot. Today, the intent of the service appears to be "multiple use," or the use of the forests for many purposes.

Preservation of a forest to be lumbered involves careful planning. If the trees are all mature, then "block," or "strip," cutting is done (Fig. 14-3). This procedure involves the removal of a large block, or swath, of trees from the forest. The surrounding trees soon reseed the cutover area. Occasionally, several lone mature trees are left standing in the vacancy to hasten the reforestation. This procedure works well with

Figure 14–3
Block cutting of Douglas fir in Mt. Hood National Forest. (Courtesy of U.S. Forest Service. Photo by F. Flack.)

trees whose seedlings thrive in full sun, such as the Douglas fir. Seedlings that require shade are best reseeded in areas that are "selectively cut." In this process the old trees are logged first, and the young trees are left to mature.

Careful forestry techniques are important during lumbering. Great care must be exercised that the immature trees are not damaged. Also, diseased and damaged immature trees are best removed before the logging operation. Old limbs and debris that possibly could feed a fire are also best removed. Beside the cutting and maintenance of the trees, management of forest fires, or diseases, and of pests is also important to prevent timber loss (Fig. 14–4). Forest fires can be of great help in developing some forested areas. For example, jack pine cones will not drop their seeds unless exposed to fire. Pure stands of longleaf pine in the Southeast occur mainly because periodic fires sweep the area and burn off the deciduous brush which would compete with the more resistant pine. The Ponderosa pine and the Douglas fir also depend on fire for development.

Yet forest fires usually prove destructive. For example, in the United States, in 1954, 8.8 million acres were destroyed by fire; in 1961, 3 million acres; and in 1962, 4.1 million acres (Fig. 14–5).

Most fires are thought to be man-caused, but lightning is another major cause. For example, in Idaho in July of drought-plagued 1966, lightning fires were a daily occurrence, and in one day 50 such fires were cited! Statistics show that lightning frequently causes fires in other heavily forested areas. Alaska is covered with a fairly young growth of trees, supposedly the product of frequent and recent major fires.

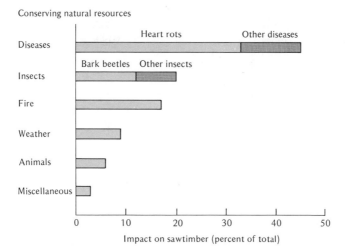

Figure 14-4
Impact of destructive agencies on saw timber as percentage of total impact on the United States including Alaska. (U.S. Forest Service data.)

Figure 14-5
Thousands of acres of forest are destroyed each year by fire. (Courtesy of U.S. Forest Service. Photo by J. Hughes.)

Pests, mainly insects, are another major threat to forests and in recent years have been more destructive than fire and diseases combined.

Insects damaged the equivalent of 5.4 billion board feet in 1962. Several insects are especially devastating. The pinebark beetle girdles the Ponderosa pine, the spruce budworm defoliates spruce, the pine weevil kills white pine, and the larch sawfly attacks larch.

Control of insects is not easy. Death of one or another developmental stage of the insect by removing tree bark from fallen trees can usually be assured, although the process is time-consuming. Burning of the bark is also effective. Spraying with insecticides is also extensively practical, but dangerous levels of residues have curbed the use of this technique in some areas. Maintaining the forest in a healthy state is probably the best prevention of insect infestation.

Diseases are also a problem. In 1962, nearly 4.0 billion board feet of saw timber were lost to tree diseases. For example, chestnut blight, a parasitic fungus imported from the Far East in the early 1900s, has destroyed the tree in this country. No effective control of the disease has been found.

Blister rust, another imported disease, has attacked the white pine in the East and the sugar pine and western white pine in the West. This fungal disease uses the gooseberry bush as an intermediate host. Consequently, control usually involves eradication of these plants. The Dutch elm disease, another fungal killer, has essentially eradicated the American elm from the central hardwood forests. No control has yet been found for this disease, although intensive research continues. Unfortunately, before controls are found, this lovely shade tree, which once characterized many towns in the United States and supplied beautifully grained wood for paneling, may be extinct.

Forests in the future

In February 1965 the U.S. Forest Service published *Timber Trends in the United States,* a report concerning the timber demands and supplies through 2000 A.D. Predictions were based on an annual population growth rate of 1.5 percent annually and a total U.S. population of 325 million people by 2000 A.D.

The needs will most likely be very great. For example, the total use of construction materials, including lumber, was projected to nearly double the consumption in 1961. The need for pulpwood for paper and paper board manufacture will also increase greatly, from 42.3 million tons in 1962 to 115.5 million tons in 2000 A.D.

The pertinent question is: Can the forest resources in the U.S. meet the demands in the future decades? Already approximately one-fourth of the original forested land has been eliminated for agricultural and other purposes. Of the remainder, 509 million acres are commercial forest that can be counted on to produce wood or wood materials now and in the future. The production of wood materials from this area must more than meet the demand if a wood shortage is to be prevented. At the moment the growth exceeds the harvest, so a crisis is not imminent.

Yet insects and diseases and other destructive forces decrease the amount of usable growing wood. Consequently, as the demand for timber increases, additional lands will probably be put into trees, more efficient methods to increase the productivity of the lands already cultivated will be investigated, and attempts to control insects and diseases probably will be intensified.

However, as the population grows and the year 2000 A.D. nears, the United States may not be able to meet the demands. Already a great amount of wood is imported from Canada and will continue to be. But as the 2.5 acres per person of commercial forest dwindles to the projected 1.6 acres per person, even imports may not meet the demand. By 2000 A.D., 21 billion cubic feet of wood will be needed, a 75 percent increase over the 1962 demand of 12 billion cubic feet.

Control of productivity is hampered by the ownership of the forested land. The majority of the forests, 59 percent, is owned by small private landholders (Fig. 14–6). Hence, much forest is not managed properly to yield maximally. The 28 percent of forest in government control and the 13 percent controlled by forest companies is managed fairly well, but in no way can proper management of less than one-half of the nation's forests meet future demands.

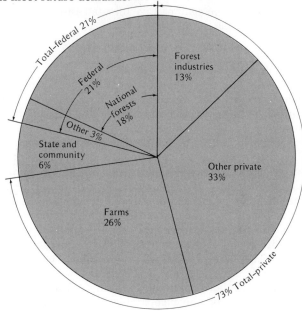

Figure 14–6
Owners of commercial forest land in the United States. (U.S. Forest Service data.)

Forests in other countries

In this section most attention was paid to forest resources in the United States. This is primarily because there are extensive data on U.S. forests. However, a few comments can be made concerning the forest reserves in other countries. Canada has a wealth of timber; approximately 619 million acres are classed as "productive" forest land. Furthermore, 407 million acres are listed as "accessible." The great amount of wood imported into the United States in 1962, approximately one-half of the total Canadian cut, will no doubt increase.

Some underdeveloped areas, such as Asia, South America, and Africa, are known to have significant resources (estimated at 5.2 billion acres) which may be tapped. Other nations may also develop the forestry industry as the year 2000 A.D. approaches. The U.S. Forest Service states in *Timber Trends in the United States* that, "although information on the world's forest resources is scanty, there is undoubtedly a huge potential flow of wood products from tropical forests," although controversy exists over the extent of renewability of extensively cut tropical forests.

Europe also has ample forest resources. Germany, for example, has pioneered in forest management practices and benefits from magnificent small stands. So do the British Isles and the Scandinavian countries. The total forested acreage in Scandinavia and western Europe, exclusive of the British Isles, is approximately 185 million acres. Testimony to the excellent management of this forested reserve, moreover, is a yield rate greater than that of the United States.

Wildlife and wilderness

With the first step of the pioneering European onto American soil came the germ of environmental disaster. The motives that drove the settlers to these shores and the values carried over from the European economy would, several centuries later, create a vast and complex dilemma to be known in the 1970s as the environmental crisis.

Although not considered seriously at the time, the first symptom of the difficulty was the disagreement with the Indians. Steward L. Udall, former Secretary of the Interior, in his book *The Quiet Crisis,* describes the attitudes of the natives toward their land, a point of view in conflict with that of the immigrants, which helps account for the Indian eradication. "The most common trait of all primitive peoples is a reverence for the life-giving earth, and the native American shared this elemental ethic: The land was alive to his loving touch, and he, its son, was brother to all creatures. His feelings were made visible in medicine bundles and dance rhythms for rain, and all his religious rites and land attitudes savored the inseparable world of nature and God, the Master of Life. During the long Indian tenure the land remained undefiled save for scars no deeper than the scratches of cornfield clearings or the farming canals of the Hohokams on the Arizona desert."

Such an ethic was met broadside by land-hungry settlers. Englishmen, especially, valued land ownership. Was not the history of their native country with united boundaries scarred with frequent battles over land ownership? Early settlers wasted little time in clearing woods to plant crops and built fences around them as the fertile farm area became more in demand. Shortly, the ethic of the Indians proved a hindrance to the expansion and development of the New World.

Today, the environmental problems encountered in this country to a major extent rest on possession and exploitation of the land. That this is the basic thinking of most Americans is especially understandable in view of the abundant and rich resources the pioneers found.

Unfortunately, such exploitation is no longer synonymous with life-giving adjustment to the environment. In fact, as pointed out earlier, such an ethic appears now to threaten man's existence. Needed now is an "ecosystem approach" to a knowledge of what is necessary for man to live harmoniously with nature.

Ample evidence testifies to his inability to do this in the past, such as his reckless use of many of the resources described in this chapter. His abuse of wildlife and wilderness also indicates his careless attitude. For example, in 1951, 2,368 vertebrate wildlife species existed in the United States, including 670 species of mammals, 811 birds, 149 reptiles, and 138 amphibian and 600 freshwater fish, according to L. Wing in *Practice of Wildlife Conservation* (John Wiley & Sons, Inc., New York, 1951). (The additions of Alaska and Hawaii in 1959 added more species to the list.) However, when the country was first discovered, more species than this existed. In fact, in the world as a whole during the last 2,000 years, probably close to 106 species or subspecies of mammals alone have become extinct. Of these, 27 were indigenous to North America.

Man in most cases was the cause, because of excessive hunting, sometimes owing to the introduction of guns to those groups who previously hunted without them, destruction of habitats, or the introduction of foreign animals into new habitats, with the eventual elimination of the native forms. Whereas exploitation of mineral or energy resources has the obvious consequence of total depletion, destruction of wildlife is often more subtle.

Yet the loss of an animal species is a sad and serious event. This is because an individual or corporate attitude which allows the extinction of animals indicates a philosophic stance toward life in general — including human life — which, in the long run, is suicidal. The reckless extermination of the whale or the passenger pigeon is evidence of little understanding of the complexity of the interrelationships of organisms and environment as product of a long history of evolution. Destruction of species indicates little thought of the organism's unique position, or niche, in the environment and the effect of its life functions on other organisms.

David Ehrenfeld puts it another way in *Biological Conservation* (Holt, Rinehart and Winston, Inc., New York, 1970): "The widespread and often preventable destruction of natural communities and the subsequent general loss of species foreshadows and often is accompanied by severe environmental problems for the human inhabitants of the region, regardless of technology." Thus, the main concern is that lack of respect for the survival of living things shows a lack of understanding or concern with the survival of man.

Unfortunately, preservation of wildlife is often only defended on esthetic or spiritual grounds, two valid but not easily persuasive reasons. After all, beauty is in the eye of the beholder, and who is to measure the loss in man's spiritual dimension when the last dodo was clubbed to death?

Fortunately, other reasons for preservation, such as recreation, have been influential. Today, many organizations, both governmental and private, exist to preserve wildlife and the areas in which they live. Of course, the increase in the human population, the advancements in technology, and the development of urban areas constantly threaten the groups' effectiveness. Yet many significant steps have been taken. The first major attempts of preservation in the country occurred from 1850 on, although as early as 1718 Massachusetts enacted a closed hunting season on deer, and by the time that the American Revolution began, all but one of the 13 colonies had closed seasons on some game species. However, a major step toward preservation of national resources occurred in 1864 when the designer of New York's Central Park, Frank Olmsted, persuaded Congress to approve a bill to preserve Yosemite Valley (which became a national park 26 years later). In 1872 President Ulysses Grant designated Yellowstone National Park; in 1909 President Theodore Roosevelt set aside Yosemite National Park.

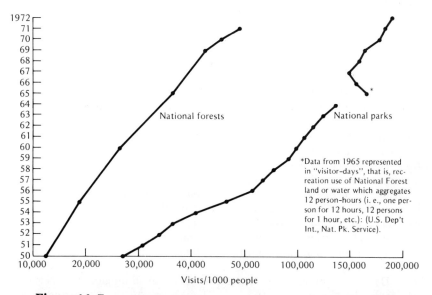

Figure 14–7

Use of national forests and national parks since 1950. (Data from U.S. Department of Commerce and U.S. Department of Interior, National Park Service.)

By 1916 the National Park Service was created to manage the new National Park System. The purpose of the parks was coming under new definition, too, for although originally the scenic areas were considered for recreation, by 1916 concern was expressed for the preservation of "the natural beauty" they contained. The purpose of the 32 national parks continues to be examined. To illustrate, a decision was reached in the 1960s to set some areas aside exclusively for wildlife and others primarily for people. Such designations necessarily must be made, the Park Service decided, in view of incidents such as the attacks of grizzly bears on tourists, no doubt provoked in part by the increasingly crowded conditions in the parks.

The use of the parks in the future will probably be even more regulated. The service more and more realizes how altered the parks' natural attractions become as their popularity and the national population increase (Fig. 14–7). Noel D. Eichhorn of the Conservation Foundation elaborates on the problem: "Large numbers of visitors can have a considerable impact on an area. They are noisy and many animals are frightened away. They feed the more appealing of those remaining and make beggars and pests of them. In and near camp grounds underbrush disappears. The soil becomes compacted and there is no regeneration — of grain or forests. Sewerage must be disposed of and streams may become polluted. Candy wrappers, Kleenex, empty beer cans, and yellow film boxes decorate the ground and the lower branches of the trees." Recently, automobile traffic was banned in Yosemite Park.

Figure 14–8
The Great Cypress Swamp typifies the unique habitats in the Everglades National Park. (Courtesy of U.S. Department of Interior, National Park Service.)

In addition to the establishment of the national parks, another step in the preservation of undisturbed areas occurred when Congress passed the Wilderness Act in 1964. This act preserved millions of wilderness acres throughout the country from any exploitation or development. In these areas the natural plants and animals will exist undisturbed for future generations. Several private agencies are constantly working toward preservation of wildlife and wilderness. The Nature Conservancy, the National Wildlife Society, and the Sierra Club are all examples of organizations committed to preservation of what appear to be natural conditions.

Special wildlife refuges have also been created. The first was Pelican Island in Florida in 1903. Now over 300 refuge areas exist, many set up with the express intent of preserving the breeding or nesting areas of threatened birds. Refuges, plus agreements such as the Convention for the Protection of Migratory Birds in the United States and Canada of 1916, help to preserve many species and areas from destruction.

Preservation of wildlife and wilderness areas on an international scale has also been attempted. For example, in 1940 the Convention on Nature Protection and Wildlife Preservation in the American Republics defined important conservation measures to be practiced between the United States and other consenting American republics. In 1948 the International Union for the Protection of Nature (IUPN) was initiated with representatives of 33 countries. Renamed in 1957 the International Union for Conservation of Nature and Natural Resources (IUCN), its purposes include inventory of threatened animal species around the world and agitation for action to preserve them.

Causes of species extinction

Despite these and other attempts to preserve species and their habitats, many animals and birds are still threatened with extinction. As the number of people increase on earth and more area and materials are used for the maintainance of today's technological developments, an increasing number of plant and animal species may soon be lost. According to Robert Silverberg in *The Auk, the Dodo, and the Oryx,* between 1801 and 1850 only two kinds of mammals are known to have become extinct. However, between 1851 and 1900, 31 species of mammals and numerous birds were exterminated, and between 1901 and 1944, many more added to the list. Today, about 600 mammal species are seriously threatened with extinction in the near future.

Habitat change

Several factors account for the increasing loss of animal species. David Ehrenfeld suggests in *Biological Conservation* that ecosystem alteration, hunting, the demands of institutions of various sorts, predator control, and superstition are the main causes of extinction. Many examples could be cited which would illustrate these. Ehrenfeld mentions an example of ecosystem alteration: development of industry and agriculture, at least in part to feed the growing human population, has strained greatly the natural water reserves in this area in the southwestern United States, and many fish species have been lost there.

Another excellent example is the destruction of the Everglades, southern Florida's unique ecosystem (Fig. 14–8). Ehrenfeld describes this unique area in *Biological Conservation* (Holt, Rinehart and Winston, Inc., New York, 1970):

> That part of Florida south of Lake Okeechobee is the only part of the continental United States that has a nearly subtropical climate, and its dominant biological community, the Everglades, is so complex that its most important biotic relationships are only beginning to be understood. The Everglades were formed and molded by the moderate climate and, equally important, by the high summer rainfall and run-off from Lake Okeechobee. When Lake Okeechobee overflowed, the waters crept south in a broad, shallow sheet, supporting what has aptly been called a river of tall sawgrass, and forming teardrop-shaped islands that became covered with a variety of trees, including mahogany and a number of other tropical species. Southwest of the lake there was an extensive region of cypress swamp whose quiet waters were dyed black by organic acids leached from the trees. In the south, sawgrass gradually gave way to mangrove, and fresh water to brackish, in an area that was a remarkably rich spawning ground for fish and invertebrates.

Today, the area is seriously threatened, mainly because the complicated and fundamental water system in the Everglades' ecosystem had been tampered with. The east-west Tamiami Canal built in 1928 was intended to drain the area for farming. Three other canals later were built as water conservation measures to intercept the slow and constant water flow from Lake Okeechobee to Florida's tip and the Everglades National Park. As a result of the canals, the park frequently does not have enough water. Consequently, fish, bird, and mammals once common to this unique area are gone, or threatened with extinction.

Such losses are especially tragic because of the unique and abundant wildlife there. John Harte and Robert Socolow describe the flora and fauna in *Patient Earth* (Holt, Rinehart and Winston, Inc., New York, 1971):

> The park, the third largest in the country after Yellowstone and Mount McKinley, contains an abundance and variety of wildlife to be seen nowhere else in the United States. Perhaps most impressive are the anhingas, sometimes referred to as water turkeys or snake birds, and the large wading birds, including the roseate spoonbill, the great white heron, the wood ibis (actually a stork, the country's only stork), the white ibis, and the limpkin. So productive are the soils and the waterways in the park that these and

over 300 other species of birds are supported here, in some cases in great density. Although the mammals, fish, and reptiles are somewhat more elusive than the large wading birds, they are no less exotic; such species as the alligator, the porpoise, the Virginia whitetailed deer, the manatee, or seacow, and even the rare panther, or mountain lion, find their niche in the Everglades ecosystem.

The plant communities too — such as the junglelike hardwood forests, the cypress swamps, the sawgrass marshes, and the mangrove swamps — are unlike those found anywhere else in the United States. In short, the park is teeming with the plant and animal life of a tropical ecological community.

Sadly, signs of distressed species are obvious. The wood ibis, the only stork in North America, only successfully nested one season between 1961 and 1967.

Hunting

Hunting with guns for food, for fur or feathers, or for sport has helped eliminate a great number of species. Without guns it is unlikely that species such as the passenger pigeon, Carolina parakeet, heath hen, or California grizzly would have been destroyed so suddenly.

Certainly, the well-documented destruction of the American bison illustrates how effectively a species can be decimated with guns (Fig. 14-9). Introduction of Sharp's .45 caliber center-fire rifle in 1880 — along with the railroad — helped to reduce the original continental herd of 80 million or more bison in 1700 to a mere 1,090 in 1893. Most of the

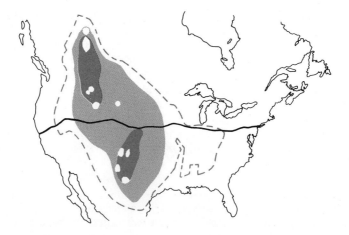

Figure 14–9
Range of the North American buffalo before 1800 (dotted line), approximately 1850 (slashed line), approximately 1875 (dark areas), and today (white areas).

decimation occurred in only 70 years — between 1820 and 1889 — and notably, in the last decades. Although many bison were killed for food or as an Indian control measure, many were killed for sport (Fig. 14–10). Today, as then, sports hunting is a major enterprise for many. When regulated, shooting for fun helps maintain the size of population of some animals. Deer, for example, in many states would outgraze their ranges and seriously upset ecosystems if an annual hunting season was not in effect. Unfortunately, hunting for sport in some states or countries has been uncontrolled or unwisely controlled. Consequently, extinction of some wildlife is imminent. The polar bear and wolf in Alaska are examples of species increasingly threatened because of a lack of protective legislation.

Some animals in recent history have been decimated for reasons other than fun or food. The craze to decorate women's fashions with ostrich or egret plumes helped reduce the population of these birds. The incredible demand for ostrich feathers — 160 tons of ostrich feathers sold in France alone in 1912 — even led to ostrich farms. Fortunately, before the bird could be exterminated, the fad waned.

The fate of the alligator is questionable, too, because poachers seek its skin for leather goods. The whale, whose products are used in margarine, cosmetics, detergents, candles, and lubricating oils, and some

Figure 14–10
Buffalo bones left on Great Plains after slaughter of these animals at the end of nineteenth century. (Brown Brothers.)

cats, such as the snow leopard or ocelot, whose furs are used in fashions, are also threatened with extinction. Hopefully, steps will be taken so that all these animals will be preserved.

Institutions

Another reason for the decimation of some species is their collection for institutions such as museums, zoos, and medical institutions. The number of animals imported annually, presumably for these institutions as well as for pets, is very large. The U.S. Fish and Wildlife Service reported that in 1967 about 74,300 mammals, 203,000 birds exclusive of parrot types and canaries, 405 reptiles, 137,000 amphibians, and 27 million fish were imported. More than 65 wild primates alone were imported in 1967, presumably to be used mainly for medical research. Such an exodus of animals from their natural habitats may well seriously reduce many to the point of species extinction.

Superstition

Many animals around the world are killed because of superstitious beliefs. V. Ziswiler cites several examples: the rhinoceros, whose ground horn is used as a aphrodisiac or made into goblets because it supposedly can detect poisoned drinks; and the tiger, whose ground bones hold medicinal value.

The biology of extinction

The *Red Data Book,* a compilation of species threatened with extinction sponsored by the International Union for Conservation of Nature and Natural Resources, lists 277 species and races of mammals, approximately 65 gravely threatened, and more than 300 birds, including 60 in extreme danger.

The criteria on which these animals and birds are included in the list vary greatly. Nevertheless, Ehrenfeld in *Biological Conservation* lists several criteria, any one of which can decimate a population. He cites predatory habit, the large size of individuals, a narrow habitat tolerance, value for fur or other products, value for market or sport, restricted distribution, migration across international boundaries, intolerance of man's presence, reproduction in a few large aggregations, a long gestation, few young per litter, maternal care required, and behaviorly nonadaptiveness to technological advances. All mammals and birds in the *Red Data Book* appear to meet one or more of these criteria.

Some of the animals listed are already feared to be very close to extinction. This is because their population is now so small that they may not be able to rally. Some populations in the past have not been able to enlarge once reduced below a certain size. A famous and sad example is the passenger pigeon. In 1800 this bird was considered the most plentiful of all bird species! By 1880 only several billion birds survived, and by 1914 the last passenger pigeon died in the Cincinnati Zoo.

Reasons for its rapid extinction are not clear — certainly man and his guns greatly reduced the population in the early 1800s. According to Ziswiler in *Extinct and Vanishing Animals*, other reasons later in the century included the destruction of the American forests, which were the birds' nesting areas, absence of appropriate breeding stimuli in the smaller flocks (paired birds in zoos never reproduced), difficulty in locating a mate, an intensification of predation as the rates of predator to prey increased in the bird's final few years, and an inability for small flocks accustomed to large aggregations to establish feeding and resting areas. Ehrenfeld suggests that another possibility may be an inability for the small populations to compensate for normal reductions through diseases, weather changes, and other disasters.

Some threatened species do not appear affected by small population size. The miraculous recovery of the sea otter is an example. Once a very abundant species, slaughter to obtain the thick pelts reduced its number alarmingly fast (Fig. 14–11). In 1856 the Russian-American Company sold 118,000 skins, but in 1885 the company sold only 8,000, and in 1910 only 400. The species was soon believed extinct. However, a few pairs survived and increased the species to considerable size once again in a short time. Laws now protect the sea otter from the threat of extinction again. Other seal species have undergone similar histories.

Unfortunately, all species cannot be depended upon to rally as did the sea otter. Consequently, some of the species now on the endangered list will become extinct. To the person who realizes that any living

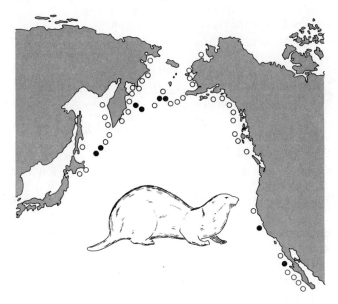

Figure 14–11
Former and present distribution of the sea otter on northern Pacific coasts. (From V. Ziswiler, *Extinct and Vanishing Animals*, Springer-Verlag New York, New York, 1967.)

creature is a product of the intricacies of evolutionary process working through millenia, the extinction of any organism provokes sentiments such as that of President Theodore Roosevelt: "When I hear of the destruction of a species I feel as if all the works of some great writer had perished."

Summary

In this chapter we have examined man's exploitation of the living resources of the earth. We have looked at forests, their distribution, kinds, uses, management, and future. We have examined attitudes toward wildlife and wilderness areas, especially the history of attempts to preserve natural areas and their inhabitants, and the factors causing extinction of animals. Throughout the chapter is that main theme: The rules by which man relates to the environment are similar to those which rule other species. Man's careless use of nature suggests a lack of concern for himself. An erosion of the quality of human life could be one result of human carelessness. Extinction of human life could be another.

Supplementary readings

Allen, S. W., and J. W. Leonard. 1966. *Conserving Natural Resources: Principles and Practices in a Democracy,* 3rd ed. McGraw-Hill Book Company, New York. 432 pp.

Black, J. D. 1968. *The Management and Conservation of Biological Resources.* F. A. Davis Company, Philadelphia. 339 pp.

Crocker, T. D., and A. J. Rogers, III. 1971. *Environmental Economics.* Dryden Press, Hinsdale, Ill. 150 pp.

Darling, F., and N. D. Eichhorn. 1967. *Man and Nature in the National Parks: Reflections on Policy,* 2nd ed. Conservation Foundation, Washington. 86 pp.

Darling, F., and J. P. Milton, eds. 1966. *Future Environments of North America.* Doubleday & Company, Inc., (Natural History Press), Garden City, N.Y. 767 pp.

Dasmann, R. F. 1972. *Environmental Conservation,* 3rd ed. John Wiley & Sons, Inc., New York. 473 pp.

Ehrenfeld, D. W. 1970. *Biological Conservation.* Holt, Rinehart and Winston, Inc., New York. 226 pp.

Farb, P. 1963. *The Forest.* Life Nature Library, Time Inc., New York. 192 pp.

Hays, S. P. 1959. *Conservation and the Gospel of Efficiency: The Progressive Conservation Movement, 1890–1920.* Harvard University Press, Cambridge, Mass. 297 pp.

Henkin, H., M. Merta, and J. Staples. 1971. *The Environment, the Establishment, and the Law.* Houghton Mifflin Company, Boston. 223 pp.

Highsmith, R. M., Jr., J. G. Jensen, and R. D. Rudd. 1969. *Conservation in the United States,* 2nd ed. Rand McNally & Company, Chicago. 407 pp.

Huberty, M. R., and W. L. Flock, ed. 1959. *Natural Resources,* 2nd ed. McGraw-Hill Book Company, New York. 556 pp.

McCormick, J. 1959. *The Living Forest.* Harper & Row, Publishers, Inc., New York. 127 pp.

Nash, R. 1967. *Wilderness and the American Mind,* rev. ed. Yale University Press, New Haven, Conn. 300 pp.

Philip, H. R. H., Duke of Edinburgh, and J. Fisher. 1970. *Wildlife Crisis.* Cowles Publications, New York. 256 pp.

Schorger, A. W. 1955. *The Passenger Pigeon: Its Natural History and Extinction.* University of Wisconsin Press, Madison, Wis. 424 pp.

Schwartz, W., ed. 1969. *Voices for the Wilderness.* Ballantine Books, Inc., New York. 366 pp.

Smith, F. E. 1966. *The Politics of Conservation.* Pantheon Books, Inc., New York. 338 pp.

Smith, R. L. 1974. *Ecology and Field Biology,* 2nd ed. Harper & Row, Publishers, Inc., New York. 850 pp.

Udall, S. L. 1963. *The Quiet Crisis.* Holt, Rinehart and Winston, Inc., New York. 209 pp.

U.S. Department of Agriculture. 1949. *Trees: Yearbook of Agriculture.* Government Printing Office, Washington, D.C.

U.S. Department of Agriculture, Forest Service. 1965. *Timber Trends in the United States,* Forest Research Report 17. Government Printing Office, Washington, D.C. 235 pp.

U.S. Department of Interior. 1968. *The A-B-Seas of Desalting.* Government Printing Office, Washington, D.C.

Ziswiler, V. 1967. *Extinct and Vanishing Animals,* rev. Engl. ed. Springer-Verlag, New York. 133 pp.

Chapter 15

I only went out for a walk and finally concluded to stay out until sundown, for going out, I found, was really going in.

John Muir

Approaching solutions to environmental problems

Sooner or later the most difficult question related to environmental problems must be asked: How are they to be solved? Any solutions are difficult to come by because of the complex and tangled causes of the problems. However, whatever the specific environmental difficulty, insight into its solution may be gained by examining what may be historical and cultural reasons for our contemporary environmental crisis.

Historical causes of the crisis in the west

A historian, Lynn White, Jr., believes that the Judaeo-Christian traditions are mainly responsible for our environmental problems today. He points out that the Biblical directive is for man to "go and dominate over nature," that is, to set himself above the rest of the natural world. This view eventually led to a definition of man as an animal with a soul, a unique organism separate from mundane nature. Very early this supernatural view of man clashed with animistic views of the world, which interpreted nature as the habitat and even manifestation of spirits and, thus, holy and deserving reverence.

Such a view was institutionalized by science. Science originated as the means to know God better through a study of His wonders. Consequently, many religious characteristics became characteristics of science — faith, fervor, dedication, even worship. White claims that the Judaeo-Christian philosophy has seriously, often negatively, influenced western culture, as manifested, for example, in environmental problems. White claims the lack of feeling for the "holiness" in nature accounts for its easy despoilation in the wake of the new religion: scientific and technological activity.

Many people do not agree that the influence of the western religion has caused our problems. Lewis Moncrief acknowledges the influence of the Judaeo-Christian viewpoint in creating our environmental difficulties, but claims that other forces also helped shape it. He points out that environmental problems occur in parts of the world that *do not* have the western religious heritage. Instead of religion, Moncrief prefers to look elsewhere for causes of the environmental troubles: capitalism, with its science and technology, and the democratization of much of western society in the late seventeenth and eighteenth centuries, both of which, in turn, led to an increase in wealth, an increase in population, individual ownership of resources, urbanization, and eventual environmental degradation. In this view, religion or even science and technology are only indirectly responsible for our problems. Specific events responsible are the revolutions in the late 1700s or later, which freed the individual to demand more services and goods, including the ownership of land, a necessity for a "good life" transferred to America with the immigrants. The "inexhaustible" frontier which met the settlers further encouraged an attitude that has led to our national environmental problems.

Today, in the United States, perhaps as a result of the national tradition for faith in a mass of people who are democratically determining policy, a personal moral direction regarding the treatment of natural resources is lacking. There is little or no recognition that the problems are most directly related to the daily activities of the individual in society. Consequently, little is accomplished toward preserving the environment. Social institutions appear unable to activate, even though democratically given the power to "handle the situation." (The failure of the Environmental Protection Agency to carry out stringent controls on air pollution may be an example.) Countries under a different political regime, such as the USSR, also have difficulty managing environmental problems. The historical cause of the problem (consider the effect of the Russian Revolution) may be similar, however. In addition, the United States and other developed countries place great faith in the ability of technology to solve any problem, including environmental ones, sooner or later.

Thus, the forces that led to the belief in the individual's acting independently and democratically to solve his problems in countries with abundant resources and a developing technology may have indirectly thrust us into a monumental crisis with the environment.

The crisis in the developed countries

Beside the opinions of White and Moncrief, others, such as Erich Fromm in *The Sane Society,* suggest factors that may have helped bring about the environmental crisis. But no matter what the specific causes, today it is upon us in many unpleasant and threatening manifestations, and solutions are hurriedly being sought. The important question now is: What action can be taken? Whatever the choice, it will be greatly influenced by economic, legal, and political considerations.

Economic considerations

Traditionally in the United States, the rise in the gross national product (GNP) for a year has been the measure of economic growth. However, some developments are not revealed in the GNP, such as the irrevocable exploitation of the natural resources and the loss of natural beauty. As the GNP rises and greater material comforts are available to the consumer, a quality of life may be insidiously eroded. P. R. Ehrlich and A. H. Ehrlich say in *Population, Resources, Environment: Issues in Human Ecology* (W. H. Freeman and Company, San Francisco, 2nd ed., 1970):

> More important than what the GNP is, however, is what it is not. It is not a measure of the degree of freedom of the people of a nation. It is not a measure of the health of a population. It is not a measure of the state of depletion of natural resources. It is not a measure of the stability of the environmental systems upon which life depends. It is not a measure of security from the threat of war. It is not, in sum, a comprehensive measure of the quality of life.

Despite any unpleasant consequences of a rising GNP, a high personal income is usually a very specific benefit. Easily available capital usually means satisfaction of immediate needs. Such needs are predominantly material ones, which, however desirable, do not necessarily provide for long-range happiness. Still, the demand for increasingly higher wages is ever with us. Hence, any solutions to the environmental problems will probably lie in maintaining our income as high as possible and reallocating our natural resources over a long period of time so that they are not exhausted. In other words, what is needed is a "balance" between the meeting of immediate human needs and the general maximizing of the long-range satisfactions of man.

The development of a balance between short- and long-range goals appears to be an economic problem. Most appropriate, then, is a close look at our economy, with the hope of discerning certain factors which are presenting an easy solution to the crisis.

A major realization is that it is, at least at present, more profitable to "abuse" the environment than to preserve it. For example, consider the pollution produced by solid wastes, specifically nonreturnable bottles or cans. The economy oriented to maximizing profit encourages the lowest-cost methods for waste disposal — an admission of many beverage companies that package their products in these bottles. To buy back the bottles is not worth the effort. As a consequence, nonreturnable bottles and cans swell the solid waste to be disposed of in dumps, let alone that scattered around the countryside.

Beside the emphasis on profit, another important reason for environmental abuse in this country is the significant role of the consumer in the capitalistic economy. Society has evolved an attitude of acceptance of the needs of the consumer, no matter of what kind. Thus, what the consumer wants (or what Madison Avenue convinces him he wants), he gets, despite any accumulating adverse effects on the society as a whole. For example, it is felt to be an individual right to own and drive a car if certain standards are met, despite whatever effect the exhaust might have on others.

A third cause for environmental problems comes from catering to the individual's needs without holding him or the producer accountable for all by-products. Economists call this "dissociation of costs and benefits." The steel mill, for example, can emit its noxious gasses into the atmosphere directly because (at least until recently) steel products are in great demand, the residents in the area need work, and the atmosphere is large enough to assimilate the wastes. However, when many steel mills build in the area, along with other industries, air pollution becomes a problem, affecting even the transient family driving through the region. The costs have become greatly unrelated to the benefits, although clearly, the polluting industries are acting in the most economically sensible manner.

Disassociation of costs and benefits is a major ingredient in environmental problems, and one most difficult to manage, mainly because it is intrinsic in the capitalistic economy. It appears that any attempt to solve the environmental problems demands that the producer or consumer be expected to assume the responsibility for the adverse effects of filling a consumer need. Such adverse effects are often called "externalities" by economists.

What are the ways by which the benefits and costs can be brought together, and the externalities of producing and consuming eliminated? In any proposed solution, one fact stands out clearly — whatever the action, some party must *pay*. This point is too often not realized by environmental "do-gooders."

The question more specifically is: *Who* shall pay? Some say the consumer, but others say the producer, because he is the immediate source of pollution. Unfortunately, this places the producer in a difficult economic position. He has two choices: raise his prices, or lower his profits. The first alternative may drive his customers to a competitor who does not subscribe to paying for his share of polluting. The second proposal can have various effects. The shareholders may become unhappy with the decrease in profit they receive, the producer may not be able to meet costs of making the product, or the producer may curtail services to the customers, which, in turn, may drive them away.

To ask who pays is to point out an unfortunate aspect of our economy — the definition of a profit for the producer does not include externalities. An example is the recent agitation to install exhaust filter devices on cars. Car manufacturers clearly stated that such additions could have been standard equipment in a new car long ago, but to keep the price down, among other reasons, the filters were omitted. Now, to sell cars with the devices as the law requires will increase new car prices. Many prospective buyers received such news unhappily, feeling that car manufacturers should assume the costs themselves. It would have been very desirable to have long had a definition of "profit" that included externalities.

The need for filter devices on cars illustrates a second means for absorbing externalities — let the consumer bear the burden of cleaning up the environment. Removal of lead from gasoline is an example. Most new cars run best with the addition of lead. Lead-free gasoline really means the substitution of another ingredient for the lead, which then leads to a boost in price, which the consumer bears.

Whether it is the producer or consumer who pays, the externalities are reduced as the responsibilities are more internalized. To further encourage internalizing the costs, government subsidies or tax relief have also been suggested. For example, suggestions have been made for internalizing the costs of a growing population. These include eliminating the personal income tax exemption for dependent children, subsidizing sterilization, and initiating new taxation rates that favor couples without children, couples who adopt children or have few children, and unmarried taxpayers.

Another serious proposal is to tax polluters according to the amounts and types of pollutants they add to the environment. A major obstacle to the taxation of pollutors is the difficulty in assessing the quantities and qualities of pollutants. It has been successfully done in some areas. For example, in the Ruhr Valley in Germany, a very dense industrial area, industries and municipalities are taxed according to the amounts and kinds of pollutants ejected into the Ruhr River. Such taxation encourages a tightening of procedures producing pollution since it obviously becomes a profit-making issue. Taxes conceivably would be added to other products believed to be potentially dangerous to the environment, such as insecticides or herbicides. Of course, such taxes often solve little because, in the final analysis, consumers still pay.

Beside the increase in consumer prices or loss of profits to the producers, another attempt to internalize the costs of environmental destruction has been the establishment of environmental quality standards, which evoke fines if violated. To date, this is the course followed by local and federal governments, mainly because it is the easiest to administer. However, some difficulties occur because of the cumbersome and inefficient nature of bureaucratic institutions and because of the difficulty in assessing the values striven for. In terms of dollars and cents, what is a non-polluted stream with good fishing worth? Or the songbird populations? Or a forest of redwoods? The problem of proper assessment also confronts a major assumption on which most environmental clean-up or preservation is based — that the benefits, whatever their nature, always exceed the costs. Such an assumption may itself be inaccurate.

Legal considerations

Any economic approach to environmental problems makes clear one main point: to clean up the environment or to prevent its destruction requires money. Someone must pay sooner or later. A legalistic approach to solving environmental problems touches on the same point but is more concerned with the manner in which individuals or corporations have been held accountable for environmental damages.

Most legal action in environmental problems centers on the internalizing of externalities. Generally, this has been accomplished in three ways: litigation, legislative limits on the creation of externalities, and negotiation. Most often litigation occurs between private parties, with little or no concern for the environment as a whole, over a personal and limited incident. For example, a farmer who sues a food processing plant upstream for emitting noxious wastes that have poisoned his cattle may be little concerned about any broader problem than his dead cows.

It is difficult today to expect the mass of private litigations to yield a consistent pattern for legal action which could be applied to environmental problems of the society as a whole. The private incidents are too limited and present precedents that are too incoherent and inconsistent. Furthermore, legal decisions that involve individual litigants generally only recompense that individual rather than others similarly affected. In other words, "class action" in law suits that are originated by private litigation is generally not allowed.

In the main, private litigations have rested on two laws — nuisance law and trespass law. *Nuisance law* prohibits any use of property that interferes with the use and enjoyment of another's property. The pollution emitted upstream which poisons cattle downstream could be substance for a law suit based on nuisance law; so could excessive noise or exhaust pollutants from a vehicle.

In general, nuisance law suits have some innate difficulties. One is the difficulty of proving who specifically is interfering with the use and enjoyment of another's property. In an environmental problem such as air pollution damage, proof is especially difficult to obtain. However, a suit could be filed against all persons or corporations who contribute to the problem, with the burden on them to demonstrate how much pollution or its quality each contributes.

Another problem is "prescriptive right," the right to continue polluting or otherwise damaging another's property if such acts began long ago and have never been challenged. Usually, it is held that the privilege to continue depends on the circumstances or reasons for the act.

Other restrictions surround the invocation of nuisance law. Sometimes, for example, only the owner of the property immediately next to the property of the pollutor can bring suit; also, normally, only individuals speaking for their private interests are allowed to invoke the nuisance law. Recently, however, the courts have recognized organizations speaking against environmental destruction.

Invocation of nuisance law has solved many environmental problems, yet its restrictions prevent it from becoming a panacea. Even if the strictures were lessened, it still is limited, because it is usually brought to bear *after* the environmental damage has been done. This will probably remain the single most serious limitation of all nuisance lawsuits.

A second legal form of litigation is the trespass law. To *trespass* is to invade uninvited an owner's or occupier's property. Hunting on another's land is trespassing, but so may be the damage caused to the vegetation on a person's land by waste gasses blown over it. Some courts have upheld such a definition, although others have not, mainly because an intervening force, such as wind or water, is usually indicted in such an event.

In addition to litigation, standards have been incorporated as a second means of legal control of environmental problems. State and federal controls on water, air, and land pollution already exist, and recently, penalties have been imposed on those individuals or groups who violate the standards. Again, problems revolve around the enforcement of the standards: What are proper standards? How much time should be allowed for a firm to solve its pollution problems? Should the government assume any of the costs of the firm needed to solve the problem? Despite these difficult questions, a great number of laws now stand to control environmental damage. Examples of national laws include the Air Quality Act of 1967, the Water Pollution Control Act of 1965, and various state laws which have been described in detail in previous chapters.

Future legal developments

Today, effective control of environmental problems is difficult because of the discrepancies in the laws and their enforcement in different areas. For example, one state may have stringent regulations to control air pollution, while a neighboring state does not. The economic implications are clear — industry will migrate into the area where more capital gain is realized. Such inequities may some day provoke more unison of restrictions and, hence, more uniform control of environmental problems. Thus, in the future new laws may be developed to handle the problems.

Within states, more effective management of environmental problems may come with the sectioning of the state into areas, accommodating geologic, hydrologic, and even industrial differences. Another development that may help resolve the problems may be an examination of private property ownership. The government may have to intercede — purchase interest in natural resources, for example, as already done to preserve the national parks. Finally, constitutional law may have to be changed to include the right of every individual to a healthful and livable environment.

Political considerations

In a democracy, no discussion of environmental problems is complete without a brief look at political aspects of meeting the problem. In most cases, despite what appear to be limits of an individual's ability to influence the economics or the laws, hope lies in the individual's opportunity to vote. Laws and economic values can, thus, be changed.

Nevertheless, to enact environment-saving laws with the vote is difficult. F. M. Potter, Jr., points out several problems. Government control involves a great deal of red tape and a lack of clear communication among the hierarchical governmental levels. As a result, directives become muddled through the chain of command, delaying action or making it ineffective. Furthermore, government agencies frequently compete because their influence overlaps, so they vie for budgets and prestige. Finally, government agencies are mutually protective, tending to "look out" for each other if one is in danger of being exposed or criticized.

Congress, too, has limitations in solving environmental problems, according to Potter. The Congressional committees are frequently composed of senators or representatives whose membership is best explained by reasons other than their expertise or even interest in the committee's purpose. Sometimes, too, their interest in local concerns dims their vision of the broader national interests. And, too, as with the branches and bureaus of government, action in the Congress is slow.

Yet, despite these shortcomings, the power of the individual vote may well be the source of hope in effecting relatively rapid change in the aspects of our society presently allowing environmental decay.

Ethical considerations

Very often those concerned about environmental problems beg for a change in society's point of view concerning values by which we live. Their conviction is that if enough people significantly change their world view, so that their habits, activities, and thinking were other than those producing the environmental difficulties today, our problems would begin to be solved. Such a claim is not without substance. Certainly, the significant religions in the world operate on this assumption, as do many political systems. The pertinent question is: What beliefs should be changed? Or: Which beliefs now held are aggravating the environmental problems?

It may be that the very characteristics innate in America's capitalistic society breed difficulty. Certainly, the disassociation of costs and benefits already described is a predictable result of a system that relies on monetary profit for viability. But other problems may result also. Erich Fromm, the humanist psychologist, points out many such problems which may lead to an unhealthy society because they damage people psychologically. Fromm cites some characteristics of our society that may be looked to as sources of problems: monetary profit, the rapid rate of technological change, mass production and mass consumption, and the development of a depersonalizing management hierarchy.

Whatever the reasons, some characteristics of our way of life are causing troubles. Scott Paradise, Executive Director of the Boston Industrial Mission, believes strongly that American priorities must certainly change. The national ideology may be no longer adequate — in view of the growing population — to handle the stresses now visible. According to him, values in our society that need adjustment include the following: Man is thought of as the source of all value. That is, in a kind of medieval view, the value of anything is relative, determined by its significance to mankind. A corollary is that the universe exists for man alone. As a consequence, man's governmental systems, his "state," exist mainly to facilitate the exploitation of environment for individual and corporate wealth and power. Thus, man's prime concern is to produce in order to consume, a process that increases endlessly. Any possibility of running out of resources is unreal, because the natural resources are believed inexhaustible; or if they become sparse, man will "remake" the environment to satisfy his needs — this attitude Paradise calls the "bulldozer mentality."

Paradise suggests that these values need to be changed. He pleads for the following: Man should be valued more highly than other creatures, but he is not to ignore, neglect, or abuse the others. In fact, he needs to become a guardian of the earth to prevent despoilation, even so far as to make the purpose of his "state," or government, to supervise a planning process that would prevent environmental impairment. Man is more than merely a producer to consume. He is "freed" to become vitally involved in nonconsumption activities of a more "spiritual" nature, such as the arts, or the interrelationships of people, activities that should take precedence over others. Furthermore, resource depletion is reduced because of the careful manner in which the earth is used, and the relationship of man to the natural environment becomes nondestructive to himself or the environment. Thus, man meets his short-range goals as well as preserves a legacy for future generations.

The change in an ethic, belief, or ideology which form the basis for action is clearly the goal for which Paradise pleads. How to achieve the goal is a vital question which he does not answer. Very likely, in our democratic country the only method to achieve such ends is through the ability of some to convince others of the long-range folly of our acts, so that a large group with political power can transform the country's values.

Helping the underdeveloped countries

In many ways, the problems of the United States are similar to those confronting other developed countries (D.C.s). Solutions to the problems in other D.C.s are probably similar. The environmental problems in underdeveloped countries (U.D.C.s) present an entirely different syndrome. Consequently, proposed solutions to their problems are drastically different.

The U.D.C.s' problems have been described in detail earlier. In brief review, among their important characteristics are a low average level of education, a low GNP, rural living for a majority of people, and poor medication. Most important is the interlinking of these factors to cause the U.D.C.s' unhappy condition. If their state is to improve, the chain must be broken, some link or links destroyed.

Another important point to keep in mind is the necessity for the D.C.s to aid the U.D.C.s to break free of the crippling bonds that hold them down. Because of the overwhelming social, political, and economic limitations, no U.D.C. can be hopeful of solving its problems alone. Yet in the past, despite the apparent best intentions of the D.C.s, inadequate aid has been forthcoming. For example, although the United Nations declared that 1960 to 1970 was to be the "Development Decade" for the U.D.C.s and that each D.C. was to contribute 1 percent of its GNP to the U.D.C.s, little was donated. Besides, too often money that has been given to the U.D.C.s has been mismanaged or diverted into projects sensible to the economy of a D.C. but not vital to a U.D.C., with its starving and homeless people. Furthermore, some projects, such as the Aswan Dam, sponsored by the USSR, are of questionable ecological wisdom.

How can the D.C.s best help? It appears unreasonable to suppose that the U.D.C.s can be aided economically to the point of transforming them into a D.C., nor does this seem wise. The drain on the world's resources by the industrialized D.C.s in order to maintain their standards of living is already exhausting natural resources around the world. Environmental limitations prevent many additional countries of such development. Consequently, the U.D.C.s' developmental route may lie with their agriculture. New crops — even enough to export — accompanied by increased facilities for storage, distribution, and processing may be the areas that could best benefit from aid from the D.C.s. Consequent developments in the U.D.C.s would probably include improved roads, better educated farmers, more machine and fertilizer factories, and more employment. Most likely, such developments would also be accompanied by increased attempts to curb population growth with whatever help the D.C.s could provide.

The improvement of living conditions in the U.D.C.s faces several major difficulties. Most U.D.C.s are burdened with huge populations now. Hence, they cannot ease into an economy that will develop parallel to a growing population. Furthermore, U.D.C.s today are faced with economic competition with D.C.s. Such a situation did not confront the D.C.s when they were developing. And, too, the governments of U.D.C.s too often are working against a great many inhabitants of the countries. Haiti and Spain are ruled by dictatorships who do little to raise their GNPs or living standards. Homeless and hungry people are common.

The D.C.s also present complications. Too long have the economies of most D.C.s been based on exploitation. Such an economical tradition must be altered if the U.D.C.s are to enjoy "a piece of the pie." Recall that the United States, for example, uses 32 percent of all the world's minerals and has only 6 percent of the world's population. Such consumption cannot possibly be transferred to other countries and, in order to aid the U.D.C.s economically in the long run, must at some time be curtailed here and in other D.C.s. Other attitudinal changes are also necessary. The United States is notorious for lack of regard for the cultural differences of the countries it aids, as pointed out well in Burdick and Lederer's *The Ugly American*. An intensified concern over what is best for the country aided, and not what is best as interpreted by the D.C. giving the aid, is essential. Projects must be developed that take into consideration the uniqueness of the U.D.C.'s cultural traditions and values.

Some suggestions have been made as to how the United States can help to improve environmental conditions in the U.D.Cs. For example, in December 1970, the Office of Science and Technology reported on the manner in which the United States could aid in protecting the world environment. In *Protecting the World Environment in Light of Population Increase,* ten recommendations were made under the auspices of three administrative arrangements. Under the United Nations, the United States was urged to call for a review of national and international programs in view of a declared international environmental policy, develop plans for a global monitoring system, and suggest that international air and water quality standards be set. Within the governmental agencies responsible for foreign affairs, it was suggested that efforts be made to strengthen international arrangements for forecasting weather and measuring climate modification, that the administration of the Agency for International Development reflect a stated concern for environmental problems in its activities, that U.S. agencies involved with U.S. business concerns overseas urge the business institutions to take into account the effect on the local environment of their operations, and that self-help plans to aid foreign countries include environmental and conservation aspects of agricultural development. Suggestions were also made on the domestic front to "broaden the base for international cooperation," that is, to further expand research and environmental management programs. Although these suggestions do not influence overpopulation or lack of adequate food, they do indicate the kinds of activity that can be undertaken to enable a D.C. to aid U.D.C.s.

Sadly, the abetting of the problems of the U.D.C.s is perhaps one of the most difficult and challenging problems of our time. Historically, it usually has not been clear how one nation can aid another. How *individuals* of a D.C. can influence another country is less clear, but in democratically run countries perhaps the vote of the individual can place responsible citizens in office so that our foreign policy reflects this global concern.

What can I do?

Eventually, in a discussion of solutions to the environmental dilemma, the individual, convinced of the need for action and willing to commit himself to doing his part, asks: What can *I* do? No question could be more appropriate as this book nears completion. Clearly, any change in economics, in laws, or in an ethic rests in the involvement of the individual. Unfortunately, in these early stages of confrontation of environmental problems, an individual is too often overwhelmed by the magnitude of the problems and their entwinement in "the system," the large superstructures of power seen in our state and national governments, our legal institutions, and even our schools.

Foremost, a person must have faith in his ability as an individual to be effective in provoking change. The societal dropout or the counterculture member may well delight in his return to a simple and natural living style, but in the long run his influence is most likely minimal. Effecting change demands involvement. The alarmist "concerned" about environment difficulties only to the point of thinking about them is not enough, although it may be better than not thinking about them at all. Change comes only with action.

"How to act," then, is the question. Too often, the individual responds by involving himself singly and impotently, because the complications of the environmental crisis obscure opportunities for the most effective involvement. Personal action is too often weak and dissipated. Strength lies in groups' rallying together to influence law or economics.

Many examples could be cited of environmental action of limited effectiveness. For example, to carry a glass jug to the food store into which newly bought milk can be poured from a disposable carton at the checkout counter may effectively emphasize a way to reduce solid waste by urging that more products be sold in returnable bottles instead of disposable cartons. But without at least equal time and effort spent discussing the concern with the store officials, the milk company, and possibly agitating for legislation to handle the problem, probably little will be accomplished toward reducing the amount of solid waste. To merely pour milk from container to container without actively relating the deed to broader concerns is a misinterpretation of the role of an environmentally concerned citizen. However, this criticism is not to say that small, individual demonstrations are not without benefit. Certainly, they are, but to practice them without a broader interpretation of the problem is a mistake and in the long run probably ineffective.

Thus, very likely the first requirement for the environmentally conscious individual is to align himself with a group or groups. The advantages are several but certainly include the strength in numbers provided by organizations. Membership in the groups also offers opportunity to talk over problems with others and, thus, the opportunity for a clarification of the facts surrounding an issue and the most effective mode of attack to solve it. Obviously, allegiance to local groups as well as at a level more widespread is an advantage.

The second main requirement is to be informed. Membership in groups helps, but enrollment in environmental courses in the area high school, college, or university and the reading of the numerous books and other materials now available are obvious sources of information. In all cases it is essential to maintain an objective and open point of view. The environmental problems have tangled causes, and hence, any solutions are not easily and simply discerned or effected. As an example, consider DDT. Advantages of its use must be compared to its disadvantages before a decision regarding its use can be made. Beware of alarmists — they may be correct, but maintain your rationale and objectivity; then act. Familiarize yourself with basic biology, economics, or law so as to make valid decisions for action.

Third, the environmentally minded individual must be alert and imaginative. Many environmental disasters have been allowed to occur, because no one noticed their development. The proper questions at the appropriate times may uncover obvious threats to the environment which could use your group's attention. Solutions to problems are best found in the same manner, with an alert and imaginative outlook. Any list of "eco-tactics," specific details as to how to proceed, is of limited value because every situation in a community needs a particular solution that demands imagination. Many of our environmental problems occur today because citizens were not alert enough earlier to sense impending trouble. Too many will remain unsolved because inadequate imagination is brought to the issue.

In the long view any solutions to environmental problems, however, will come only when everyone reevaluates his position in nature's "web of life." Kenneth Cauthen, theologian, says it well in *Christian Biopolitics: A Credo and Strategy for The Future* (Abingdon Press, Nashville, 1971.):

> *At this point it must be recognized that the interdependence of all life in relationship to the planetary environment places special obligations and limitations on the present generation. There is one world, one human family, one interrelated web of life woven on the spherical skin of the earth. Plants, animals, and men share a common environment. It is imperative not only that nations and races learn to live in peace with justice for all but that we also learn how to relate ourselves to our natural surrounding in such a way as to stay alive and prosper. We cannot afford to continue to make war, to tolerate oppression, to allow the gap between the rich and the poor to persist. But neither can we indiscriminately and indefinitely plunder the planet for its resources, overpopulate it with people, and pollute our air and water without paying the terrible consequences in human misery. If we are to have a future at all, we must at least learn the elementary requirements of biological survival.*

Supplementary readings

Anderson, W., ed. 1970. *Politics and Environment: A Reader on Ecological Crises.* Goodyear Publishing Co., Inc., Pacific Palisades, California. 362 pp.

Baldwin, M., and J. K. Page, Jr., eds. 1970. *Law and the Environment.* Walker & Company, New York. 432 pp.

Black, J. N. 1970. *The Dominion of Man: The Search for Ecological Responsibility.* Edinburgh University Press, Edinburgh, Scotland. 169 pp.

Cailliet, G. M., P. Y. Setzer, and M. S. Love. 1971. *Everyman's Guide to Ecological Living.* Macmillan Publishing Co., Inc., New York. 119 pp.

Caldwell, L. K. 1970. *Environment: A Challenge for Modern Society.* Doubleday & Company, Inc. (Natural History Press), Garden City, N.Y. 292 pp.

Commoner, B. 1971. *The Closing Circle.* Alfred A. Knopf, Inc., New York. 326 pp.

Cooley, R. A., and G. Wandesforde-Smith, eds. 1970. *Congress and the Environment.* University of Washington Press, Seattle, Wash. 277 pp.

Dales, J. H. 1968. *Pollution, Property and Prices: An Essay in Policy-making and Economics.* University of Toronto Press, Toronto. 111 pp.

Darling, F. F., and J. P. Milton, eds. 1966. *Future Environments of North America.* Doubleday & Company, Inc. (Natural History Press), Garden City, N.Y. 767 pp.

Dasmann, R. F. 1968. *A Different Kind of Country.* Macmillan Publishing Co., Inc., New York. 276 pp.

Davies, J. C., III. 1970. *The Politics of Pollution.* Pegasus, Indianapolis, Ind. 231 pp.

Disch, R. ed. 1970. *The Ecological Conscience: Values for Survival.* Prentice-Hall, Inc., Englewood Cliffs, N.J. 206 pp.

Dorfman, R., and N. S. Dorfman, eds. 1972. *Economics of the Environment.* W. W. Norton & Company, Inc., New York. 426 pp.

Ehrlich, P. R., and A. H. Ehrlich. 1970. *Population, Resources, Environment: Issues in Human Ecology,* 2nd ed. W. H. Freeman and Company, San Francisco. 509 pp.

Ellul, J. 1967. *The Technological Society.* W. H. Freeman and Company, San Francisco. 449 pp.

Fromm, E. 1965. *The Sane Society.* Holt, Rinehart and Winston, Inc., New York. 370 pp.

Garvey, G. 1972. *Energy, Ecology, Economy.*
W. W. Norton & Company, Inc., New York. 235 pp.

Graham, F., Jr. 1966. *Disaster by Default: Politics and Water Pollution.*
M. Evans & Co., Inc., New York. 256 pp.

Hardin, G. 1968. "The Tragedy of the Commons." *Science,* vol. 162, pp. 1243–1248.

Herfindahl, O. C., and A. V. Kneese. 1965. *Quality of the Environment: An Economic Approach to Some Problems in Using Land, Water, and Air.*
The Johns Hopkins Press, Baltimore, Md. 96 pp.

Huth, H. 1957. *Nature and the American: Three Centuries of Changing Attitudes.*
University of California Press, Berkeley, Calif. 250 pp.

Istock, C. E. 1971. "Modern Environment Deterioration as a Natural Process." *International Journal of Environmental Studies,* vol. 1, pp. 151–155.

Jarrett, H., ed. 1966. *Environmental Quality in a Growing Economy.*
The Johns Hopkins Press, Baltimore, Md. 173 pp.

Landau, N. J., and P. D. Rheingold. 1971. *The Environmental Law Handbook.*
Ballantine Books, Inc., New York. 496 pp.

Leopold, A. 1966. *A Sand County Almanac.*
Oxford University Press, New York. 269 pp.

Mitchell, J. G., and C. L. Stallings. 1970. *Ecotactics: The Sierra Club Handbook for Environmental Activists.*
Pocket Books, New York. 288 pp.

Moncrief, L. W. 1970. "The Cultural Basis for Our Environmental Crisis." *Science,* vol. 170, pp. 508–512.

Montgomery, J. C. 1971. "Population Explosion and United States Law." *Hastings Law Journal,* vol. 22, no. 3, pp. 629–659.

Murphy, E. F. 1971. *Man and His Environment: Law.*
Harper & Row, Publishers, Inc., New York. 168 pp.

Myrdal, G. 1970. *Challenge of World Poverty: A World Anti-poverty Program in Outline.*
Pantheon Books, Inc., New York. 518 pp.

Odum, H. T. 1971. *Environment, Power, and Society.*
John Wiley & Sons, Inc., New York. 331 pp.

Ramsey, W., and C. Anderson. 1972. *Managing the Environment: An Economic Primer.*
Basic Books, Inc., Publishers, New York. 302 pp.

Rathlesberger, J., ed. 1972. *Nixon and the Environment: The Politics of Devastation.*
Taurus Communications, New York. 279 pp.

Ridgeway, J. 1970. *The Politics of Ecology.* E. P. Dutton & Co., Inc., New York. 222 pp.

Roos, L. L., Jr., ed. 1971. *The Politics of Ecosuicide.* Holt, Rinehart and Winston, Inc., New York. 404 pp.

Sax, J. L. 1970. "Environment in the Courtroom." *Saturday Review,* Oct. 3, pp. 55–57.

Sax, J. L. 1971. *Defending the Environment: A Strategy for Citizen Action.* Alfred A. Knopf, Inc., New York. 252 pp.

Sprout, H., and M. Sprout. 1971. *Toward a Politics of the Planet Earth.* Van Nostrand Reinhold Company, New York. 499 pp.

Stone, C. D. 1974. *Should Trees Have Standing?* W. Kaufmann, Los Altos, Calif. 128 pp.

Study of Critical Environmental Problems (SCEP). 1970. *Man's Impact on the Global Environment.* MIT Press, Cambridge, Mass. 319 pp.

U.S. National Goals Research Staff. 1970. *Toward Balanced Growth: Quantity and Quality.* Government Printing Office, Washington, D.C.

Ward, B., and R. Dubos. 1972. *Only One Earth: The Care and Maintenance of a Small Planet.* W. W. Norton & Company, Inc., New York. 225 pp.

Watt, K. E. F. 1968. *Ecology and Resource Management: A Quantitative Approach.* McGraw-Hill Book Company, New York. 450 pp.

White, L., Jr. 1967. "The Historical Roots of Our Ecologic Crisis." *Science,* vol. 155, pp. 1,203–1,207.

Index

Abortion: and birth control, 217, 230–32; legal aspects of, 231–32; model law, 230
Acetylcholine, and nerve impulses, 131
Acheulian tool industry, 87–8
Adenosine triphosphate (ATP), molecular structure, 39
Adhesion, of water molecules, 303
Adrenal gland, and population growth, 198
Africa: birth control, 216; as man's original environment, 87
Aggression, and territoriality, 151–52, 175–76, 178, 200
Agricultural Revolution: and ecosystem, 57; and man's survival, 157–59
Air Quality Act of 1967, 295, 427
Air pollution. *See under* Pollution
Aldehydes, 288, 289
Algae: blue-green, fossils of, 33; as food source, 265–66; growth curve of, 196; and pollution, 335, 337
Allen's rule, 114
Altitude, and yarrow plants, 7, 8, 9, 17
Aluminum, 384
Amebae, movements of, 124
Amino acids, 35, 36, 38, 238, 240, 249–50, 266, 267–68
Anabolism, 37
Ancestors, common, of organisms, 19, 24–7
Anemia, hypochromic, 244
Animals: adaptation to cold climates, 114–15; first domesticated, 59; and fluorides, 285; and ozone, 290; and realms of earth, 97–105; selective breeding, 159; simple, behavior of, 123–29; waste from, 314
Apes, evolutionary development, 63–4
Aquatic life, and oil pollution, 320
Ariboflavinosis, 245
Arsenic as pesticide, 342
Art of Cro-Magnon man, 80–1
Asia: arable land, 260; food production increase, 271
Aswan Dam, 254–55
Atmosphere: changes in, 286–88; formation of, 32
Atomic energy, 377–80
ATP. *See* Adenosine triphosphate, 39
Australian realm, 100
Australopithecines, 65–73: comparison of faces, 68
Australopithecus: and *Homo erectus,* 75–6; *africanus,* 66, 72, 87; *robustus,* 72
Automobile(s): electric, 292–93; and pollution, 278, 279, 285, 288, 291–93, 295
Autotroph(s): as first plant kingdom, 41; as food producers, 51–2
"Baby boom," 184–85
Bacteria, and water pollution, 312
Barriers, natural, and realms of earth, 97, 100–01, 102–03
Bees, behavior of, 129–31
Behavior: determining survival, 120–53; learned and instinctive, defined, 136; social, 145–52
Bergmann's rule, 114

Beri-beri, 245
Biafra, and kwashiorkor, 245, 246
Bilateral symmetry, development of, 45
Biochemical oxygen demand (BOD), 312
Biocides, 342–63
Biogeography, 26, 86
Biological evolution, 4, 10, 24–7, 62–82
Biomes, 105–09: altitudinal and latitudinal, 106
Biotic potential, of population, 193
Birds: and DDT, 355, 358–60; and oil pollution, 320
Birth control: clinics, 212, 214; history of, 212–14; methods of, 218–35; research, 212, 215
Birth rates, 164–65: and population change, 192; United States, 184–86
Bison, destruction of, 414–15
Blindness, river, 257–58
Body, human, and water, 304–05
Botanical pesticides, 342, 347
Brain: effects of hunger, 248–49; evolution of, 132–35; parts of, 135–36
Brain capacity: of *Homo sapiens,* 79; of Neanderthal man, 77; of Peking man, 74
Brazil: and fertilizers, 252; increased food needs, 243
British Isles, biography of, 95–8
Bronze, 385–86
"Bulldozer mentality," 429
Calories, 240, 241
Cancer, and DDT, 360
Carbon, cycling of, in ecosystem, 53
Carbon compounds as pollutants, 285–88
Carbon dioxide, 386–88
Carbon monoxide, 285–86, 292, 296
Carboxy-hemoglobin, 286
Catabolism, 37
Cave paintings of Cro-Magnon man, 80–1
Cell(s): evolution of, 41; origin of, 37, 40; relaying information, 131–32; structure of, 40
Cereal grains: first domesticated, 157, 158; as staple diet, 239–40
Cerebrum, areas of, 134, 135–36
Change: as basic factor, 4–28; two types of, 10
Chemicals: as pollutants, 318–22; synthetic organic, 328, 348–49, 363–64
Chick edema, 346–47, 348
Chimpanzees, studies of, 141–42
China: birth control, 216; irrigation in, 254, 255; population, 176
Chloracne, 346
Chlorella, 266, 267
Cholinesterase, and nerve impulses, 131–32
Chromosomes, 39
Cigarette smoke, 286
Circulatory disease, and smog, 276, 278
Class actions, and environment, 426
Clean Air Act of 1963, 295
Clones, production of, 9
Coacervates, 34, 37, 39
Coal, 369–70

437

438 Index

Coelenterate, nervous system of, 125, 126
Color, and birth rates, 185–86
Coloring, protective, 17–8, 26–7, 114
Communication: among bees, 129–31; development of, 121; of early man, 73, 74
Community succession, 206–07
Competition, and population density, 203–05
Competitive exclusion principle. See Gause's principle
Composting, 331
Compounds in earth's original atmosphere, 32–3
Comstock Law, 212, 215
Conceptualization ability of Cro-Magnon man, 80, 81
Conditioning, 140: operant, 142–44
Conservation: agreements, 412; forest, 403–06; soil, 392, 394
Consumer, and environmental crisis, 424, 425, 426
Continental drift hypothesis, 102
Contraceptives, 224–30: future, 233–34; for men, 234
Coaling towers, 325, 327
Copper, 385
Corn, new varieties, 269–70
Corridor, as animal migration route, 104
Cost-benefits dissociation, and environmental crisis, 424, 426, 429
Crankcase blow-by, 291
Cretinism, 243
Cro-Magnon man, 68, 79–81, 89
Crops: and air pollution, 282, 284; ancient, 159; losses from pests, 349; new, 267–72; and soil erosion, 391; yields, and biocides, 350–51, with fertilizer, 251–53, increased, 271
Cultivation, increased, and food problem, 259–62
Cultural anthropology, 121, 145
Cycling of water, 300, 301
Dams, 253–54
Dances of bees, 129–30
Darwin's finches, 94–5, 204–05
DDT, 343, 348, 349, 351, 352–62: and population growth, 170
Death rates, 164–65: from malnutrition, 238, 247; and population change, 192; reduction of, 170–71; smog, 276; United States, 182–84
Deer, decimation of, 190, 191, 195, 201, 202
Defoliation, 346, 347
Demes, 15
Dependents, D.C.s and U.D.C.s, 173
Desalinization, 258
Desert: adaptation in, 305; biome, 112–13
Detergents as pollutants, 308, 315–16, 321
Developed countries (D.C.s), 162: aid to U.D.C.s, 430–32; environmental crisis, 423; family planning in, 214–16; and fertilizers, 252; vs. underdeveloped, 160–68, 169, 175
Disease, and pesticides, 351
Donora, Pa., smog, 276, 284, 288
Drought, 387
Dust bowl, 386–88

Earth, primitive, Miller's simulation, 34
Earthworms, 127
Ecological diversity, 364
Ecological niche. See Gause's principle
Ecology, defined, 4
Economics, and environmental crisis, 423–26
Ecosystem(s), 32, 47–58: definition of, 49; flow of energy, 52–3; flow of matter, 52; "foreign" organism in, 48; nature of, 49–52
Ectoparasites, 202
Electrical power plants, and pollution, 285, 322
Emphysema, 277–78
Endoparasites, 202
Energy: from burning trash, 330; dissipation of, 52–5; for metabolism, 240; sources of, 368–80
Engram, 144
Environment: and ancient man, 77; early (before life), 32–3; and evolution of cell, 38; man and, 93; and population growth, 161–62; new, and changes in organisms, 18; organisms and, 4, and population change, 191–93; and realms, 100–05; and tools, 87–9
Environmental crisis, 408: historical causes, 422–23; solutions to, 422–36
Environmental resistances, 193ff.
Enzymes, origin of, 38, 39
Erosion, soil, 390–92: forests and, 403
Estrogens, 223–24
Ethics, and environment, 423–30
Ethiopian realm, 99
Eutrophy of lakes, 335
Everglades, destruction of, 413
Everglades National Park, 411
Evolution: biological, 4, 10, 24–7; of environment into life, 41; significance to man, 27–8; synthetic theory of, 10–12, 15–17, 25–6, 27, 32
Exhaust controls, automobile, 291, 292, 425
Externalities, economic, 424–25, 426
Extinction, 192, 409: causes of, 412–16; biology of, 416–18
Eye(s): compound, of insects, 128, 129; diseases of, 244
Family, beginnings of, 123
Famines, 159–60
Ferroalloy metals, 384
Fertile Crescent region, 157, 158, 159
Fertilizer(s): and food problem, 251–3; and hardpan, 257; as pollutants, 315
Filter, as animal migration route, 104
Fish: and DDT, 356–58; and heat, 323, 324; instinct in, 138; and lake entrophy, 335; and water pollution, 312
Fish farming, 264
Fish meal, as protein source, 263–64
Fishing industry, expansion of, 262–63
Flatworms, nervous system of, 127, 129
Flukeworm, blood, 257, 258–59, 260
Fluorides, as pollutants, 285
Follicle-stimulating hormone (FSH), 223, 224, 226
Fontechevade man, 78

Index

Food chain, 50, 52: and pesticides, 354–55
Food problem, global, 240–43: solutions to, 249–58
Food(s): basic groups, 238–40, 242; fortified, 249–50; losses of, 262; new, 250–51
Food supplies, world, 238–73
Food web, 51, 52
Forest fires, 404, 405
Forests, 398–408: distribution of, 400–02; future of, 406–07; management, 403–06
Fossil fuels, 368–74
Fossils: dating of, 102: of man's ancestors, 62–81
Fruit fly, genetic research and, 18–19
Fungi, as food source, 265–67
Fungicides, 343
Fusion reactions, as energy source, 379–80
Fusion torch, 331
Galápagos Islands, biogeography of, 93–5
Ganglia, 127, 132
Gas, natural, 372–73
Gause's principle, 204
Genes, 39: and sexual reproduction, 43–4; theory of, 12–5
Genetic drift, 13
Genetic recombination, 13
Geological events, and distribution of organisms, 101
Geology, and synthetic theory of evolution, 19, 20–3
Geothermal energy, 377
Glaciation: *Homo erectus* in, 89; and Neanderthal man, 77
Glacier hypothesis of evolution, 77
Goiter, 243
Grain-breeding programs, 267–72
Grassland biomes, 107–08, 109
Greenhouse effect, 286–88
Gross national product (GNP), 423, 430
Group living among animals, 149
Habitat change, and extinction, 412–14
Habituation, 140
Hardpan from irrigation, 256–57
Heat: buffering by water, 304; from burning trash, 330; as pollutant, 322–28
Herbicides, 345–47
Heterotrophs, 41, 50–1: and autotrophs, 50–2
Homeostasis of ecosystem, 47
Homo erectus, 68, 73–6: migration of, 88–9
Homo neanderthalis. See Neanderthal man
Homo sapiens, 78–81
Hormones: long-lasting, 233; in oral contraceptives, 226; and ovulation, 223–24, 226
Horse(s): ancient and modern, 110; biogeography of, 107, 109–12
Hunger: and areas of world, 241; "hidden," 240
Hunting: by Cro-Magnon man, 79, 80, 81; excessive, 409, 414
Hunting culture, and man's social origins, 121–22
Hydra, nerve net of, 126
Hydrocarbons, 278, 288, 291, 292, 295, 321: as insecticides, 348–49
Hydrogen, structure of, 301–02

Hydrogen bond, 302–04
Hypothalamus and water in body, 305
Ice, 303
India: and birth control, 215; and fertilizers, 252, 262; food losses in, 262; increased food needs, 243; irrigation in, 255; pests in, 349
Indians, American, vs. settlers, 408
Individual, and environmental crisis, 432–33
Industrialization and population growth, 162
Industrial pollutants, 293–94
Insecticides, 347–49: damage from, 202
Insects: and forests, 405–06; instinctive behavior, 136; nervous apparatus, 127
Insight, as learning, 140–41
Instinct, vs. learning, 136–44
Intelligence, 141: tests of, and malnutrition, 249
Intrauterine device (IUD), 228–29, 234
Iodine, and goiter, 243
Ireland, fauna of, 96–7
Iron: as catalyst, 38; deficiency, 244; supply of, 381–84
Irrigation and food problems, 253–58
Islands, oceanic vs. continental, biogeography of, 95
Isolation: behavioral, 15; and development of horses, 110; geographical, 14; mechanical, 15; and natural selection, 13–15; seasonal, 15
Japan: birth control, 216–17; and fertilizers, 252; population, 176
Java man. See *Pithecanthropus erectus*
J curve, of population growth, 195–96
Judaeo-Christian traditions, and environmental problems, 422–23
Judaism: and abortion, 232; and birth control, 213
Kaibab Plateau, 190, 191, 195, 201, 202
Korea, birth control in, 216
Krakatoa Island and primary succession, 207, 209
Kwashiorkor, 245
Lake Michigan, and natural succession, 207
Land environment, transition organisms, 44–5
Land fill, 331–33
Land, potentially arable, 260
Land use, 399
Language, beginnings of, 121–22
Laterite, 261
Latitude, and races, 90
Lead as pollutant, 290–91
Learning: categories of, 140; and instinct, 139; programmed, 144
Legal action, and environmental crisis, 426–28
Lemmings, 198
Life: beginnings of, 33–6; distribution of, 86–117; diversification of, 41, 46–7; human, pesticide threat to, 359–61; quality of, and population growth, 182; and water, 300–01, 304–05
Light waves, and communication of bees, 130

Livestock wastes as pollutants, 314–15
London smog, 276, 278–79, 284, 288
Los Angeles smog, 278–80, 288
Lumber industry, 402–03
Luteinizing hormone, 223, 224, 226
Lysine: added to food, 250; deficiency, 249–50
Macroconsumers, 51
Malaria, effects of DDT, 170, 351
Malnutrition: mental effects of, 247; See also Nutritional diseases
Mammals, of realms of world, 98
Man: ancestors of, 62–81; distribution of, 87–93; and ecosystems, 55–8; evolution of, 71; significance of evolution, 27–8; supernatural view of, 422, 429
Marasmus, 246
Marriage, origins of, 122–23
Mating behavior, 146–47
Matter-energy reactions, and beginning of life, 33
Medical programs, and population, 161, 169–71
Membranes, development of, 37
Memory, nature of, 144–45
Mental effects, of malnutrition, 245, 247–50
Mercantilism, Malthus and, 178, 180
Mercury poisoning, 343, 345
Mesoderm, 45
Metabolism, 37
Metals, scrap, 333
Methemoglobinemia, 313
Mexico, and grain-breeding, 269–70
Microconsumers (saprotrophs), 51
Microspheres, 36, 38
Middens, 156
Middle East, birth control in, 216
Migration: of animals, 103–05; environment and, 90–3; and mixing of populations, 13; and population change, 192; and population density, 173, 175
Minerals: in diet, 238; exhaustion of, 380–86; as pollutants, 318–22; predicted lifetimes, 383
"Missing link," 65–6
Mitosis, beginnings of, 42–3
Molecules, coding, 39
Molybdenum, 386
Mongoloids, migrations of, 90, 93
Monkeys, differentiation from apes, 63
Moral restraint, as Malthusian population check, 180
Moth, peppered, color change, 5, 6, 8, 9, 17–8
Mousterian tool industry, 89
Multicellularity, development of, 42–3
Mutationism, 13
Natural succession. See Succession
Neanderthal man, 68, 76–8: tools of, 89; two types of, 77
Nearctic realm, 99
Nematocysts, 125, 126
Neotropical realm, 99
Nerve chain, simple. See Reflex arc
Neurons, and behavior, 131
New York City, air pollution, 281
Niacin deficiency, 245

Nickel, 386
Night blindness. See Nyctalopia
Nitrogen compounds: as pollutants, 285, 295, 313; cycling of, in ecosystem, 52, 54
Nuclear reactor, 377–80
Nuisance laws and pollution, 426–27
Nutritional diseases, 243–49
Nutritional requirements, 238–40, 242
Nyctalopia, 244
Ocean(s): as food source, 262–64; formation of, 33, 300; oil and, 320
Oil, as pollutant, 318–22
Oldowan tool industry, 87
Olefins, 288, 289
Oligopithecus, 63
Oparin's hypothesis, 33–6
Oral contraceptives, 226–28: side effects, 228
Organisms: disease-causing, in water pollution, 313; nutritional relationships, 50–2
Oriental realm, 100
Osteomalacia, 244
Overfishing, 264
Ovulation: and hormones, 223–24, 226; and rhythm method, 219, 222
Oxidation pond, 308–09
Oxygen: introduction of, 41, 42; structure of, 301–02
Ozone, 289–90: layer, 42
Pacific Islands, migration paths, 91
Pair bonding, 122, 147, 149, 150
Pakistan: birth control, 215; and fertilizers, 252; increased food needs, 243; irrigation in, 255; pests in, 349
Palearctic realm, 97, 98, 99
Paleocene epoch, and beginning of man, 62
Paleontology, 19
Paraffins, and yeast culture, 267
Paramecium: competition of species, 203, 204; movements of, 124–25
Paranthropus robustus, 67
Parasites and disease, 257–58
Parasitism, and population density, 202, 203
Particulates, as pollutants, 282–83, 287, 293, 315, 316
Parks, national, 410–11
Pecking order, 149
Peking man, 74
Pellagra, 245, 249
Penicillin, and population growth, 170
Peptide bond, formation of, 35
Permafrost, 113, 114
Peru, and fishing industry, 263
Pest control, natural, 361–63
Pest damage, costs of, 349
Pesticides, 328, 342–61: advantages, 249–52; disadvantages, 352–61
Petroleum, 370–72
Phermones, 137
Phosphates as pollutants, 315–16, 336
Phosphorus, cycling of, in ecosystem, 55
Photochemical products as pollutants, 288–89
Photosynthesis, origin of, 41–2
Physiological control, internal, and population growth, 197–99
Pig, miniature, 16

Pithecanthropus erectus, 73, 74
Pithecanthropus pekinesis, 74
Plague, and population size, 160
Plankton, J curve growth, 195
Plant nutrients as pollutants, 313–16
Plants: and air pollution, 283–85, 289, 290; pasture, changes in, from heavy grazing, 6, 8, 9, 17, 18
Pleistocene overkill, 93
Politics, and environmental crisis, 428
Pollutants: air, 278, 279, kinds and effects of, 282–91; water, 311–28
Pollution: air, 276–97, control of, 291–96, cost of, 280–82, 284; and fertilizers, 253; in national parks, 411; of ocean, 264; taxation and, 425–26; thermal, 322–28; as threat to survival, 28; used-vehicle, 332–33, 334; water, 300–39
Population change (of organisms), 190–96
Population control (human), 212–35: of Malthus, 180
Population density, 173–77
Population growth (human), 156–87: calculation of, 165–66; D.C.s compared to U.D.C.s, 168, 169; resistance to, 196–205; and wildlife, 410, 412
Population(s): age distribution, 171, 172; animal, characteristics of, 190–209; evolution of, 10–28, evidence of, 17–27; size, and gene changes, 13
Predators: in pest control, 362, 363, 364; and population density, 201–02
Prehistoric men, skulls, compared with ape, 80
Prey-predator relationship, 195, 200–02: and pesticides, 354
Primates, evolution of, 67
Proconsul, 63, 64, 65
Progesterone, 224
Propliopithecus, 63
Prosimians, 62
Protein molecules, in first step in evolution, 34
Proteinoids, and early life processes, 36
Protein(s), 238, 240: in algae and fungi, 266–67; beverages, 251; deficiency of, 245, 248, 249; new sources of, 250–51
Protestants: and abortion, 232; and birth control, 214
Protozoans, 124
Pulp and paper industry, 403
Pyramids of plant-animal relationships, 54, 55, 56
Races (of man): development of, 89–93; distribution, 92; as subpopulations, 15
Radioactive materials: 377–80; as pollutants, 316–17
Rain forest, tropical, 115
Ramapithecus, 64
Realms (of earth), 97–105: distribution of organisms in, 105–09
Recycling of waste, 329–31, 333, 381
Reflex arc, 132
Refuges, wildlife, 412
Reinforcement, in conditioned response, 142, 143, 144
Religion, and birth rates, 186

Religionists: and abortion, 232; and birth control, 213–14, 217; and Java man, 73
Renaissance, 160
Research: agricultural, international, 270; birth control, 212, 215
Resistance (to population growth): density-dependent, 197–205; density-independent, 196–97
Resources, nonrenewable, 368–95
Respiration, cellular, effect of oxygen, 42
Respiratory diseases, and air pollution, 277, 282, 283
Riboflavin: deficiency, 245; in enriched foods, 249
Ribonucleic acid (RNA), and memory, 144–45
Rice: new strains of, 268, 270; polished, and beri-beri, 245
Rickets, 244
RNA. *See* Ribonucleic acid
Rodenticides, 343
Rocks: metamorphic, 33; sedimentary, 33
Roman Catholicism: and abortion, 232; and birth control, 214, 217, 222
Salt deposits from irrigation, 256, 257
Schistosomiasis, 257, 258–59, 260, 261
Screwworm fly, 350, 361, 363
Sea water, desalting of, 258
Selection, natural, 12, 31, 203, 354: and early ape men, 72
Self-regulation, importance in ecosystem, 49
Sewage: as pollutant, 336; treatment of, 306–09
Sex roles, differentiation of, 122
Sexual reproduction, evolution of, 43–4
Shale, as source of oil, 373–74
Shock disease and population density, 198–99
Shrew tree, 62, 63
Sigmoid growth curve, 193–95, 201
Silt: in dams, 256; in Mississippi Valley, 316, 317
Sinanthropus pekinesis. See Peking man
Skinner box, 142, 143
Smog, 288: legal control of, 294–96; lethal effects of, 276–78
Snails, and schistosomiasis, 257, 259, 260
Social origins of man, 121–23
Socioeconomic status, and birth rates, 186
Soil: preservation of, 390–92; structure of, 388–90; use of, 386–95
Soil Conservation Act, 394
Solar energy, 374–75
Solid Waste Disposal Act, 331
South America: birth control policies, 217; fauna of, 104; pests in, 349
Soybeans, 250, 266
Sparrow: English, population growth, 193; song, distribution of subspecies, 14
Speciation, 15
Species: endangered, 412, 414–18; new, production of, 12–18
Starvation, threat of, 243; preventive measures, 249–58
Steinheim man, 78
Sterilization, 229–30

442 Index

Strip mining, 370
Stromatolites, 33
Structural developments and behavior, 120, 125
Subspecies, 15
Succession: and population change, 206–09; primary, 207, 209
Sulfhydryl groups, 38
Sulfur compounds as pollutants, 283–85, 293–94, 296
Sunlight, and pigmentation, 90, 244
Supersitition, and extinction, 416
Survival: of the fittest, 11–12; and understanding of environment, 27–8; and use of resources, 368; and water, 304–05
Swanscombe man, 78
Sweepstakes route, of animal migration, 104
Taboos: diet, 247; sex, 122–23
Taiga forest, 106–07, 108
Taiwan, birth control in, 216
Tapeworm, life history of, 203
Tapiola, 327
Tarpans, 112
Tarsier, 62, 64
Taung baby, 66
Taxonomy, 19, 24, 86; of ape-man fossils, 70–1, 74
Technology and environment, 276, 422
Teeth: of horses, effect of environment, 111–12; monkey, ape, and man, 64, 66
Temperate deciduous forests, 105–06, 107
Temperature: animal adaptation to, 114–15; and development of races, 90
Territoriality, 149–52; and population density, 199–200
Thermal inversion, 279, 281, 296
Thiamine, in enriched foods, 249
Threonine, in fortified foods, 250
Thyroid gland, disease of, 243
Tidal power, 376–77
Time scale, geological, 19, 20–3
Tin, 385–86
Token learning, 143
Tools, 87–9: and man's survival, 156; of Australopithecines, 70, 72; of Peking man, 74
Trees, diseases of, 355, 356, 406
Trespass law, 427
Trial and error: in learning, 140; and paramecium, 125
Tryptophane deficiency, 250
Tuna, mercury ingestion by, 345
Tundra, 113–15
Tungsten, 386
Underdeveloped countries (U.D.C.s), 163; crop yields, 261; environmental crisis, 430–32; family planning, 215–18; and fertilizers, 252–53; food problem, 241–43, 272; and fortified foods, 250; nutritional diseases, 244, 245, 247; population crisis, 169–74
United States, population growth, 181–86
Urban living, and population problems, 181
Utopianism, Malthus and, 178, 180
Vaccination as contraceptive, 234
Values and the environment, 428–30
Vanadium, 386
Vegetation: and names of biomes, 105; vertical stratification, 115
Vertebrates: bone similarities, 25; evolution of brain, 133, 135; embryonic similarity, 26; instinctive behavior, 137–39
Vitamin D and skin pigmentation, 90
Vitamins: deficiencies, 244, 245; in diet, 238, 239, 240
Vehicle pollution, 291–93: used vehicles, 332–33, 334
Wallacea, 100
Warbler species, and Gause's principle, 205
War(s): and population, 160; and territoriality, 152
Waste disposal, 305–09, 313: cost of, 329; and environmental crisis, 424; methods of, 331; and pollution, 328–33
Wastes: radioactive, 379; as water pollutants, 311–12
Water: demand, 309–11; effects of temperature, 323; molecule, 302–04; as power source, 375–77; properties, 301–04; shortage, 311; stratification, 324–25, 326
Water-dependent organisms, 44, 45
Water Quality Act of 1965, 328, 427
Water supply, 300–39: desert, 112–13; grasslands, 108; tropical rain forest, 115; world, 258
Wells, tube (irrigation), 255–56
Whaling industry, death of, 264–65
Wheat: crop yields, 269; evolution, 158; new varieties, 270
Wilderness Act of 1964, 412
Wildlife: exploitation of, 408–12; preservation of, 410, 412
Wood, uses of, 402–04
World population, 160–63; U.N. predictions, 163–64
Worms, structure and behavior of, 125, 127
Xerophthalmia, 244
Yarrow plants, height changes in, and altitude, 6, 7, 8, 17, 18
Yeasts, as food source, 265, 267
Yellowstone National Park, 410
Yellowstone Timberland Reserve, 403
Yosemite National Park, 410
Young, care of, 147–48
Youth, increased numbers of, 171–73
Zinc, 384
Zinjanthropus boisei, 69, 72
Zygote, formation of, 43